Wireless Multimedia Communications

Convergence, DSP, QoS, and Security

Wireless Multimedia Communications

Convergence, DSP, QoS, and Security

K.R. Rao
Zoran S. Bojkovic
Dragorad A. Milovanovic

CRC Press
Taylor & Francis Group
Boca Raton   London   New York

CRC Press is an imprint of the
Taylor & Francis Group, an **informa** business

CRC Press
Taylor & Francis Group
6000 Broken Sound Parkway NW, Suite 300
Boca Raton, FL 33487-2742

© 2009 by Taylor & Francis Group, LLC
CRC Press is an imprint of Taylor & Francis Group, an Informa business

No claim to original U.S. Government works
Printed in the United States of America on acid-free paper
10 9 8 7 6 5 4 3 2 1

International Standard Book Number-13: 978-0-8493-8582-7 (Hardcover)

Visit the Taylor & Francis Web site at
http://www.taylorandfrancis.com

and the CRC Press Web site at
http://www.crcpress.com

Contents

Preface

Multimedia communication is one of the major themes in today's information communication technologies. It has many facets, from multimedia networking to communications systems, signal processing, and security.

The rapid growth of interactive multimedia applications, such as video telephones, video games, and TV broadcasting, has resulted in spectacular progress of wireless communications. However, the high error rates and the stringent delay constraints in wireless systems still significantly impact applications and services. On the other hand, the development of more advanced wireless systems provides opportunities for proposing novel wireless multimedia protocols and new applications and services that can take maximum advantage of the systems.

The impressive evolution of mobile networks and the potential of wireless multimedia communications pose many questions to operators, manufacturers, and scientists working in the field. The future scenario is open to several alternatives: thoughts, proposals, and activities of the near future could provide the answer to the future trends of the wireless world.

Wireless mobile communications may not only complement the well-established wireline network, they may also become serious competition in years to come.

The perspective of today's information society calls for a multiplicity of devices including Internet Protocol (IP)-enabled home appliances, vehicles, personal computers, and sensors, all of which are globally connected. Current mobile and wireless systems and architectural concepts must evolve to cope with these complex connectivity requirements. Research in this truly multidisciplinary field is growing fast. New technologies, new architectural concepts, and new challenges are emerging.

This book reflects the latest work in the field of wireless multimedia communications, providing both underlying theory and today's design techniques by addressing aspects of convergence, quality of service (QoS), security, and standardization activities.

BOOK OBJECTIVES

Anyone who seeks to learn the core wireless multimedia communication technologies, concerning convergence, QoS, and security, will need this book. The practicing engineer or researcher working in the area of wireless multimedia communication is forced to own a number of different texts and journals to ensure satisfactory coverage of the essential ideas and techniques of the field. Our first objective for the book is to be the source of information on important topics in wireless multimedia communications, including the standardization process. Another of the book's objectives is to provide a distillation from the extensive literature of the central ideas and primary methods of analysis, design, and implementation of wireless multimedia communications systems. The book also points the reader to the primary reference sources that give details of design and analysis methods.

Finally, the purpose of the book is not only to familiarize the reader with this field, but also to provide the underlying theory, concepts, and principles related to the power and practical utility of the topics.

ORGANIZATION OF THE BOOK

Following an Introduction, Chapter 2 provides an overview of the key convergence technologies to offer many services from the network infrastructure point of view. After a short presentation of the next generation network architecture, we deal with convergence technologies for third generation (3G) networks. This chapter also reviews technologies for 3G cellular wireless communication systems. Next, the 3G wideband code-division multiplex access (WCDMA) standard, which has been enhanced to offer significantly increased performance for packet data broadcast services, is presented. Challenges in the migration to fourth generation (4G) mobile systems conclude this chapter.

Chapter 3 surveys wireless video that has been commercialized recently or is expected to go to market in 3G and beyond mobile networks. We present a general framework that takes into account multiple factors (source coding, channel resource allocation, error concealment) for the design of energy-efficient wireless video communication systems. This chapter also reviews rate control in streaming video over wireless. We continue with a short presentation of content delivery technologies, and conclude with the H.264/AVC standard in the wireless video environment, together with a video coding and decoding algorithm, network integration, compression efficiency, error resilience, and bit rate activity.

Chapter 4 seeks to contribute to a better understanding of the current issues and challenges in the field of wireless multimedia services and applications. This chapter begins with real-time IP multimedia services, including evolution from short to multimedia message services. Extended IP multimedia system (IMS) architecture is provided, too. After that, we examine the current IMS policy control to update the significant changes in the core network. A number of service delivering platforms that have already been developed and commercially deployed are discussed.

Chapter 5 summarizes specifications for wireless networking standards that support a broad range of applications: wireless local area networks (WLANs), wireless personal area networks (WPANs), wireless metropolitan area networks (WMANs), and wireless wide area networks (WWANs). After a short presentation of IEEE 802.X standards, we deal with WLAN link layer standards. Wireless asynchronous transfer mode LAN, together with the European Telecommunication Standard Institute (ETSI) BRAN HIPERLAN standard, is included, as well. This chapter also reviews WPAN devices and Bluetooth. We continue with an overview of WMANs. The emphasis is on the IEEE 802.16 network arrangement, medium access control protocol, as well as on the orthogonal frequency division multiplexing (OFDM) physical (PHY) layer protocol.

Chapter 6 focuses on advances in wireless video. We start by introducing error robustness support using the H.264/AVC standard that makes it suitable for wireless video applications. Also, error concealment and limitation of error propagation are considered. Next, we move to error resilience video transcoding for wireless communications. We provide an overview of the error resilience tools including benefits according to category (localization, data partitioning, redundant coding, and concealment-driven techniques). This chapter highlights recent advances in joint source coding and optimal energy allocation, including joint source-channel coding and power adaptation. Finally, multipath transport is analyzed, together with general architecture for multipath transport of video streams. This is a promising technique for efficient video communications over ad hoc networks.

Chapter 7 concentrates on cross-layer wireless multimedia design. Then, we describe cross-layer architecture for video delivery over a wireless channel, and continue with a cross-layer optimization strategy. Using this strategy, information is exchanged between different layers, while end-to-end performance is optimized by adapting to this information at each protocol layer. A short overview of cross-layer design approaches for resource allocation in 3G CDMA network is also provided. After that, we move to the problem of cross-layer resource allocation for integrated voice/data traffic in wireless cellular networks.

The goal of Chapter 8 is to find a large audience and help stimulate further interest and research in mobile Internet and related technologies. First, related protocols for mobile Internet are presented and analyzed. Next, we describe IP mobility for cellular and heterogeneous mobile networks. We continue with scalable application-layer mobility protocols. We also review mobility and QoS. A network architecture analysis for seamless mobility services concludes this chapter.

The book ends with Chapter 9, which is devoted to evaluation of future 4G networks. 4G is a very promising generation of wireless communications that will change people's lives in the wireless world. After a discussion including migration to 4G mobile systems, as well as beyond 3G and toward 4G networks, we speak about 4G technologies from the user's perspective. The emphasis is on heterogeneous system integration and services. After that we present an all-IP 4G network architecture. Then, we outline the issues concerning QoS for 4G networks. Next, we continue with security in 4G networks, together with infrastructure security and secured handover between heterogeneous networks. Network operators' security requirements conclude the chapter.

Each chapter has been organized so that it can be covered in 1 to 2 weeks when this book is used as a principal reference or text in a senior or graduate course at a university. It is generally assumed that the reader has prior exposure to the fundamentals of wireless communication systems. The book can be also very useful for researchers and engineers dealing with wireless multimedia communication systems.

The references are grouped according to the various chapters. Special efforts have been taken to make this list as up to date and exhaustive as possible.

A major challenge during the preparation of this book was the rapid pace of development. Many specific applications have been realized in the past few years. We have tried to keep pace by including many of these latest developments. Finally, we hope that this book will provide readers with a valuable tool and resource when working in wireless communications.

Acknowledgments

We hope that this book will provide readers with a valuable tool and resource when working in wireless multimedia communications generally and, in particular, when dealing with convergence, digital signal processing, quality of service, and security. Many years of work resulted in these wireless multimedia communication pages, but also in many lifetime friendships among people all around the world. Thus, it is a pleasure to acknowledge the help received from colleagues associated with various universities, research labs, and industry. This help was in the form of technical papers and reports, valuable discussions, information, brochures, the review of various sections of the manuscript, and more.

Sincere and special thanks are due to the following people:

Ling-Gee Chen, National Taiwan University, Institute of Electrical Engineering, Taipei, Taiwan

Jae-Jeong Hwang, Kunsan National University, School of Electronic and Information Engineering, Kunsan, Korea

Valeri Mladenov, Technical University–Sofia, Faculty of Informatics, Sofia, Bulgaria

Fernando Pereira, Lisbon Technical University, Department of Electrical and Computer Engineering of *Instituto Superior Technico*, Lisbon, Portugal

Jurij Tasic, Faculty of Electrical Engineering, Ljubljana, Slovenia

Special thanks go to MSc Bojan Bakmaz, University of Belgrade, Serbia. He was instrumental in the final preparation of this book.

List of Acronyms

1G:	first generation
2G:	second generation
3G:	third generation
3GPP:	Third Generation Partnership Project
3GPP2:	3G Partnership Project 2
4G:	fourth generation
AAA:	authentication, authorization, and accounting
AAAC:	authentication, authorization, accounting, and charging
AAL:	ATM adaptation layer
AC:	access category
ACF:	association control function
ACH:	association control channel
ACK:	acknowledgment
ACL:	asynchronous connectionless
AES:	advanced encryption standard
AF:	application function
AF:	assured forwarding
AFC:	access feedback channel
AIFS:	arbitrary interframe space
AIPN:	all-IP network
AIR:	adaptive intrarefresh
AKA:	authentication and key agreement
ALT PHY:	alternative physical layer
AM:	aggregation module
AMC:	adaptive modulation and coding
AMPS:	advanced mobile phone service
AMPS:	Association of Radio Industries and Broadcasting
AN:	access network
ANSI:	American National Standardization Institute
AP:	access point
APA:	adaptive power allocation
API:	application programming interface
APP:	application layer
AR:	access router
ARM:	advanced router mechanisms
ARQ:	automatic repeat request
AS:	application server
ASDL:	asymmetric digital subscriber line
ASIC:	application-specific integrated circuit
ASO:	arbitrary slice ordering
AT:	access time

ATM:	asynchronous transfer mode
BC:	broadcast channel
BCC:	broadcast control channel
BCH:	Bose-Chandhuri-Hocqnenghem (code)
BER:	bit error rate
BGCF:	breakout gateway control function
BLER:	block error rate
BoD:	bandwidth-on-demand
BPSK:	binary PSK
BRAN:	broadband radio access network
BRDN:	broadband railway digital network
BS:	base station
BSC:	base station controller
BSS:	basic service set
BSS ID:	BSS identifier
BTS:	base transceiver station
BTS:	base transceiver system
BW:	broadband wireless
BWA:	broadband wireless access
CA:	composition agreement
CAC:	call admission control
CAI BIOS:	common air interface basic input–output system
CAMEL:	customized application for mobile network enhanced logic
CARD:	candidate access router discovery
CBP:	coded block pattern
CCK:	complementary code keying
CDMA:	code-division multiple access
CEPT:	European Conference of Postal and Telecommunications
CID:	connection identifier
CIP:	cellular IP
CIR:	committed information rate
CL:	convergence layer
CM:	centralized mode
CN:	core network
CN:	corresponding node
CoA:	core of address
CoP:	care-of-port
CORBA:	common object requests broker architecture
CP:	common part
CP:	content provider
CPDU:	control PDU
CPS:	common part sublayer
CQI:	channel quality indicator
CRC:	cyclic redundancy code
CS:	circuit switched
CS:	convergence sublayer

CSCF:	call state control function
CSCF:	call/session control function
CSI:	channel state information
CSMA/CA:	carrier sense multiple access with collision avoidance
CSMA/CD:	carrier sense multiple access with collision detection
CT:	context transfer
CTAP:	channel time allocation period
CTS:	clear-to-send
CW:	contention window
DAB:	digital audio broadcast
DAD:	duplicate address detection
DARPA:	Defense Advanced Research Projects Agency
DBA:	dynamic bandwidth allocation
DCC:	DLC connection control
DCCH:	dedicated control channel
DCF:	distributed coordinated function
DCS:	dynamic channel selection
DCT:	discrete cosine transform
DDCA:	distributed dynamic channel allocation
DECT:	digital enhanced cordless telecommunications
DHA:	dynamic home agent
DHCP:	dynamic host configuration protocol
DHT:	distributed hash table
DiffServ:	differentiated services
DIFS:	DCF interframe space
DL:	downlink
DLC:	data link control
DLFP:	DL frame prefix
DM:	direct link mode
DMA:	dynamic mobility agent
DoS:	denial of service
DP:	data partitioning
DPDCH:	dedicated physical data channel
DRC:	data rate control
DS:	direct sequence/distributed system
DSL:	digital subscriber line
DSSS:	direct sequence spread spectrum
DVB:	digital video broadcasting
EAP:	Extensible Authentication Protocol
EC:	error control
ECN:	explicit congestion notification
EDCA:	enhanced distributed channel access
E-DCH:	enhanced dedicated channel
EDGE:	enhanced data services for GSM evolution
EEP:	equal error protection
EF:	expedited forwarding

EGPRS:	enhanced GPRS
ELN:	explicit loss notification
EMIP:	enhanced mobile IP
EMS:	enhanced messaging service
EREC:	error resilience entropy coding
ERP:	enterprise resource planning
ESS:	enhanced service set
ETSI:	European Telecommunications Standards Institute
EUL:	enhanced uplink
EV-DO:	evolution data–optimized
FA:	foreign agent
FACH:	forward access channel
FBC:	flow-based charging
FBWA:	fixed broadband wireless access
FCC:	Federal Communications Commission
FCH:	frame channel
FCH:	frame control channel
FCH:	frame control header
FCS:	frame check sequence
FDD:	frequency division multiplex
FDMA:	frequency division multiple access
FEC:	forward error correction
FFT:	fast Fourier transform
FH:	frequency hopping
FHO:	fast handover
FHSS:	frequency hopping spread spectrum
FM:	frequency modulation
FMO:	flexible macroblock ordering
FMWC:	fixed mobile wireless convergence
FN:	foreign network
FPLMTS:	Future Public Land Mobile Telecommunications System
FT:	frequency time
FTP:	File Transfer Protocol
GCoA:	global care of address
GDR:	gradual decoding refresh
GEO:	geosynchronous satellites
GFA:	gateway foreign agent
GGSN:	gateway GPRS support node
GIS:	Geographical Information System
GMSC:	gateway mobile switching center
GPRS:	General Packet Radio System
GPS:	Global Positioning System
GSM:	Global System for Mobile
GTP:	GPRS Tunneling Protocol
GW:	gateway
HA:	home agent

H-ARQ:	hybrid automatic repeat request
HCCA:	HCF controlled channel access
HCF:	hybrid coordination function
HDR:	high data rate
HDTV:	high definition television
HEC:	header extension code
HIPERLAN:	high performance LAN
HiperMAN:	high performance metropolitan area network
HLR:	home location register
HMIP:	hierarchical MIP
HN:	home network
HRD:	hypothetical reference decoder
HS:	high-speed
HSCSD:	high-speed circuit-switched data
HSDPA:	high-speed downlink packet access
HS-DSCH:	high-speed downlink shared channel
HSS:	home subscription service
HTML:	Hypertext Mark-Up Language
HTTP:	Hypertext Transport Protocol
IBSS:	independent BSS
IC:	integrated circuit
ICANN:	Internet Corporation for Assigned Names and Numbers
IDMP:	intradomain mobility arrangement protocols
IDR:	instantaneous decoding refresh
IDS:	intrusion detection systems
IEC:	interactive error control
IEC:	International Electrotechnical Commission
IEEE:	Institute of Electrical and Electronics Engineers
IEEE-SA:	IEEE Standards Association
IETF:	Internet Engineering Task Force
IIS:	intelligent interface selection
IMS:	IP multimedia subsystem
IMT:	international mobile telecommunication
IP:	Internet Protocol
IPTV:	Internet television
IPv6:	Internet Protocol version 6
IR:	infrared
IrDA:	infrared data association
ISDN:	integrated services digital network
ISI:	intersymbol interference
ISM:	industrial, scientific, and medical
ISO:	International Organization for Standardization
ISP:	Internet service provider
ISUP:	ISDN signaling user part
ISUP:	ISDN user part
ITU:	International Telecommunication Union

ITU-R:	ITU–Radio-Communication Standardization Sector
ITU-T:	ITU–Telecommunication Standardization Sector
JSC:	joint source channel
JVT:	Joint Video Team
LAN:	local area network
LBS:	location-based services
LCH:	link control channel
LCoA:	local care of address
LDR:	low data rates
LEO:	low-earth orbit
LLC:	logical link control
LMDS:	local multipoint distribution service
LOS:	line of sight
LPD:	low probability of detection
LPI:	low probability of interception
LTE:	long-term evolution
LU:	location update
M2M:	machine-to-machine
MA:	mobile agent
MAC:	medium access control
MAL:	mobility abstraction layer
MAN:	metropolitan area network
MANET:	mobile ad hoc networking
MAP:	mobile anchor point
MB:	macroblock
MBMS:	multimedia broadcast multicast services
MBWA:	mobile broadband wireless access
MC:	multicarrier
MC-CDMA:	multicode CDMA
MDC:	multiple description coding
MEO:	medium-earth orbit
MGCF:	media gateway control function
MGCP:	Media Gateway Control Protocol
MGW:	media gateway
MIHO:	mobile initiated handover
MIMO:	multiple input multiple output
MIP:	mobile IP
ML:	maximum likelihood
MM:	measurement module
MMDS:	multichannel multipoint distribution service
MMS:	multimedia messaging service
MMSC:	MMS center
MMSE:	MMS environment
MMS-IOP:	MMS Interoperability Group
MN:	mobile node
MNO:	mobile network operator

MPEG:	Motion Picture Experts Group
MRF:	media resource function
MRFC:	multimedia resource function controller
MRFP:	multimedia resource function processor
MS:	mobile station
MSC:	mobile switching center
MSDU:	MAC service data unit
MSME:	MAC sublayer management entity
MSS:	multiple subscriber stations
MT:	mobile terminal
MTC:	mobile terminal controller
MTSO:	mobile telephone switching office
MVB:	multivehicle bus
m-WLAN:	moving wireless LAN
NACK:	negative acknowledgment
NAL:	network abstraction layer
NALU:	NAL unit
NAPT:	network address port translation
NAT:	network address translate
NDS:	network domain security
NEMO:	networks in motion
NGN:	next generation network
NIHO:	network initiated handover
NLOS:	non-line of sight
NMS:	Network Management System
NRT:	non-real time
NS:	network service
NVUP:	network view of the user profile
OAM:	operations and management
OAMandP:	operations, administration, maintenance, and provisioning
OBEX:	object exchange
OFDM:	orthogonal frequency division multiplexing
OMA:	Open Mobile Alliance
OS:	operating system
OSA:	open service access
OSI:	open systems interconnection
OSM:	Office of Spectral Management
P2P:	peer-to-peer
PA:	paging agent
PA:	performance attendant
PAN:	personal area network
PBX:	private branch exchange
PC:	personal computer
PCC:	policy and charging control
PCF:	point coordination function
PCRF:	policy and charging rules function

PCS:	personal communication system
P-CSCF:	proxy–call state control function
PDA:	personal digital assistant
PDC:	personal digital cellular
PDF:	policy decision function
PDM:	packet division multiplex
PDN:	packet data network
PDP:	Packet Data Protocol
PDU:	packet data unit
PDU:	protocol data unit
PEF:	policy enforcement function
PHS:	personal handyphone system
PHY:	physical (layer)
PIFS:	PCF interframe space
PLCP:	Physical Layer Convergence Protocol
PLMN:	public land mobile network
PLR:	packet loss rate
PM:	performance manager
PMD:	physical medium dependent
PMP:	point to multipoint
PNC:	piconet controller
PNM:	personal network management
PNNI:	private network-to-network interface
PoC:	push-to-talk over cellular
POS:	personal operating space
PPP:	Point-to-Point Protocol
PPS:	picture parameter sets
PS:	packet switched
PSC:	parameter set concept
PSD:	power spectral density
PSK:	phase shift keying
PSNR:	peak signal-to-noise ratio
PSTN:	public switched telephone network
PTT:	push-to-talk
QAL:	QoS abstraction layer
QAM:	quadrature amplitude modulation
QoS:	quality of service
QoSB:	QoS broker
QoSM:	QoS manager
QP:	quantization parameter
QPSK:	quadrature PSK
RAFC:	random access feedback channel
RAN:	radio access network
RBCH:	radio broadcast channel
RCH:	random channel
RCPC:	rate-compatible punctured convolutional (code)

RD:	Reed-Solomon (code)
RF:	radio frequency
RFC:	request for comments
RG:	radio gateway
RLC:	radio link controller
RNC:	radio network controller
RNS:	radio network subsystem
RP:	referent point
RP:	register and proxy
RR:	register and redirect
RRC:	radio resource control
RSs:	redundant slices
RSVP:	Resource Reservation Protocol
RT:	real time
RTCP:	Real-Time Control Protocol
RTG:	receive/transmit transition gap
RTP:	Real-Time Protocol
RTS:	request-to-send
RTSP:	Real-Time Streaming Protocol
RTT:	round-trip time
RVLC:	reversible variable length code
SA:	subnet agent
SACK:	selective acknowledgment
SAE:	system architecture evolution
SAMP:	Scalable Application Layer Mobility Protocol
SAP:	service access point
SBLP:	service-based local policy
SC:	security context
SC:	selective combining
SC:	single carrier
SCIM:	service capability interaction manager
SCO:	synchronous connection–oriented
SCP:	service control point
S-CSCF:	serving–call state control function
SDMA:	space–division multiple access
SDP:	Session Description Protocol
SDP:	service delivery platform
SDR:	software defined radio
SDU:	service data unit
SEI:	supplement enhancement information
SG:	study group
SGSN:	service GPRS support node
SGW:	signaling gateway
SINR:	signal-to-interference-plus-noise ratio
SIP:	Session Initiation Protocol
SLA:	service level agreement

SLF:	subscription local (locator) function
SMS:	short message service
SMTP:	Simple Mail Transfer Protocol
SOA:	Session Description Protocol
SOAP:	Simple Object Access Protocol
SONET:	synchronous optical network
SP:	service provider
SPN:	service provider network
SPS:	sequence parameter sets
SS:	spread spectrum
SS:	subscriber stations
SS7:	signaling system no. 7
SSCS:	service specific convergence sublayer
SSI:	source significance information
S-T:	space–time
STC:	short transport channel
STM:	synchronous transfer mode
TAG:	Technical Advisory Group
TBF:	temporary block flow
TBTT:	target beacon transmission time
TC:	time code
TCAP:	transaction capability application part
TCP:	Transmission Control Protocol
TD:	time division
TD CDMA:	time division CDMA
TDD:	time division duplex
TDM:	time division multiplex
TDMA:	time division multiple access
TE:	terminal
TFI:	temporary flow identity
TFRC:	TCP-friendly rate control
TG:	task group
TIA:	Telecommunication Industry Association
TIMIP:	terminal independent MIP
TISPAN:	telecoms and Internet converged services and protocols for advanced networks
TPC:	transmit power control
TTC:	Telecommunications Technology Council
TTG:	transmit/receive transition gap
TTI:	transmission time interval
UAC:	user agent client
UAS:	user agent server
UBCH:	user broadcast channel
UDCH:	user data channel
UDP:	User Datagram Protocol
UE:	user equipment

UEP:	unequal error protection
UMA:	unlicensed mobile access
UMCH:	user multicast channel
UMTS:	Universal Mobile Telecommunication System
UNII:	unlicensed national information infrastructure
UP:	uplink
UPDU:	user PDU
URI:	user resource identifier
USAP:	user service access point
USIM:	UMTS subscriber identity module
UT:	user terminal
UTRA:	UMTS terrestrial radio access
UTRAN:	UMTS radio access network
UWB:	ultrawideband
VAS:	value-added service
VCC:	voice call continuity
VCEG:	Video Coding Expert Group
VCL:	video coding layer
VFIR:	very fast infrared
VHO:	vertical handover
VLC:	variable-length code
VLC:	variable length coding
VLR:	visitor location register
VLSI:	very large scale integration
VMSC:	visitor mobile switching center
VoIP:	voice over IP
VPN:	virtual private network
W3C:	World Wide Web Consortium
WAN:	Wireless area net
WAP:	Wireless Applications Protocol
WARC:	World Administrative Radio Conference
WATM:	wireless ATM
WCDMA:	wideband CDMA
WDM:	wavelength division multiplex
WEP:	wired equivalent privacy
WG:	working group
WiFi:	wireless fidelity
WiMAX:	worldwide interoperability for microwave access
WISP:	wireless Internet service provider
WLAN:	wireless local area network
WLL:	wireless local loop
WMAN:	wireless MAN
WPAN:	wireless PAN
WS:	Web services
WSDL:	Web Services Description Language
WWAN:	wireless WAN

WWRF: Wireless World Research Forum
XML: Extensible Mark-up Language

Overview

Taking into account speculation on the future of wireless multimedia communications, in this introductory section, we start with a brief overview. It includes the evolution and vision of wireless communications together with technical challenges. Cellular systems in operation today are presented very briefly. The design details of these systems are constantly evolving with new systems emerging and old ones going by the wayside. Finally, wireless spectrum allocation is outlined.

EVOLUTION OF WIRELESS COMMUNICATIONS

Wireless communication is a fast-growing part of the dynamic field of electronic communication. The term wireless has come to mean nonbroadcast communication, usually between individuals who very often use portable or mobile equipment. Wireless mobile communication may not only complement the well-established wireline network, it may also become a serious competitor in years to come.

The first wireless systems developed in the preindustrial age transmitted information over line-of-sight distances. The early communication networks were replaced first by the telegraph network and later by the telephone. A few decades after the telephone was invented, the first radio transmission and radio communications were born. Radio technology advanced rapidly to enable transmission over large distances with better quality, less power, and smaller, cheaper devices, thereby enabling public and private radio communications, television, and wireless networking.

Early radio systems transmitted analog signals. Today, most radio systems transmit digital signals composed of binary bits, where the bits are obtained directly from a data signal or by digitizing an analog signal. A digital radio can transmit a continuous bit stream or it can group the bits into packets. The latter type of radio is called a packet radio and is characterized by burst transmission: the radio is idle except when it transmits a packet. The packet networks continue to be developed for military use. Packet radio networks also found commercial application in supporting wide-area wireless data services. These services, first introduced in the early 1990s, enable wireless data access (including e-mail, file transfer, and Web browsing) at fairly low speeds, on the order of 20 kbps. A strong market for these wide-area wireless data services never really materialized, due mainly to their low data rates, high cost, and lack of killer applications. These services mostly disappeared in the 1990s, supplanted by the wireless data capabilities of cellular telephones and wireless local area networks.[1]

The introduction of wired Ethernet technology in the 1970s steered many commercial companies away from radio-based networking. Ethernet 10 Mbps data rate far exceeded anything available using radio, and companies did not mind running cables within and between their facilities to take advantage of these high rates. In 1958, the FCC (Federal Communications Commission) enabled the commercial development of wireless local area networks (LANs) by authorizing the public use of the ISM (industrial, scientific, and medical) frequency bands for wireless LAN products. The ISM band was very attractive to wireless vendors because they did not need to obtain an FCC license to operate in this band. However, the wireless LAN systems could not interfere with the primary ISM band users, which forced them to use a low-power profile and an inefficient signaling scheme. Moreover, the interference primary users experienced within this frequency band was quite high. As a result these initial wireless LANs have very poor performance in terms of data rates and overage. This poor performance, coupled with concerns about security, lack of standardization, and high cost, resulted in weak sales. Few of these systems were actually used for data networking; they were relegated to low-tech applications like inventory control. The current generation of wireless LANs, based on the family of IEEE 802.11 standards, has better performance. Despite the data rate differences, wireless LANs are becoming the preferred Internet access method in many homes, offices, and campus environments due to their convenience and freedom from wires. However, most wireless LANs support applications such as e-mail and Web browsing that are not bandwidth intensive. The challenge for future wireless LANs will be to support many users simultaneously with bandwidth-intensive and delay-constrained applications such as video. Range extension is also a critical goal for future LAN systems.

Cellular systems exploit the fact that the power of a transmitted signal falls off with distance. Thus, two users can operate on the same frequency at spatially separate locations with minimal interference between them. This allows very efficient use of cellular spectrum, so that a large number of users can be accommodated.[2] The explosive growth of the cellular industry took almost everyone by surprise. Throughout the late 1980s, as more and more cities became saturated with demand for cellular service, the development of digital cellular technology for increased capacity and better performance became essential.

The second generation (2G) of cellular systems, first deployed in the early 1990s, was based on digital communications. The shift from analog to digital was driven by its higher capacity and the improved cost, speed, and power efficiency of digital hardware. Although 2G cellular systems initially provided mainly voice services, these systems gradually evolved to support data services such as e-mail, Internet access, and short messaging. Unfortunately, the great market potential for cellular phones led to a proliferation of 2G cellular standards: three different standards in the United States alone, and other standards in Europe and Japan, all incompatible. The fact that different cities have different, incompatible standards makes roaming throughout the United States and the world using one cellular phone standard impossible. Moreover, some countries have initiated service for third generation (3G) systems, for which there are also multiple incompatible standards. As a result of the standards proliferation, many cellular phones today are multimode: they incorporate multiple digital standards to facilitate nationwide and worldwide roaming, and possibly the first generation (1G) analog standard as well, because only this standard provides universal coverage throughout the United States.

Satellite systems are typically characterized by the height of satellite orbit, low-earth orbit (LEOs at roughly 2,000 km altitude), medium-earth orbit (MEOs at roughly 9,000 km altitude), or geosynchronous orbit (GEOs at roughly 40,000 km altitude).

The geosynchronous orbits are seen as stationary from the Earth, whereas the satellites with other orbits have their coverage area change over time. Geosynchronous satellites have large coverage areas, so fewer satellites are necessary to provide wide-area or global coverage. However, it takes a great deal of power to reach the satellite, and the propagation delay is typically too large for delay-constrained applications like voice. These disadvantages caused a shift in the 1990s toward lower-orbit satellites. The goal was to provide voice and data service competitive cellular systems. However, the satellite mobile terminals were much bigger, consumed much more power, and cost much more than contemporary cellular phones, which limited their appeal. The most compelling feature of these systems is their ubiquitous worldwide coverage, especially in remote areas or third-world countries with no landline or cellular system infrastructure. Unfortunately, such places do not typically have large demand or the resources to pay for satellite service. As cellular systems became more widespread, they took away most revenue that LEO systems might have generated in populated areas. With no real market left, most LEO satellite systems went out of business.[3,4]

A natural area for satellite systems is the broadcast environment. Direct broadcast satellites operate in the 12 GHz frequency band. These systems offer hundreds of TV channels and are major competitors to cable. Satellite-delivery digital radio has also become popular. These systems, operating in both Europe and the United States, offer digital audio broadcasting at near-CD quality.

VISION OF WIRELESS COMMUNICATIONS

The vision of wireless communications supporting information exchange between people or devices is the communications frontier of the next few decades, and much of it already exists in some form. This vision will allow multimedia communication from anywhere in the world using a small handheld device or laptop. Wireless networks will connect palmtop, laptop, and desktop computers anywhere within an office building or campus, as well as from the corner café. In the home, these networks will enable a new class of intelligent electronic devices that can interact with each other and with the Internet, in addition to providing connectivity between computers, phones, and security/monitoring systems. Such smart homes can also help elderly and disabled individuals with assisted living, patient monitoring, and emergency response. Wireless entertainment will permeate the home and any place that people congregate. Video teleconferencing will take place between buildings that are blocks or continents apart, and these conferences can include travelers as well.

Wireless video will enable remote classrooms, remote training facilities, and remote hospitals anywhere in the world. Wireless sensors have an enormous range of both commercial and military applications. Commercial applications include monitoring of fire hazards, hazardous waste sites, stress and strain in buildings and bridges, carbon dioxide movement, and the spread of chemicals and gases at a disaster site. The wireless sensors self-configure into a network to process and interpret sensors measurements, and then convey this information to a centralized control location. Military applications include identification and tracking of enemy targets, detection of chemical and biological attacks, support of unmanned robotic vehicles, and counterterrorism. Finally, wireless networks enable distributed control systems, with remote devices, sensors, and actuators linked together via wireless communication channels. Such networks enable automated highways, mobile robots, and easily reconfigurable industrial automation.

The described applications are the components of the wireless vision. A question often arises: What, exactly, does wireless communication represent? This complex topic could be segmented into different applications, systems, or coverage regions.

Wireless applications include voice, Internet access, Web browsing, paging and short messaging, subscriber information services, file transfer, video teleconferencing, entertainment, sensing, and distributed control. Systems include cellular telephone systems, wireless LANs, wide-area wireless data systems, satellite systems, and ad hoc wireless networks. Coverage regions include in-building, campus, city, regional, and global. The question of how best to characterize wireless communications along these various segments has resulted in considerable fragmentation in the industry, as evidenced by the many different wireless products, standards, and services offered or proposed. One reason for this fragmentation is that different wireless applications have different requirements. Voice systems have relatively low data rate requirements (around 20 kbps) and can tolerate a fairly high probability of bit error (bit error rates, or BERs, of around 10^{-3}), but the total delay must be less than around 30 ms or it becomes noticeable to the end user. On the other hand, data systems typically require much higher data rates (1 to 100 Mbps) and very small BERs (the target BER is 10^{-8} and all bits received in error must be retransmitted) but do not have a fixed delay requirement. Real-time video systems have high data rate requirements coupled with the same delay constraints as voice systems, whereas paging and short messaging have very different low data rate requirements and no delay constraints. These diverse requirements for different applications make it difficult to build one wireless system that can efficiently satisfy all these requirements simultaneously. Wired networks typically integrate the diverse requirements of different applications using a single protocol. This integration requires that the most stringent requirements for all applications be met simultaneously. Although this may be possible on some wired networks, with data rates on the order of gigabits per second and BERs on the order of 10^{-12}, it is not possible on wireless networks, which have much lower data rates and higher BERs. For these reasons, at least in the near future, wireless systems will continue to be fragmented, with different protocols tailored to support the requirements of different applications.

The exponential growth of cellular telephone use and wireless Internet access have led to great optimism about wireless technology in general. The 1G cellular wireless mobile systems were analog and were based on frequency-division multiplex (FDM) technology. Limited to the technologies of that time, most phones were large, placed in a briefcase-sized case, and permanently installed in a vehicle. Based on the number of vehicles that might need a phone and the number of people who could afford to pay, it was once projected by some that the cellular industry would see only limited growth. Indeed, the growth of cellular subscribers was moderate before the 1980s.

By the end of the 1980s, however, advances in semiconductor technologies provided a vital boost to the cellular mobile industry. Using application-specific integrated circuits (ASICs), the size of the phone shrank to a small handset. It turned out that this small technical evolution led to a major revolution for the cellular mobile industry for at least two reasons. First, the industry's consumer base was changed from the number of vehicles to the number of people, which is a much larger base. Second, the function of the phones was also changed from being able to call from a vehicle to being able to call from anywhere. This greatly increased people's desire to have a phone and therefore significantly increased the penetration rate.

The second boost for the cellular industry came from the introduction of the 2G digital technology standards, including GSM (Global System for Mobile Communications), IS-136 (time-division multiple access, TDMA), IS-95 (code-division multiple access, CDMA), and personal digital cellular (PDC). Digital technology has not only improved voice quality and services, but more importantly, has significantly reduced the cost of handset and infrastructure services, leading to further acceleration of the industry's growth.[5] Extension of the 2G system is introduced in 2.5G systems packet service enhancement.

Moving forward to the twenty-first century, further acceleration of growth was widely anticipated. 3G systems have been deployed,[6] and while 3G systems significantly improve the spectral efficiency and possibly the cost of the system, a more profound feature is the significant improvement of data and multimedia capabilities. Although this feature seems to be a mere evolution from a technical viewpoint, its potential lies in the promotion of communications not only from person to person, but also from person to machine and from machine to machine. Similar to the expansion from vehicles to people in 2G systems, this expansion is hoped to lead to a significant increase in the user base, as the number of machines can be an order of magnitude larger than the number of people.

In addition to the possible user base expansion, growth is further fueled by the significant increase of the user penetration rate. Increasingly, more functions are being built into the wireless mobile handset. One example is the combination of the phone with a personal digital assistant (PDA). This combination, with a user-friendly interface, can greatly improve convenience for a user over that provided by a wireline phone. When further combined with improved voice quality, reliability, multiple functionality, and reasonable air price, it can indeed become a serious competitor to the wireline phone system.[7]

Furthermore, the continuous success of mobile communication systems, as well as its consequences in terms of the need for better quality of service (QoS), more efficient systems, and more services, promises a transmission rate of up to 2 Mbps, which makes it possible to provide a wide range of multimedia services including videotelephony, paging, messaging, Internet access, and broadband data. However, it is expected that there will be a strong demand for multimedia applications which require higher data rates above 2 Mbps in cellular systems, especially in the downlink, where mobile users will enjoy high-speed Internet access and broadcast services. To offer such broadband packet data transmission services, a 3.5G wireless system has been introduced. This system is expected to achieve higher performance with a peak data rate of about 10 Mbps. High-speed packet transmission is possible by time-sharing a commonly used data channel among access users, called high-speed downlink shared channel (HS-DSCH). The 3.5G system relies on new technologies that make it possible to achieve such a high data rate. These new technologies include adaptive modulation and coding, hybrid automatic repeat request, fast cell selection, and fast packet scheduling. A packet scheduler controls the allocation of channels to users within the coverage area of the system by deciding which user should transmit during a given time interval. Therefore, to a large extent, it determines the overall behavior of the system.[8]

Fourth-generation (4G) beyond 3G wireless networks promises much higher overall data throughput and many more diverse services than current networks.[6] All-IP (Internet Protocol) wireless has emerged as the most preferred platform for 4G wireless networks. In such networks different access systems are integrated on an all-IP-based network, including network and internetworking of different systems with the backbone. The design of a future wireless networking architecture has to take into account the fact that the dominant load in 4G wireless networks will be high-speed, content-rich, burst-type traffic, which already poses a great challenge to all existing wireless networking technologies deployed in current networks. Many research activities have been carried out to design suitable architecture for 4G wireless networks, which should be efficient, adaptive, flexible, and scalable, and should also be able to work harmoniously with different network technologies (including currently deployed networks) and accommodate heterogeneous networking applications. Additional research activities are required to investigate how to realize smooth migration and seamless interoperability between the legacy networks and future 4G wireless networks. The architecture of 4G wireless networks should efficiently address the constraints and problems existing in the currently deployed wireless networking platforms, such as rigid network structure, low overall bandwidth efficiency, strictly interference-limited capacity, difficulty performing rate-matching algorithms, lack of flexibility to implement cross-layer network design, and so forth. The research on the next generation wireless networking involves many cutting-edge research topics, such as cross-layer joint optimization design, quality of service assurance, dynamic network resource allocation, ad hoc or mesh network routing algorithms, heterogeneous networking, cooperative network detection, vertical/horizontal network service integration, and so on.

ELEMENTS OF COMMUNICATION SYSTEMS

The most basic possible wireless system consists of a transmitter, a receiver, and a channel, usually a radio link, as shown in Figure 0.1. Because radio cannot be used directly with low frequencies such as those in a human voice, it is necessary to superimpose the information content onto a higher frequency carrier signal at the

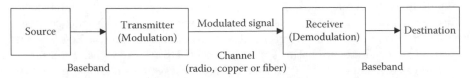

FIGURE 0.1 Elements of a single communication system.

transmitter, using a process called modulation. The use of modulation also allows more than one information signal to use the radio channel by simply using a different carrier frequency for each. The inverse process, demodulation, is performed at the receiver in order to recover the original information.

The information signal is also sometimes called the intelligence, the modulating signal, or the baseband signal. An ideal communication system would reproduce the information signal exactly at the receiver, except for the inevitable time delay as it travels between transmitter and receiver, and except, possibly, for a change in amplitude. Any other changes constitute distortion. Any real system will have some distortion, of course; part of the design process is to decide how much distortion, and of what type, is acceptable.

Figure 0.1 represents a simplex communication system. The communication is one way only, from transmitter to receiver. Broadcasting systems are like this, except that there are many receivers for each transmitter.

Most of the systems we discuss in this book involve two-way communication. Sometimes communication can take place in both directions at once. This is called full-duplex communication. An ordinary telephone call is an example of full-duplex communication. It is quite possible (though perhaps not desirable) for both parties to talk at once, with each hearing the other. Figure 0.2 shows full-duplex communication. Note that it simply doubles the previous figure—we need two transmitters, two receivers, and, usually, two channels.

Some two-way communication systems do not require simultaneous communication in both directions. An example of this half-duplex type of communication is a conversation over citizens' band (CB) radio. The operator pushes a button to talk and releases it to listen. It is not possible to talk and listen at the same time, as the receiver is disabled while the transmitter is activated. Half-duplex systems save bandwidth by allowing the same channel to be used for communication in both directions. They can sometimes save money as well by allowing some circuit components in the transceiver to be used for both transmitting and receiving. They do sacrifice some of the naturalness of full-duplex communication, however. Figure 0.3 shows a half-duplex communication system.

The full- and half-duplex communication systems shown so far involve communication between only two users. Again, CB radio is a good example of this. When

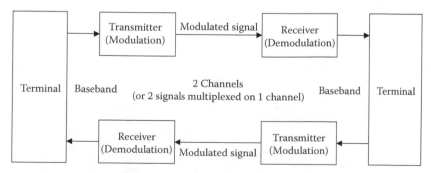

FIGURE 0.2 Full-duplex communication system.

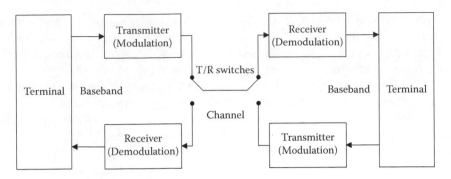

FIGURE 0.3   Half-duplex communication system.

there are more than two simultaneous users, or when the two users are too far from each other for direct communication, some kind of network is required. Networks can take many forms, and several are examined in this book. Probably the most common basic structure in wireless communication is the classic star network, shown in Figure 0.4.

The central hub in a radio network is likely to be a repeater, which consists of a transmitter and receiver, with their associated antennas, located in a good position from which to relay transmissions from and to mobile radio equipment. The repeater may also be connected to wired telephone or data networks. The cellular and personal communication system (PCS) telephone systems have an elaborate network of repeater stations.

The architect of any wireless communications system must overcome three unique and fundamental obstacles that do not affect wireline systems:

- The physical channel is completely nonengineerable. The wireless communication system (transmitters and receivers) must be designed around the natural or *given* characteristics of the radio channel.

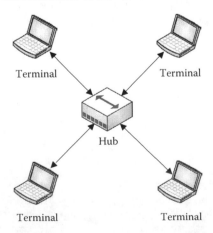

FIGURE 0.4   Star wireless network.

- The channel is, so to speak, nonclonable. In a wireline network, the growth in traffic between point A and point B can be addressed, at a cost, by adding new identical transmission facilities between those nodes: more fiber, more coax, more copper. In a wireless network, this option does not exist. The total available physical bandwidth between points A and B is finite.
- The wireless system architect must allow for the fact that the signal will experience significant destructive interactions with other signals (including images of itself) during transmission. These effects are produced, in part, by the physical channel itself.

TECHNICAL CHALLENGES

Many technical challenges must be addressed to enable the wireless applications. These challenges extend across all aspects of the system design. Computers process voice, image, text, and video data, but breakthroughs in circuit design are required to implement the same multichip, lightweight, handheld device. Because consumers do not want large batteries that frequently need recharging, transmission and signal processing in the portable terminal must consume minimal power. The signal processing required to support multimedia applications and networking functions can be power-intensive infrastructure-based networks, such as wireless LANs and cellular systems, which place as much of the processing burden as possible on fixed sites with large power resources. The associated bottlenecks and single points-of-failure are clearly undesirable for the overall system.

Energy is a particularly critical resource in networks where nodes cannot recharge their batteries sensing applications. Network design to meet the application requirements remains a big technological hurdle. The finite bandwidth and random variations of wireless channels also require robust applications that degrade gracefully as network performance degrades.

As the signal propagates through a wireless channel, it experiences random fluctuations in time if the object's transmitter surroundings are moving, due to changing reflections and attenuation. Thus, the channels appear to change randomly with time, which makes it difficult to design reliable system performance. Security is also more difficult to implement in wireless systems. The analog cellular systems have no security, and one can easily overhear conversations by scanning the analog cellular frequency band. All digital cellular systems implement some level of encryption. However, with enough knowledge, time, and determination, most methods can be cracked and, indeed, several have been compromised. To support applications like electronic commerce and credit card transactions, the wireless network must be secure against such listeners.

Wireless networking is also a significant challenge. The network must be able to locate a given user wherever that user is among billions of globally distributed mobile terminals. It must then route a call to that user as it moves. The finite resources of the network must be allocated in a fair and efficient manner based on user demands and ranges of locations. Moreover, there currently exists a tremendous infrastructure of wired networks—the telephone system, the Internet, and fiber-optic cable—which should be used to connect wether into a global network. Interfacing between wireless and wired networks is a difficult problem.

The layers in a wireless system include the link or physical layer, which handles bit transmissions over the communications medium; the access layer, which handles shared access to the communications medium; the transport layers, which route data across the network and ensure end-to-end connectivity; and the application layer, which dictates the end-to-end data rates and delay constraints associated with the application. Although a layering methodology reduces complexity and facilitates modularity and standardization, it leads to inefficiency and performance loss due to the lack of a global design optimization.

Wireless links can exhibit very poor performance along with user connectivity and network topology changes over time. In fact, the very notion of a wireless link is somewhat fuzzy due to the nature of radio propagation and broadcasting. The dynamic networks must be optimized for this channel and must be robust and adaptive to its variations, as well as to network dynamics. Thus, these networks require integrated and adaptive protocols at all layers, from the link layer to the application layer. This cross-layer protocol design requires interdisciplinary expertise in communications, signal processing, and network theory and design.

WIRELESS SYSTEMS IN OPERATION TODAY

The design details of current wireless systems are constantly evolving with new systems emerging and old ones going by the wayside. We focus mainly on the high-level design aspects of the most common systems.

Cellular systems provide two-way voice and data communication with regional, national, or international coverage. Cellular systems were initially designed for mobile terminals inside vehicles with an antenna on the vehicle roof. Today these systems have evolved to support lightweight handheld mobile terminals operating inside and outside buildings at both pedestrian and vehicle speeds.

The basic premise behind cellular system design is frequency reuse, which exploits the fact that signal power falls off with distance to reuse the same frequency spectrum at spatially separated locations. Specifically, the coverage area of a cellular system is divided into nonoverlapping cells where some set of channels is assigned to each cell. This same channel set is used in another cell some distance away, as shown in Figure 0.5, where C_i denotes the channel set used in a particular cell. Operation within a cell is controlled by a centralized base station. The interference caused by users in different cells operating on the same channel set is called intercell interference. The spatial separation of cells that reuse the same channel set, the reuse distance, should be as small as possible so that frequencies are reused as often as possible, thereby maximizing spectral efficiency. However, as the reuse distance decreases, intercell interference increases, due to the smaller propagation distance between interfering cells. Since intercell interference must remain below a given threshold for acceptable system performance, reuse distance cannot be reduced below some minimum value. In practice, it is quite difficult to determine this minimum value because both the transmitting and interfering signals experience random power variations due to the characteristics of wireless signal propagation. To

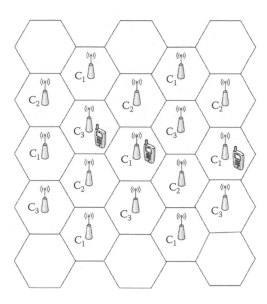

FIGURE 0.5 Cellular systems.

determine the base station placement, an accurate characterization of signal propagation within the cells is needed.

Cellular systems in urban areas mostly use smaller cells with base stations close to street level transmitting at much lower power. These smaller cells are called microcells or picocells, depending on their size. Smaller cells occur for two reasons: the need for higher capacity in areas with high user density size, and the cost of base station electronics. A cell of any size can support roughly the same number of users if the system is scaled accordingly. Thus, for a given coverage area a system with many microcells has a higher number of users per unit area than a system with just a few macrocells. In addition, less power is required at the mobile terminals in microcellular systems because the terminals are closer to the base stations. However, evolution to smaller cells has complicated network design. Mobiles traverse a small cell more quickly than a large cell, and therefore handoffs must be processed more quickly. It is also harder to develop general propagation models for small cells, because signal propagation in these cells is highly dependent on base station placement and the geometry of the surrounding reflectors.[9]

All base stations in a given geographical area are connected via a high-speed communications link to a mobile telephone switching office (MTSO), as shown in Figure 0.6. The MTSO acts as a central controller for the network, allocating channels within each cell, coordinating handoffs between cells when a mobile traverses a cell boundary, and routing calls to and from mobile users. The MTSO can route voice calls through the public switched telephone network (PSTN) or provide Internet access. A new user located in a given cell requests a channel by sending a cell's base station over a separate control channel. The request is relayed to the MTSO, which accepts if a channel is available in that cell. If no channels are available, then the call request is rejected. A call handoff is initiated when the base station

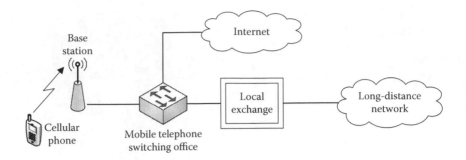

FIGURE 0.6   One example of cellular network architecture.

or the mobile in a given cell detects that the received signal power for that call is approaching a given minimum threshold. In this case the base station informs the MTSO that the mobile requires a handoff, and the MTSO then queries surrounding base stations to determine if one of these stations can detect that mobile signal. If so, then the MTSO coordinates a handoff between the base stations. If no channels are available in the cell with the new base station, then the handoff fails and the call is terminated. A call will also be dropped if the signal strength between a mobile and its base station drops below the minimum threshold needed for communication due to random signal variations.

1G cellular systems used analog communications, primarily because these systems were designed in the 1960s, before digital communications became prevalent. 2G systems moved from analog to digital due to the many advantages of digital. The components are cheaper, faster, smaller, and require less power. Voice quality is improved due to error correction coding. Digital systems also have higher capacity than analog systems because they can use more spectrally efficient digital modulation and more efficient techniques to share the cellular spectrum. They can also take advantage of advanced compression techniques and voice activity factors. In addition, encryption techniques can be used to secure digital signals against eavesdropping.

Spectral sharing in communication systems, also called multiple access, is done by dividing the signaling dimensions along the time, frequency, and/or code space axes. In frequency-division multiple access (FDMA), the total system bandwidth is divided into orthogonal frequency channels. In time-division multiple access (TDMA), time is divided orthogonally and each channel occupies the entire frequency band over its assigned time slot. TDMA is more difficult to implement than FDMA because the users must be time synchronized. However, it is easier to accommodate multiple data rates with TDMA because multiple time slots can be assigned to a given user. Code-division multiple access (CDMA) is typically implemented using direct-sequence or frequency-hopping spread spectrum with either orthogonal or nonorthogonal codes. In direct sequence each user modulates its data sequence by a different sequence, which is much faster than the data sequence. In the frequency domain, the narrowband data signal is convolved with the wideband signal, resulting in a signal with a much wider bandwidth than the original data signal. In frequency hopping the carrier frequency is used to modulate the narrowband data. This results in a modulated signal that hops over different carrier frequencies. Typically, spread spectrum signals are superimposed onto each of the distinct signals by separately decoding each spreading sequence. However, for nonorthogonal codes users within a cell interfere with each other (intracell interference) and codes that are reused in other cells cause intercell interference. Both the intracell and intercell interference power is reduced by the spreading gain of the code. Moreover, interference in spread spectrum systems can be further reduced through multiuser detection and interference cancellation.

Efficient cellular system designs are interference limited; that is, the interference dominates the noise floor because otherwise more users could be added to the system. As a result, any technique to reduce interference in cellular systems leads to an increase in system capacity and performance. Some methods for interference reduction in use today or proposed for future systems include cell sectorization, directional and smart antennas, multiuser detection, and dynamic resource allocation.

The 1G cellular systems in the United States, called the AMPS (Advanced Mobile Phone Service), used FDMA with 30 kHz FM-modulated voice channels. Many of the 1G cellular systems in Europe were incompatible, and the Europeans quickly converged on a uniform standard for 2G digital systems called GSM. The GSM standard uses a combination of TDMA and slow frequency hopping with frequency-shift keying for the voice modulation. In contrast, the standard activities in the United States surrounding 2G digital cellular provoked a raging debate on spectrum-sharing techniques, resulting in several incompatible standards. In particular, there are two standards in the 900 MHz cellular frequency band: IS-54, which uses a combination of TDMA and shift-keyed modulation; and IS-95, which uses direct-sequence CDMA with binary modulation and coding.[10,11]

All 2G digital cellular standards have been enhanced to support high-rate packet data systems and provide data rates of up to 100 kbps by aggregating all time slots together for a single user. This enhancement is called GPRS. A more fundamental enhancement, Enhanced Data Services for GSM Evolution (EDGE), further increases data rates using a high-level modulation format combined with forward error correction (FEC) coding. This modulation is more sensitive to fading effects, and EDGE uses adaptive techniques to mitigate this problem.

The received signal-to-noise ratio (SNR) is measured at the receiver and fed back to the transmitter, and the best modulation and coding combination for this SNR value is used. The IS-54 and IS-136 systems provide data rates of 40 to 60 kbps by aggregating time slots and using high-level modulation. This evolution of the IS-136 standard is called IS-136H (high-speed). The IS-95 systems support higher data rates using a time-division technique called high data rate (HDR).

The 3G cellular systems are based on a wideband CDMA standard developed under the auspices of the International Telecommunications Union (ITU). The standard, initially called International Mobile Telecommunications 2000 (IMT-2000), provides different data rates depending on mobility and location, from 384 kbps for pedestrian use, to 144 kbps for vehicular use, to 2 Mbps for indoor office use. The 3G standard is incompatible with 2G systems, so service providers must invest in a new infrastructure before they can provide 3G service. The first 3G systems were deployed in Japan. One reason that 3G services emerged first in Japan is the process of 3G spectrum allocation, which in Japan was awarded without much upfront cost. The 3G spectrum in both Europe and the United States is allocated based on auctioning, thereby requiring a huge initial investment for any company wishing to provide 3G service.

Motivated by the ever-increasing demand for wireless communications, the cellular network has evolved to the 3G, for example, the UMTS (Universal Mobile Telecommunication System), which is specified by the 3G Partnership Project (3GPP) and is one of the most popular 3G systems nowadays. The 3G cellular network is capable of supporting QoS critical to multimedia services, but at the expense of high complexity and implementation costs. For example, four service classes are supported in UMTS: conversational, streaming, interactive, and background services. However, the expensive radio spectrum for 3G cellular networks prohibits rapid deployment, and the low bandwidth restricts system capacity. The future 4G wireless networks need to address these existing constraints and problems effectively. Heterogeneous networking is a promising approach to accelerate the technological evolution toward 4G wireless networks. In recent years, IEEE 802.11 wireless LANs have proliferated due to a high performance-to-cost ratio. Usually, operating at license-free frequency bands, wireless LANs can occupy a much wider spectrum than the cellular system and provide data services using a simple medium access control (MAC) protocol. The complementary characteristics of the 3G cellular network and wireless LANs promote their interworking. Future mobile devices can be equipped with network interfaces to both the 3G network and wireless LANs at a reasonable price. The dual-mode mobile devices can then enjoy services in cellular wireless LAN integrated networks.[13]

4G, or beyond 3G, wireless networks promise much higher overall data throughput and much more diverse services than do current networks. All-IP wireless has emerged as the most preferred platform for 4G wireless networks. In such networks different access systems are integrated on an all-IP-based network, including interworking of different systems with the backbone. The design of a future wireless networking architecture has to take into account the fact that the dominant load in 4G wireless networks will be high-speed, content-rich, burst-type traffic, which already poses a great challenge to all existing wireless networking technologies deployed in current networks. Many research activities have been carried out to design suitable architectures for 4G wireless networks, which should also be able to work harmoniously with different network technologies (including currently deployed networks) and accommodate heterogeneous networking applications. The architecture of 4G wireless networks should effectively address the constraints and problems existing in the currently deployed wireless networking platforms, such as low overall bandwidth efficiency, strictly interference-limited capacity, difficulty performing rate-matching algorithms, lack of flexibility to implement cross-layer network design, and so forth.

WIRELESS SPECTRUM ALLOCATION

Most countries have government agencies responsible for allocating and controlling the use of the radio spectrum. In the United States, spectrum is allocated by the FCC for commercial use and by the Office of Spectral Management (OSM) for military use. Commercial spectral allocation is governed in Europe by the

European Telecommunications Standards Institute (ETSI) and globally by the ITU. Governments decide how much spectrum to allocate between commercial and military use, and this decision is dynamic depending on need. Historically, the FCC allocated spectral blocks for specific uses and assigned licenses to use these blocks to specific groups or companies. For example, in the 1980s the FCC allocated frequencies in the 800 MHz band for analog cellular phone service, and provided spectral licenses to two operators in each geographical area based on a number of criteria. Although the FCC and regulatory bodies in other countries still allocate spectral blocks for specific purposes, these blocks are now commonly assigned through spectral auctions to the highest bidder. Although some argue that this market-based method is the fairest way for governments to allocate the limited spectral resource, in addition to providing significant revenue to the government, there are others who believe that this mechanism stifles innovation, limits competition, and hurts technology adoption. Specifically, the high cost of spectrum dictates that only large companies or conglomerates can purchase it. Moreover, the large investment required to obtain spectrum can delay the ability to invest in infrastructure for system rollout and results in very high initial prices for the end user. The 3G spectral auctions in Europe, following which several companies ultimately defaulted, have provided fuel to the fire against spectral auctions.

In addition to spectral auctions, spectrum can be set aside in specific frequency bands that are free to use with a license according to a specific set of rules. The rules may correspond to specific communications standards, power levels, and so on. The purpose of these unlicensed bands is to encourage innovation and low-cost implementation. Many extremely successful wireless systems operate in unlicensed bands, including wireless LANs, Bluetooth, and cordless phones. A major difficulty of unlicensed bands is that they can be killed by their own success. If many unlicensed devices in the same band are used in close proximity, they generate much interference to each other, which can make the band unusable.

Underlay systems are another alternative to allocate spectrum. An underlay system operates as a secondary user in a frequency band with other primary users. Operation of secondary users is typically restricted so that primary users experience minimal interference. This is usually accomplished by restricting the power/hertz of the secondary users. Ultrawideband (UWB) is an example of an underlay system, as are unlicensed systems in the ISM frequency bands. The trend toward spectrum allocation for underlays appears to be accelerating, mainly due to the lack of available spectrum for new systems and applications.

Satellite systems cover large areas spanning many countries and sometimes the globe. For wireless systems that span multiple countries, spectrum is allocated by the ITU-R. The standards arm of this body, ITU-T, adopts telecommunications standards for global systems that must interoperate with each other across national boundaries.

Most wireless applications reside in the radio spectrum between 30 MHz and 30 GHz. These frequencies are natural for wireless systems because they are not affected by the Earth's curvature, require only moderate-sized antennas, and can penetrate the ionosphere. Note that the required antenna size for good reception is inversely proportional to the square of signal frequency, so moving systems to a higher frequency allows for more compact antennas. However, received signal power

TABLE 0.1

Licensed Spectrum Allocated to Wireless Systems in the United States

AM radio	535–1,605 kHz
FM radio	88–108 MHz
Broadcast TV (Channels 2–6)	54–88 MHz
Broadcast TV (Channels 7–13)	174–216 MHz
Broadcast TV (UHF)	470–806 MHz
3G broadband wireless	746–764 MHz, 776–794 MHz
3G broadband wireless	1.7–1.85 MHz, 2.5–2.69 MHz
1G and 2G digital cellular phones	806–902 MHz
Personal communications service (2G cell phones)	1.85–1.99 GHz
Wireless communications service	2.305–2.32 GHz, 2.345–2.36 GHz
Satellite digital radio	2.32–2.325 GHz
Multichannel multipoint distribution service (MMDS)	2.15–2.68 GHz
Digital broadcast satellite (satellite TV)	12.2–12.7 GHz
Local multipoint distribution service (LMDS)	27.5–29.5 GHz, 31–31.3 GHz
Fixed wireless services	38.6–40 GHz

with nondirectional antennas is proportional to the inverse of frequency squared, so it is harder to cover large distances with higher frequency signals.

Spectrum is allocated either in licensed bands, which regulatory bodies assign to specific operators, or in unlicensed bands, which can be used by any system subject to certain operational requirements. Table 0.1 shows the licensed spectrum allocated to major commercial wireless systems in the United States. There are similar allocations in Europe and Asia.

Note that digital TV is slated for the same bands as broadcast TV, so all broadcasters must eventually switch from analog to digital transmission. Also, the 3G broadband wireless spectrum is currently allocated to UHF TV stations 60 to 69 but is slated to be reallocated. Both 1G analog and 2G digital cellular services occupy the same cellular band at 800 MHz, and the cellular service providers decide how much of the band to allocate between digital and analog services. Table 0.2 shows the unlicensed spectrum allocations in the United States.

TABLE 0.2

Unlicensed Spectrum Allocations in the United States

ISM Band I (cordless phones, 1G wireless LANs)	902–928 MHz
ISM Band II (Bluetooth, 802.1 lb wireless LANs)	2.4–2.4835 GHz
ISM Band III (wireless PBX)	5.725–5.85 GHz
NII Band I (indoor systems, 802.1 la wireless LANs)	5.15–5.25 GHz
NII Band II (short outdoor and campus applications)	5.25–5.35 GHz
NII Band III (long outdoor and point-to-point links)	5.725–5.825 GHz

ISM Band I has licensed users transmitting at high power that interferes with the unlicensed users. Therefore, the requirements for unlicensed use of this band are highly restrictive and performance is somewhat poor. The U-NII bands have a total of 300 MHz of spectrum in three separate 100 MHz bands, with slightly different restrictions on each band. Many unlicensed systems operate in these bands.

1 Introduction to Wireless Networking

Wireless communications are expected to be the dominant mode of access technology in this century. In addition to voice, a new range of services such as audio, image, video, high speed data, and multimedia are being offered for delivery over wireless networks. Mobility will be seamless, realizing the concept of persons in contact anywhere, at any time. Wireless communications networks are experiencing phenomenal growth in various parts of the world.[1] Cellular services have evolved from sparse coverage and heavy terminal equipment—mainly for vehicles—to almost ubiquitous coverage with very small terminals affordable to the customer. Wireless systems are poised to play a continuing significant role in meeting the telecommunications needs of the twenty-first century. According to some forecasts, the number of mobile phone subscribers by 2010 may reach 1 billion and surpass fixed phone lines. Thus, wireless communication is likely to become the dominant mode of access technology. Also, as we consider the current and future needs of wireless, it is apparent that a new generation of wireless technologies is needed that enables the convergence of many different wireless technologies across various existing networks and is flexible enough to handle a wide range of customer demands. One example is the fourth generation (4G) systems.

Wireless systems are poised to play a continuing significant role in meeting the telecommunications needs. There is a desire to communicate simultaneously using speech, video, and data. The speed of communication is also important.

Wireless communications are expected to be the dominant mode of access technology.[2] The rapid growth of interactive multimedia applications, such as video telephones, video games, and TV broadcasting, has resulted in spectacular studies in the progress of wireless communication systems. The current third generation (3G) wireless systems and the next generation (4G) wireless systems support high bit rates.

However, the high error rates and stringent delay constraints in wireless systems are still significant obstacles for these applications and services. On the other hand, the development of more advanced wireless systems provides opportunities for proposing novel wireless multimedia protocols, applications, and services that can take the maximum advantage of these systems.

The impressive evolution of mobile networks and the potential of wireless multimedia communications pose many questions to operators, manufacturers, and scientists working in the field. The future scenario is open to several alternatives. Thoughts, proposals, and activities of the near future could provide the answer to the open points and dictate the future trends of the wireless world.

The perspective of today's information society calls for a multiplicity of devices, including Internet Protocol (IP)–enabled home appliances, vehicles, personal

2 Wireless Multimedia Communications

computers, sensors, and actuators, all of which are globally connected. Current mobile and wireless systems and architectural concepts must evolve to cope with these complex connectivity requirements. Scientific research in this truly multidisciplinary field is growing fast. New technologies, new architectural concepts, and new challenges are emerging.[3–7] A broader band knowledge spanning different layers of the protocol stack is required by experts involved in the research, design, and development aspects of future wireless networks.

Network design using the layered open systems interconnection (OSI) architecture has been a satisfactory approach for wired networks, especially as the communication links evolved to provide gigabit-per-second rates and bit error rates (BER) of 10^{-12}. Wireless channels typically have much lower data rates (on the order of few megabits per second), higher BERs (10^{-2} to 10^{-6}), and exhibit sporadic error bursts and intermittent connectivity. These performance characteristics change a network topology, and user traffic also varies over time. Consequently, good end-to-end wireless network performance will not be possible without a truly optimized, integrated, and adaptive network design. Each level in the protocol stack should adapt to wireless link variation strategies at the other layers in order to optimize network performance.

1.1   EVOLUTION OF MOBILE NETWORKS

The passage from generation to generation is not only characterized by an increase in the data rate, but also by the transition from pure circuit-switched (CS) systems to CS voice/packed data and IP core-based systems. Evolution of cellular communications from second generation (2G) to 3G is presented in Figure 1.1. 2G systems are a milestone in the mobile world. The evolution from the first generation (1G) of analog systems meant the passage to a new system, while maintaining the same offered service: voice. The success of 2G systems, which extend the traditional PSTN (public switched telephone network) or ISDN (integrated services digital network) and allow

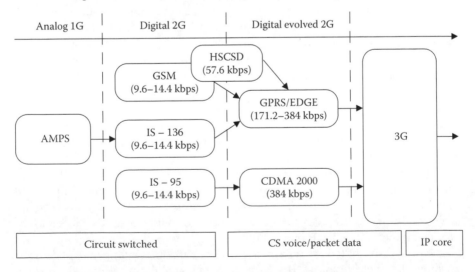

FIGURE 1.1   Evolution of cellular communications from 2G to 3G.

for nationwide or even worldwide seamless roaming with the same mobile phone, has been enormous.

Today's most successful digital mobile cellular system is GSM (Global System for Mobile Communications).[8,9] GSM is digital system in Europe. In Japan the PDC (personal digital cellular) system is operated. In the United States the digital market is divided into several systems, TDMA (time division multiple access)-based and GSM systems. This fragmentation has led to severe problems regarding coverage and service availability. Some mobile subscribers in United States and Canada still use analog AMPS (advanced mobile phone services) systems.[10,11] 2G mobile systems are still mainly used for voice traffic. The basic versions typically implement a circuit-switched service focused on voice, and only offer low data rates (9.6 to 14.4 kbps). Transitional data technologies between 2G and 3G have been proposed to achieve faster data rates sooner and at a lower cost than 3G systems. The evolved systems are characterized by higher data rates (64 to 384 kbps) and packet data mode.

HIGH-SPEED CIRCUIT-SWITCHED DATA

HSCSD (high-speed circuit-switched data) comes from the need to solve problems related to the slowness of GSM in data transmission. In fact, GSM supports data transmissions with data rates up to 9.6 to 14.4 kbps in circuit-switched mode, and the transfer on signaling channels of small size packets (up to 160 characters).

HSCSD was proposed by ETSI (European Telecommunications Standards Institute) in early 1997. The key idea is to exploit more than one time slot in parallel among the eight time slots available with a proportional increment in the data rates.[10] For example, HSCSD allows the user to access a company LAN (local area network), send and receive e-mail, and access the Internet on the move. On the other hand, HSCSD service does not effectively take advantage of the bursty nature of the traffic (e.g., Web browsing, e-mail). Channels are reserved during the connection. Furthermore, the exploitation of more time slots per user in a circuit-switched mode leads to a drastic reduction of channels available for voice users. For instance, four HSCSD users, each with four time slots assigned, prevent 16 voice users from accessing the network. Therefore, there is a need for packet-switched mode to provide more efficient radio resource exploitation when bursty traffic sources are involved.

I-MODE SERVICES

A great success in Japan has been obtained by the i-mode services, introduced in 1999, which are provided by the packet-switched communication mode of the PDC system.[11] Hence, the i-mode represents a transitional step of PDC toward 3G. The i-mode service utilizes a compact HTML (Hypertext Mark-Up Language) protocol, thus easing the interface to the Internet. Subscribers can send/receive e-mail and access a large variety of transitions, entertainment, and database-related services, as well as browsing Web sites and home pages. i-Mode is user friendly and all instructions can be managed by only 10 keys.

GPRS AND EDGE FOR GSM EVOLUTION

GPRS (general packet radio service) and EDGE (enhanced data rates for GSM evolution) have been introduced as transitional data technologies for the evolution of GSM (Figure 1.1). GPRS is the packet mode extension to GSM, supporting data applications and exploiting the already existing network infrastructure in order to save the operator's investment.

GPRS needs a mode adaptation at the radio interface level of GSM hardware. However, it adopts a new physical channel and mapping into physical resources.[11] The new physical channel is called 52-multiframe and is composed of two 2G control multiframes of voice-mode GSM. High data rates can be provided since the GPRS users can exploit more than one time slot in parallel with the possibility, contrary to the HSCSD technology which varies the number of time slots assigned to a user. The maximum theoretical bit rate of the GPRS is 171.2 kbps (using eight time slots). Current peak values are 20/30 kbps. The 52-multiframe is logically divided into 12 radio blocks of four consecutive frames, where a radio block (20 ms) represents the minimum time resource assigned to a user. If the user is transmitting or receiving big flows of data, more than one radio block can be allocated to it. The whole set of these blocks received/transmitted by a mobile terminal during a reception/transmission phase forms the temporary block flow (TBF), which is maintained only for the duration of the data transfer. The network assigns each TBF a temporary flow identify (TFI) which is unique in both directions. Contrary to the GSM, GPRS service can flexibly handle asymmetric services by allocating a different number of time slots in uplink and downlink. Time slots can be allocated in two ways:

- On demand, where the time slots not used by voice calls are allocated, and in case of resource security for voice calls (congestion), time slots already assigned to GPRS service can be deallocated.
- State in which some time slots are allocated for GPRS and they cannot be exploited by voice calls.

Another new aspect of the GPRS with respect to GSM is the possibility of specifying a quality of service (QoS) profile. This profile determines the service priority (high, normal, low), reliability, and delay class of the transmission, and user data throughput.[12,13] The radio link protocol provides a reliable link, while MAC (multiple access control) protocols control access with signaling procedures for the radio channel and the mapping of LLC (link layer control) frames into the GSM physical channels. Concerning the fixed backbone, the GPRS introduces two new network elements: service GPRS support node (SGSN) and gateway GPRS support node (GGSN). The architecture reference model is shown in Figure 1.2.[14]

The SGSN represents for the packet world what the mobile switching center (MSC) represents for the circuit world. The SGSN performs mobility management (e.g., routing area update, attach/detach process, mobile station, MS paging) as well as security tasks (ciphering of user data, authentication). GGSN tasks are comparable to those of a gateway MSC. It is not connected directly to the access network, but provides a means to connect SGSNs to other nodes of external packet data networks

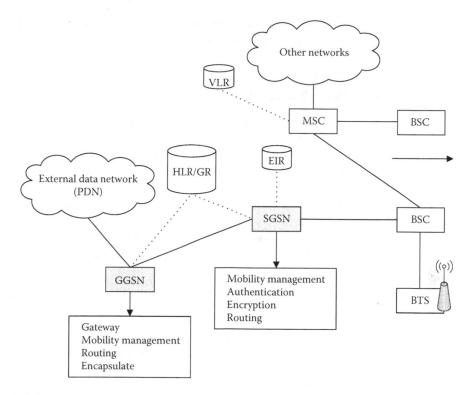

FIGURE 1.2   GSM-GPRS network architecture.

(PDNs). It also provides routing for packet coming from external networks to the SGSN where the MS is located as specified by the home location register (HLR). The new hardware boards for the BSC are called packet data units (PDUs), and their main functions are GPRS radio channel management (e.g., set-up/release); multiplexing of users among the available channels; power control; congestion control; broadcast of system information to the cells; and GPRS signaling from/to the MS, base transceiver station (BTS), and SGSN.

The evolution of the GPRS will be in the direction of improving the QoS by applying some of the concepts belonging to the 3G mobile systems (like the connection-oriented QoS), as well as using powerful coding schemes and a more efficient modulation scheme, thus providing the user with services closer to real-time services.

In the path to 3G systems, EDGE can be seen as a generic air interface for the efficient provision of higher bit rates with respect to GSM.[15,16] A typical GSM network operator deploying EDGE has a running GSM network, where EDGE can be introduced with minimal effort and costs.

EDGE uses enhanced modulation schemes with respect to GSM in order to increase the gross bit rate on the air interface and the spectral efficiency with moderate implementation complexity. Data rates up to 384 kbps using the same 200 kHz wide carrier and the same frequencies as GSM can be achieved. EDGE can be introduced incrementally, offering some channels that can switch between EDGE and

GSP/GPRS. This will relieve both capacity and data rate bottlenecks in such a way that Internet and low bit rate audiovisual services become feasible on an on-demand basis.

1.2 THIRD GENERATION SYSTEMS HISTORY

3G is now the generally accepted term used to describe the new way of mobile networks and services. 1G is used to categorize the first analog mobile systems to emerge in the 1980s, such as the AMPS and Nordic mobile telephony (NMT). These systems provided a mobile solution for voice, but have major limitations, particularly in terms of interworking, security, and quality. The next wave, 2G, arrived in the late 1980s and moved toward a digital solution which gave the added benefit of allowing the transfer of data and provision of other nonvoice services. GSM communication has been the most successful with its global roaming model. 3G is based on the developments in cellular to date, and combines them with complementary developments in both the fixed-line telecom networks and from the world of the Internet. The result is the development of a more general purpose network, which offers the flexibility to provide and support access to any service, regardless of location. These services can be voice, video, or data and combinations thereof, but as already stated the emphasis is on the service provision as opposed to the delivery technology. The motivation for this development has come from a number of main sources:

- Subscriber demand for nonvoice services, mobile extensions to fixed-line services, and richer mobile content
- Operator requirements to develop new revenue sources as mobile voice services and mobile penetration levels reach market saturation
- Operators with successful portfolios of nonvoice services now unable to sustain the volume of traffic within their current spectrum allocation
- Equipment video requirements to market new products as existing 2G networks become mature and robust enough to meet current consumer demand

3G systems can provide higher data rates, thus enabling a much broader range of services[17–21] and the following types of services have been identified:

- Basic and enhanced voice services including applications such as audio conferencing and voice mail
- Low data rate services supporting file transfer and Internet access at rates on the order of 64 to 144 kbps
- High data rate services to support high-speed packet and circuit-based network access, as well as high-quality video conferencing at rates higher than 64 kbps
- Multimedia services, which provide concurrent video, audio, and data services to support advanced interactive applications
- Multimedia services capable of supporting different quality of service requirements for different applications

Compatibility with 2G systems is one of the main goals of 3G systems. Different groups tried to unify the different proposals submitted to the International Telecommunication Union (ITU) in 1998, from ETSI for Europe, Association of Radio Industries and Broadcasting (ARIB) and Telecommunication Technology Council (TTC) for Japan, and American National Standards Institute (ANSI) for the United States.

IMT-2000

In 1985 the ITU defined the vision for a 3G cellular system, at first called Future Public Mobile Telecommunications System (FPMTS) and later renamed International Mobile Telecommunications-2000 (IMT-2000). ITU has two major objectives for the 3G wireless system: global roaming and multimedia services. The World Administrative Radio Conference (WARC'92) identified 1.885 to 2.025 MHz and 2.110 to 2.200 MHz as the frequency bands that should be available worldwide for the new IMT-2000 systems.[22]

A common spectral allocation, along with a common air interface and roaming protocol design throughout the world, can accomplish the global roaming capability. To simultaneously support new multimedia services that require much higher data rates and better QoS than only-voice services, the 3G wireless system envisages:

- Higher data rate services (up to 384 kbps for mobile users, and 2 Mbps for fixed users, increasing to 20 Mbps)
- Increased spectral efficiency and capacity
- Flexible air interfaces, as well as more flexible resource management

Following are some key features of IMT-2000 systems:

- Communications anywhere, at any time by using small lightweight pocket communicators that could be adaptable and reprogrammable. This enables the handsets to work in many environments.
- The facilities of universal personal communications, which will enable customers to roam freely between fixed and mobile networks anywhere in the world and allow them to communicate with any one, any time at any place, and in any form. Roaming will not be restricted due to coverage or geographic location.
- Access to a wide range of services besides speech, such as data, multi-media, supplementary services, roaming, and so on. An important objective of IMT-2000 is to support multimedia services, for example, real-time audio and video conference services, information retrieval services from a number of sources, and ability to surf the World Wide Web. Examples of multimedia calls are shown in Figure 1.3. A single multimedia call may involve simultaneous communication with different devices, or multiple calls simultaneously in progress from one terminal to different devices.

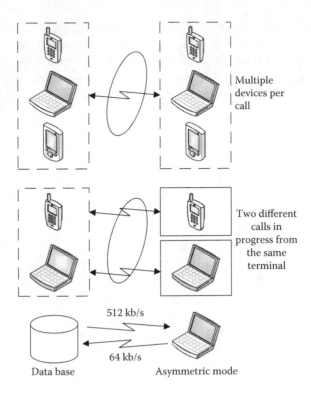

FIGURE 1.3 Examples of multimedia calls.

- A unified infrastructure will be achieved by unifying the many diverse wireless systems that exist today into a seamless radio infrastructure capable of offering a wide range of services.
- Integration of mobile and fixed networks will result in improved service integration between these networks.
- Telecommunications services for developing countries at low capital cost, that is, provision of a fixed wireless interface that supports basic POTS (plain old telephone service) and narrowband ISDN services.
- Provision of a virtual home environment. The intention is that the user receives exactly the same service regardless of his geographic location, subject to the constraints of the operating environment.

The above features are encapsulated in Figure 1.4, which was generated by the USA delegation to ITU-R TG8/1. Key concepts that are utilized in mobile communications technologies are multiple access, performance and capacity enhancement battery technology, and networking control security. Electronic commerce using a fixed network will adopt a public key methodology. On the other hand, electronic commerce across wireless networks needs higher level cryptographic systems to maintain the confidentiality, integrity, and the availability of the information network.

Universal Mobile Telecommunications System (UMTS) is the European version of IMT-2000. The UMTS Terrestrial Radio Access (UTRA) was approved by ITU

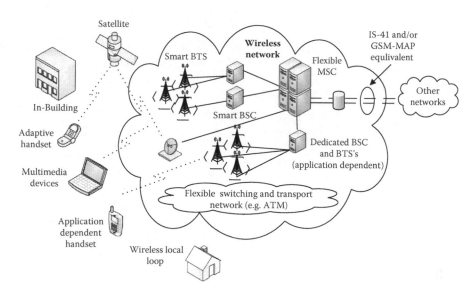

FIGURE 1.4 Multifunctional network. Reproduced with permission from M. Shafi et al. "Wireless communications in the 21st century: A perspective," *Proc. of the IEEE* 85 (October 1997): 1622–38.

in 2000.[14] Wideband CDMA (W-CDMA), supported by groups in Japan (ARIB) and Europe and background-compatible with GSM, has been selected for UTRA frequency division duplex (FDD), whereas TD-CDMA has been selected for the UTRA time division duplex (TDD).[23–25] The introduction of TDD mode is mainly because of the asymmetric frequency bands designed by ITU. Also, the asymmetric nature of the data traffic on the forward and reverse links anticipated in the next generation wireless systems (e.g., Internet applications) suggests that TDD mode might be preferred over FDD. The use of two access methods (FDD and TDD) together with the exploitation of variable bit rate techniques is an important aspect in order to fulfill the flexibility requirement.

W-CDMA supports asynchronous mode operation where reception and transmission timings of different cell sites are not synchronized. It also supports forward and reverse fast closed loop power control with an update rate of 1,600 Hz that is double with respect to the update rate.[26] In TDD mode, code, frequency, and time slot define a physical channel. In FDD mode, a physical channel is defined by its code and frequency and possibly by the relative phase. They have the following structure: a frame length of 10 ms organized in 1.5 time slots. The frame is the minimum transmission element in which the information rate is kept constant. The source bit rate can be different frame by frame, while the chip rate is always kept constant. A time slot has a duration of 10/15 ms and it is the minimum transmission element in which the transmission power is kept constant. Power control can update the transmission power level each time slot. Dual channel quadrature phase shift keying modulation is adopted on the reverse link, where the reverse link dedicated physical data channel (DPDCH) and the dedicated physical control channel are mapped to the I and Q channels, respectively. The I and Q channels are then spread to the chip rate with

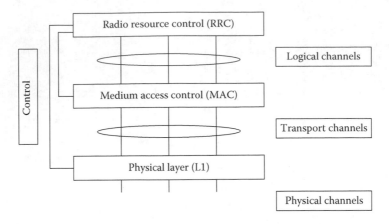

FIGURE 1.5   W-CDMA radio interface protocol architecture.

two different channelization codes and subsequently complex-scrambled by a mobile station-specific complex code. For multicode transmission, each additional reverse link DPDCH may be transmitted on the I or Q channel. Either short or long scrambling codes should be used on the reverse link. Figure 1.5 provides an overview of the radio protocols architecture of the UMTS Radio Access Network (URAN).[27]

Radio protocols can be divided into three levels: physical layer, data link layer, network layer. The link layer is divided in two sublayers: MAC and radio link control (RLC). MAC protocols provide optimized radio access for packet data transmission through the statistical multiplexing of some users on a set of shared channels. MAC protocols are crucially important in providing an effective exploitation of the limited radio resources. The RLC provides a reliable transport of the information through retransmission error recovery mechanisms. The radio resource control (RRC) is part of the network layer and it is responsible for resource management. A radio bearer is allocated by the RRC with bit rates and QoS such that the service required by upper layers can be provided with the available resources at that moment. Note that resource management is an important issue in each mobile and wireless system, but it is crucial in CDMA-based systems, and has quite different aspects with respect to the resource management in FDMA/TDMA systems like GSM. In this context, power control and radio admission control are key resource management mechanisms.[28] W-CDMA also has built-in support for future capacity enhancements such as adaptive antennas, advanced receiver structures, and downlink transmit diversity. Furthermore, to improve the spectral efficiency to the extent possible, turbo codes, capable of near Shannon limit on power efficiency, have recently been adopted by the standards-setting organizations in the United States, Europe, and Asia. For the UTRA/W-CDMA, the same constituent code is used for the rate 1/3-turbo code. Other codes are obtained by the *rate matching* process, where coded bits are punctured or repeated accordingly.[29,30]

Figure 1.6 shows the architecture of a UMTS network.[27] It consists of a core network (CN) connected with interface I to the URAN, which collects all the traffic coming from the radio stations. The URAN consists of a set of radio network subsystems

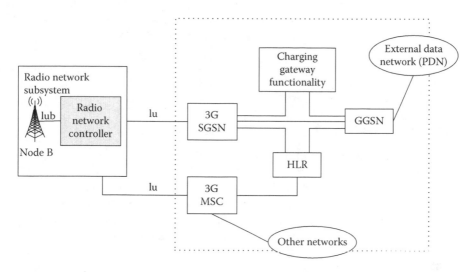

FIGURE 1.6   UMTS network architecture.

(RNSs) connected to the CN through the I interface. Each RNS is responsible for the resources of its set of cells, and each node has one or more cells. An RNS is analogous to the BSS in the GSM-GPRS architecture and consists of a radio network controller (RNC) and one or more nodes, B. The installation of a node B requires a complete replacement of the analogous BTS, since it must handle the different air interface introduced in W-CDMA. A node B is connected to the RNC through the I interface. An RNC separates the circuit-switched traffic (voice and circuit-switched data) from the packet-switched traffic and routes the former to the 3G-MSC and the latter to the 3G-SGSN. The 3G-MSC requires modifying the GPRS-MSC to handle new voice compression and coding algorithms, and it processes the circuit-switched traffic routed to it by the RNC. The MSC then sends the data to a PSTN or another public land mobile network (PLMN). The packet-switched information is routed using the IP-over-ATM protocol specified by the ATM adaptation layer 5 (AAL5). The SGSN is modified to handle AAL5 traffic but performs the same function as in GPRS. Signaling and control functions between the mobile MS and the RAN typically depend on the radio technology, whereas signaling and control functions between the MS and the CN are independent from the radio technology (i.e., access technique).

GSM Evolution to UMTS

The UMTS standard is specified as a migration from the 2G GSM standard to UMTS via the general packet radio service (GPRS) and enhanced data rates for global evolution (EDGE), as shown in Figure 1.7.[31]

The goal of 3G is to provide a network infrastructure that can support a much broader range of services than existing systems, so the changes to the network should reflect this. However, many of the mechanisms in the existing networks are equally applicable to supporting new service models, for example, mobility management. For a successful migration, the manufacturers and suppliers of new 3G equipment

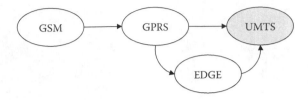

FIGURE 1.7   GSM evolution to UMTS.

understand that most licenses granted for 3G network operation will be to existing 2G operators, and thus the next step must be an evolution rather than a revolution. Operators in the main are expected to introduce GPRS functionality before taking the step to 3G. This will allow them to educate and develop the consumer market for these new services prior to making a major investment in new technology. This means that the CN will comprise the GSM circuit switched core and the GPRS packet switched core. The first release (Release 99) specification for UMTS networks is focused on changes to the RAN rather than the CN. This allows the CN to continue in function-ality, although changes will be made in areas of performance due to the higher data rates required by subscribers in the future networks. Maintaining this functionality allows the mobile network operators to continue using their existing infrastructure and progress to 3G in steps. The handover between UMTS and GSM offering worldwide coverage has been one of the main design criteria for the 3G system.

IMT-2000 STANDARDIZATION PROCESS

IMT-2000 is not a particular technology, but rather a system which should allow seamless, ubiquitous user access to services. The task is to develop a next generation network fulfilling criteria of ubiquitous support for broadband real-time and non-real-time services. The key criteria are:

- High transmission rates for both indoor and outdoor operational environments
- Symmetric and asymmetric transmission of data
- Support for circuit and packet switched services
- Increased capacity and spectral efficiency
- Voice quality comparable to the fixed line network
- Global availability, providing roaming between different operational environments
- Support for multiple simultaneous services to end users

The process is intended to integrate many technologies under one roof. Therefore, it should not be seen that wireless technologies from different regional standardiza-tion bodies, or supported by different manufacturers, are competing with each other, but rather that they can be included in the IMT-2000 family. This is evident with the development of such interworking models as wireless LAN and 3G. A major enabler of the ITU-T vision is the emergence of software defined radio (SDR). With SDR, the air interface becomes an application, which enables a single mobile device to be

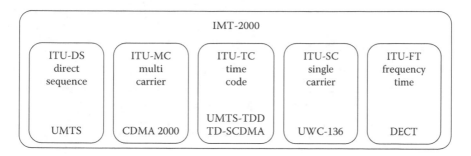

FIGURE 1.8 IMT-2000 technologies.

able to operate with a variety of radio technologies, dynamically searching for the strongest signal, or the most appropriate network to connect to. Thus far, the ITU-T has given the imprimatur of 3G to five different radio access technologies, as shown in Figure 1.8.[31]

ITU-DS is the UMTS frequency division duplex (FDD) standard; ITU-MC is CDMA-2000; and ITU-TC covers both UMTS time division duplex (TDD) and time division synchronous CDMA. The IMT-SC system, UWC-136, is the EDGE standard. The ITU-FT incorporates the European standard for cordless telephones—digital enhanced cordless telecommunications (DECT). DECT provides a local access solution which may be used, for example, in a home environment. The handset can automatically handover to a subscriber's domestic access point, providing dedicated resources. Although the integration of DECT with GSM has been standardized, it has yet to see any exposure. The development of these standards is under the control of two partnership organizations formed from a number of regional standardization bodies. The 3GPP (Third Generation Partnership Project) is responsible for UMTS and EDGE, while the 3GPP2 (Third Generation Partnership Project 2) deals with CDMA2000 (Figure 1.9). DECT is the exception to this, with its standards developed solely by ETSI.[31]

As can be seen, there is considerable overlap in terms of the bodies involved in the two organizations. The various bodies are described in Table 1.1.

1.3 EVOLVING WIRELESS MULTIMEDIA NETWORKS

In what follows, the objective is to provide a brief and yet comprehensive introduction to the evolution of wireless multimedia networks and the key technological

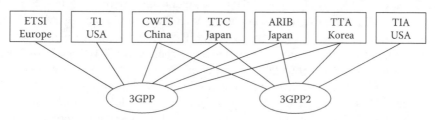

FIGURE 1.9 3G partnerships.

TABLE 1.1
Standardization Bodies

Body	Description
ETSI	The European Telecommunications Standards Institute is responsible for the production of standards for use principally throughout Europe, but standards may be used worldwide.
T1	Committee T1 develops technical standards and reports in the United States with regard to the interconnection and interoperability of telecommunications networks at interfaces with end user systems.
CWTS	The China Wireless Telecommunication Standard group has the responsibility to define, produce, and maintain wireless telecommunication standards in China.
TTC	The Telecommunication Technology Committee is a Japanese organization whose role is to contribute to the standardization and dissemination of standards in the field of telecommunications.
ARIB	The Association of Radio Industries and Businesses conducts investigations into new uses of the radio spectrum for telecommunications and broadcasting in Japan.
TTA	The Telecommunications Technology Association is an IT standards organization that develops new standards and provides testing and certification for IT products in Korea.
TIA	The Telecommunications Industry Association is the leading U.S. trade association serving the communications and information technology industries.

aspects and challenges associated with this evolution. In this context, we aim to define the appropriate framework for the emerging wireless multimedia technologies and applications.

Undoubtedly, the most widely supported evolving path of wireless networks today is the path toward IP-based networks, also known as all-IP networks. The term all-IP emphasizes the fact that IP-based protocols are used for all purposes, including transport, mobility, security, QoS, application-level signaling, multimedia service provisioning, and so on. In a typical all-IP network architecture, several wireless and fixed access networks are connected to a common core multimedia network, as illustrated in Figure 1.10. Users are able to use multimedia applications over technologies, such as WLANs, WPANs, 3G cellular such as UMTS, CDMA2000, and so forth. In this environment, seamless mobility across the different access networks is considered a key issue. Also, native, multimedia support by these networks is very important.

In the all-IP network architecture shown in Figure 1.10, the mobile terminals use the IP-based protocols defined by IETF to communicate with the multimedia IP network and perform, for example, session/call control and traffic routing. All services in this architecture are provided on top of the IP protocol. As shown in the protocol architecture of Figure 1.11, the mobile networks, such as UMTS, CDMA2000, and so on, turn into access networks (e.g., cellular voice) and are used only to support the legacy 2G and 3G terminals, which do not support IP-based applications (e.g., IP telephony). On the user plane, protocols such as RTP and RTSP are employed. On the other hand, on the control plane, protocols such as Session Initiation Protocol (SIP) and Resource Reservation Protocol (RSVP) are employed.[32]

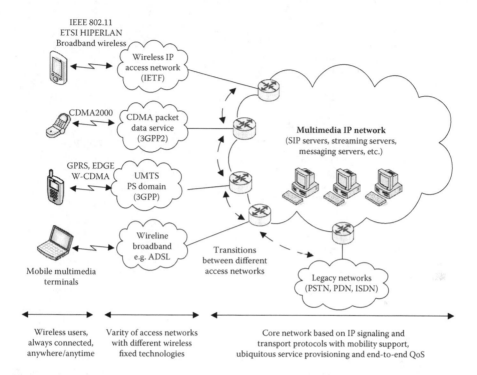

FIGURE 1.10 Multimedia IP network architecture with various access technologies. Reproduced with permission from A. Salkintrizis and N. Passas, eds. *Emerging Wireless Multimedia: Services and Technologies.* New York: John Wiley & Sons, 2005.

For the provision of mobile bearer services, the access networks mainly implement micromobility management, radio resource management, and traffic management for providing QoS. Micromobility management in UMTS access networks is

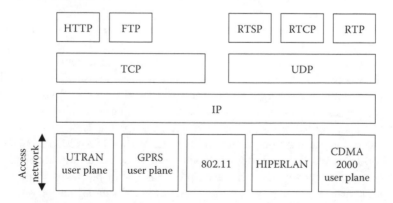

FIGURE 1.11A Simplified protocol architecture in an all-IP network architecture: *a)* user plane; *b)* control plane. Reproduced with permission from A. Salkintrizis and N. Passas, eds. *Emerging Wireless Multimedia: Services and Technologies.* New York: John Wiley & Sons, 2005.

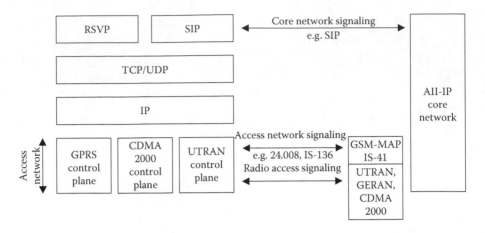

FIGURE 1.11B   (Continued).

based on GPRS Tunneling Protocol (GTP)[33] and uses a hierarchical tunneling scheme for data forwarding. On the other hand, micromobility management in CDMA2000 access networks is based on IP micromobility protocols. Macromobility, that is, mobility across different access networks, is typically based on mobile-IP, as per RFC 3344.[34]

In the short term, the all-IP network architecture would provide a new communications paradigm based on integrated voice, video, and data. You could, for instance, call a user's IP multimedia subsystem (IMS) number and be redirected to the user's web page. Here you could be presented with several options, for example, write an e-mail to the user, record a voice message, click on an alternative number to call if the user is on vacation, and so forth. You could also place an SIP call to a server and update your communication preferences, which could be in the form "only my manager can call me, all others are redirected to my web page" (or vice versa!). At the same time, you can be on a conference call briefing your colleagues about the outcome of a meeting.

At this point, it is instructive to record the key aspects of the evolution toward the wireless multimedia network architecture shown in Figure 1.10. The most important aspects relevant to the evolution toward the wireless multimedia networks follow.

Wireless networks will evolve to an architecture encompassing an IP-based multimedia core network and many wireless access networks (Figure 1.1). As discussed above, the key aspect in this architecture is that signaling with the multimedia core network is based on IP protocols (more correctly, on protocols developed by IETF) and it is independent of the access network (be it UMTS, CDMA2000, WLAN, etc.). Therefore, the same IP-based services could be accessed over any access network. An IP-based core network uses IP-based protocols for all purposes, including data transport, networking, mobility, multimedia service provisioning, and so on. The first commercial approach toward this IP-based multimedia core network is the so-called IP multimedia core network subsystem (IMS) standardized by 3GPP and 3GPP2.

The long-term trend is toward all-IP mobile networks, where not only the core network but also the radio access network are based solely on IP technology. In this approach, the base stations in a cellular system are IP access routers and mobility/ session management is carried out with IP-based protocols (possibly substituting the cellular-specific mobility/session management protocols, such as GTP).

Enhanced IP multimedia applications will be enabled in wireless networks by means of application-level signaling protocols standardized by IETF (e.g., SIP, HTTP).

End-to-end QoS provisioning will be important for supporting the demanding multimedia applications. In this context, extended interworking between, for example, UMTS QoS and IP QoS schemes is needed; or, more generally, interworking between layer-2 QoS schemes and layer-3 QoS (i.e. IP QoS) is required for end-to-end QoS provision.

Voice over IP (VoIP) will be a key technology. As discussed in Chapter 4, several standards organizations are specifying the technology to enable VoIP, for example, the ETSI BRAN TIPHON project, IETF SIP WG.

The mobile terminals will be based on software-configurable radios with capa bilities to support many radio access technologies across many frequency bands.

The ability to move across hybrid access technologies will be an important requirement, which calls for efficient and fast vertical handovers and seamless mobility. The IETF working groups, SEAMOBY and MOBILE-IP, are addressing some of the issues related to seamless mobility. Fast Mobile IP and micromobility schemes are key technologies in this area.

In a highly hybrid access environment, security will also play a key role. IEEE 802.11 Task group I (TGi) is standardizing new mechanisms for enhanced security in WLANs, and the IETF SEAMOBY group addresses the protocols that deal with (security) context transfer during handovers.

For extended roaming between different administrative domains and/or different access technologies, advanced authentication, authorization, and accounting (AAA) protocols and AAA interworking mechanisms will be implemented.

Wireless personal area networks (WPANs) will play a significant role in the multimedia landscape. WPANs have already started spreading, and they will become integrated with the hybrid multimedia network architecture, initially providing services based on Bluetooth technology, and later based on IEEE 802.15.3 high-speed wireless PAN technology, which satisfies the requirement of the digital consumer electronics market (e.g., wireless video communications between a PC and a video camera).

WLANs will also contribute considerably to wireless multimedia provisioning. WLAN technology will evolve further and will support much higher bit rates, on the order of hundreds of megabits per second.

1.4   MULTIMEDIA OVER WIRELESS

The evolutionary aspects summarized above call for several technological advances, which are coupled with new technological challenges. These challenges become even tougher when we consider the limitations of wireless environments. One of the most important challenges is the support of multimedia services, such as video broadcasting, video conferencing, combined voice video applications, and so on. The demand for high bandwidth is definitely the key issue for these services, but it is not enough. Other major requirements that should also be considered include seamless mobility, security, context awareness, flexible charging, and unified QoS support, to name but a few.[37]

Owing to the widespread adoption of IP, most multimedia services are IP based. The IP protocol, up to version 4 (IPv4), was designed for fixed networks and *best effort* applications with low network requirements, such as e-mail and file transfer, and, accordingly, it offers an unreliable service that is subject to packet loss, reordering, packet duplication, and unbounded delays. This service is completely inappropriate for real-time multimedia services such as video-conference and VoIP, which call for specific delay and loss figures. Additionally, no mobility support is natively provided, making it difficult for pure IP to be used for mobile communications. One of the benefits of version 6 of IP (IPv6) is that it inherently provides some means for QoS and mobility support, but it still needs supporting mechanisms to fulfill the demanding requirements that emerge.[37]

The efficient support of IP communications in wireless environments is considered a key issue of emerging wireless multimedia networks. The IP protocol and its main transport layer companions (TCP and UDP) were also designed for fixed networks, with the assumption that the network consists of point-to-point physical links with stable available capacity. However, when a wireless access technology is used in the link layer, it could introduce severe variations in available capacity, and could thus result in low TCP protocol performance. There are two main weaknesses of the IP-over-wireless links:

- The assumption of reliable communication links. Assuming highly reliable links (as in fixed networks), the only cause of undelivered IP packets is congestion at some intermediate nodes, which should be treated in higher layers with an appropriate end-to-end congestion control mechanism. UDP, targeted mainly for real-time traffic, does not include any congestion control, as this would introduce unacceptable delays. Instead, it simply provides direct access to IP, leaving applications to deal with the limitations of the IP best effort delivery service. TCP, on the other hand, dynamically tracks the round-trip delay on the end-to-end path and times out when acknowledgments are not received in time, retransmitting unacknowledged data. Additionally, it reduces the sending rate to a minimum and then gradually increases it in order to probe the network's capacity. In WLANs, where errors can occur due to temporary channel quality degradation, both these actions (TCP retransmissions and rate reduction) can lead to increased delays and low utilization of the scarce available bandwidth.
- The lack of traffic prioritization. Designed as a "best effort" protocol, IP does not differentiate treatment according to the kind of traffic. For example, delay sensitive real-time traffic, such as VoIP, will be treated in the same way as ftp or e-mail traffic, leading to unreliable service. In fixed networks, this problem can be relaxed with overprovisioning of bandwidth, wherever possible (e.g., by introducing high capacity fiber optics). In WLANs this is not possible because the available bandwidth can be as high as a few tens of megabits per second. But even if bandwidth was sufficient, multiple

access could still cause unpredictable delays for real-time traffic. For these reasons, the introduction of scheduling mechanisms is required for IP over WLANs in order to ensure reliable service under all kinds of conditions.

MULTIMEDIA SERVICES IN WLAN

WLAN systems are technologies that can provide very high data rate applications and individual links (e.g., in company campus areas, conference centers, airports) and represent an attractive way to set up computers networks in environments where cable installation is expensive or not feasible. These systems represent the coming together of two of the fastest-growing segments of the computer industry: LANs and mobile computing, thus recalling the attention of equipment manufacturers. This shows high potential and justifies the big attention paid to WLANs by equipment manufacturers.

Today, IEEE 802.11 WLANs can be considered as a wireless version of Ethernet, which supports best-effort service. The mandatory part of the original 802.11 MAC is called distributed coordination function (DCF), and is based on carrier sense multiple access with collision avoidance (CSMA/CA), offering no QoS guarantees. Typically, multimedia services such as VoIP or audio/video conferencing require specified bandwidth, delay, and jitter, but can tolerate some losses. However, in DCF mode, all mobile stations compete for the resources with the same priorities. There is no differentiation mechanism to guarantee bandwidth, packet delay, and jitter for high-priority mobile stations or multimedia flows. Even the optional polling-based point coordination function (PCF) cannot guarantee specific QoS values. The transmission time of a polled mobile station is difficult to control. A polled station is allowed to send a frame of any length between 0 and 2,346 bytes, which introduces the variation of transmission time. Furthermore, the physical layer rate of the polled station can change according to the varying channel status, so the transmission time is hard to predict. This makes a barrier to providing guaranteed QoS services for multimedia applications.

The rapidly increasing interest in wireless networks supporting QoS has led the IEEE 802.11 Working Group to define a new supplement called 802.11e to the existing legacy 802.11 MAC sublayer.[40] The new 802.11e MAC aims at expanding the 802.11 application domain, enabling the efficient support of multimedia applications. The new MAC protocol of the 802.11e is called the hybrid coordination function (HCF). This describes the ability to combine a contention channel access mechanism, referred to as enhanced distributed channel access (EDCA), and a polling-based channel access mechanism, referred to as HCF-controlled channel access (HCCA). EDCA provides differentiated QoS services by introducing classification and prioritization among the different kinds of traffic, while HCCA provides parameterized QoS services to mobile stations based on their traffic specifications and QoS requirements. To perform this operation, the HCF has to incorporate a scheduling algorithm that decides how the available radio resources are allocated to the polled stations. This algorithm, usually referred to as the *traffic scheduler*, is one of the main research areas in 802.11e, as its operation can significantly affect the overall system performance. The traffic scheduler is now part of the 802.11e standard, and can thus serve as a product differentiator that should be carefully designed and implemented, as it is directly connected to the QoS provision capabilities of the system. In the open technical literature, only a limited number of 802.11e traffic schedulers have been proposed so far. The current approaches for supporting IP QoS over WLANs fall into the following categories:

- Pure end-to-end. This category focuses on the end-to-end TCP operation and the relevant congestion-avoidance algorithms that must be implemented on end hosts to ensure transport stability. Furthermore, enhancements for fast recovery such as the TCP selective acknowledgment (SACK) option and NewReno are also recommended.
- Explicit notification based. This category considers explicit notification from the network to determine when a loss is due to congestion, but, as expected, would require changes in the standard Internet protocols.
- Proxy-based. Split connection TCP and Snoop are proxy-based approaches, applying the TCP error control schemes only on the last host of a connection. For this reason, they require the access point (AP) to act as a proxy for retransmissions.
- Pure link layer. Pure link layer schemes are based on either retransmissions or coding overhead protection at the link layer—that is, automatic repeat request (ARQ) and forward error correction (FEC), respectively—so as to make errors invisible at the IP layer. The error control scheme applied is common to every IP flow irrespective of its QoS requirements.
- Adaptive link layer. Finally, adaptive link layer architectures can adjust local error recovery mechanisms according to the application requirements (e.g., reliable flows vs. delay-sensitive) and/or channel conditions.

AD HOC NETWORKS AND MULTIMEDIA SERVICES IN WPANS

Many WLANs of today need an infrastructure network that provides access to other networks and includes MAC. Ad hoc wireless networks do not need any infrastructure. In these systems mobile stations may act as a relay station in a multihop transmission environment from distant mobiles to base stations. Mobile stations will have the ability to support base station functionality. The network organization will be based on interference measurements by all mobile and base stations for automatic and dynamic network organization according to the actual interference and channel assignment situation for channel allocation of new connections and link optimization. These systems will play a complementary role to extend coverage for low power systems and for unlicensed applications. A central challenge in the design of ad hoc networks is the development of dynamic routing protocols that can efficiently find routes between two communication nodes. A Mobile Ad Hoc Networking (MANET) Working Group has been formed within the IETF to develop a routing framework for IP-based protocols in ad hoc networks. Another challenge is the design of proper MAC protocols for multimedia ad hoc networks.[41]

LANs without the need for an infrastructure and with a very limited coverage are being conceived for connecting different small devices in close proximity without expensive wiring and infrastructure. The area of interest could be the personal area around the individual using the device. This new emerging architecture is indicated as WPAN. The concept of personal area network refers to a space of small coverage (less than 10 m) around a person where ad hoc communication occurs, and is also referred to as personal operating space (POS). The network is aimed at interconnecting portable and mobile computing devices such as laptops, personal digital assistants (PDAs), peripherals, cellular phones, digital cameras, headsets, and other electronics devices.[42]

Starting with Bluetooth,[43] WPANs became a major part of what we call heterogeneous network architectures, mainly due to their ability to offer flexible and efficient ad hoc communication in short ranges without the need of any fixed infrastructure.

One of the main advances for multimedia applications in WPANs is ultrawideband (UWB) communications. The potential strength of the UWB radio technique lies in its use of extremely wide transmission bandwidths, which results in desirable capabilities, including accurate position location and ranging, lack of significant fading, high multiple access capability, covert communications, and possible easier material penetration. The UWB technology itself has been in use in military applications since the 1960s, based on exploiting the wideband property of UWB signals to extract precise timing/ranging information. However, recent Federal Communication Commission (FCC) regulations have paved the way for the development of commercial wireless communication networks based on UWB in the 3.1 to 10.6 GHz unlicensed band. Because of the restrictions on the transmit power, UWB communications are best suited for short-range communications, namely, sensor networks and WPANs. To focus standardization work in this technique, IEEE established subgroup IEEE 802.15.3a, within 802.15.3, to develop a standard for UWB WPANs. The goals for this new standard are data rates of up to 110 bps at 10 m, 200 Mbps at 4 m, and higher data rates at smaller distances. Based on those requirements, different proposals are being submitted to 802.15.3a. An important and open issue of UWB lies in the design of multiple access techniques and radio resource sharing schemes to support multimedia applications with different QoS requirements. One of the decisions that will have to be made is whether to adopt some of the multiple access approaches already being developed for other wireless networks, or to develop entirely new techniques. It remains to be seen whether the existing approaches offer the right capabilities for UWB applications.

Multimedia Services over 3G Networks

Over the past few years, there have been major standardization activities undertaken in 3GPP and 3GPP2 for enabling multimedia services over 3G networks. The purpose of this activity has been to specify an IP-based multimedia core network, the IMS mentioned before, that can provide a standard IP-based interface to wireless IP terminals for accessing a vast range of multimedia services independently from the access technology. Figure 1.12 shows the IP multimedia subsystem network providing a standardized IP-based signaling for accessing multimedia services.

The interface uses the SIP specified by IETF for multimedia session control (see RFC 3261). In addition, SIP is used as an interface between the IMS session control entities and the service platforms which run the multimedia applications. The initial goal of IMS was to enable the mobile operators to offer to their subscribers

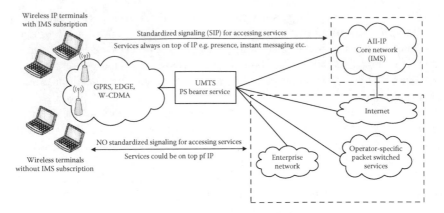

FIGURE 1.12 IP multimedia subsystem network. Reproduced with permission from A. Salkintrizis and N. Passas, eds. *Emerging Wireless Multimedia: Services and Technologies.* New York: John Wiley & Sons, 2005.

multimedia services based on and built on Internet applications, services, and protocols. Thus, IMS forms a single core network architecture that is globally available and can be accessed through a variety of technologies, such as mobile data networks, WLANs, fixed broadband (e.g., xDSL), and so on. No matter what technology is used to access IMS, the user always employs the same signaling protocols and accesses the same services.

In a way, IMS allows mobile operators to offer popular Internet-like services, such as instant messaging, Internet telephony, and so on. IMS can offer the versatility to develop new applications quickly. In addition, IMS is global (identical across 3GPP and 3GPP2), and it is the first convergence between the mobile world and the IETF world.

Multimedia messaging services are now emerging in 3G cellular networks, providing instant messaging by exploiting the SIP-enabled IMS domain. By combining the support of messaging with other IMS service capabilities, such as presence, new rich and enhanced messaging services for the end users can be created. The goal of 3G operators is to extend mobile messaging to the IMS, while also interoperating with the existing SMS, EMS, and MMS wireless messaging solutions, as well as SIP-based Internet messaging services. The SIP-based messaging service should support interoperability with the existing 3G messaging services SMS, EMS, and MMS, as well as enable development of new messaging services, such as instant messaging, chat, and so forth.

It should be possible, in a standardized way, to create message groups (*chat rooms*) and address messages to a group of recipients as a whole, as well as to individual recipients. Additional standardized mechanisms are expected, in order to create and delete message groups, enable and authorize members to join and leave the group, and also to issue mass invitations to members of the group.

HYBRID MULTIMEDIA NETWORKS

Efficient mobility management is considered to be one of the major factors toward seamless provision of multimedia applications across heterogeneous networks. Thus, a large number of solutions have been proposed in an attempt to tackle all the relevant technical issues. Of course, the ultimate goal is to realize a seamless multimedia environment. The high penetration of WLANs, and especially the higher data rates they offer, caused cellular operators to investigate the possibility of integrating them into their systems and support a wider range of services for their users. This integration led to hybrid multimedia systems. The design and implementation of these systems present many technical challenges, such as vertical handover support between, for example, WLAN and UMTS; unified AAA; and consistent QoS and security features. Despite the progress in WLAN/UMTS interworking standardization, up until now most attention has been paid to AAA interworking issues and less to mobility and QoS.[44,45]

Owing to the intrinsic differences in mobility management between UMTS and IP-based architectures, a mobility management scheme tailored to heterogeneous environments has to overcome technology-specific particulars and combine their characteristics. Based on the above scenarios, a number of proposals are aimed at offering some degree of seamless integration. Some of them focus on the general framework and the functionalities that these networks should provide for advanced mobile capabilities, using mobile IP as the basic tool for system integration. This has the advantage of simple implementation, with minimal enhancements on existing components, but at the expense of considerably larger handover execution time.

1.5 USERS' PERSPECTIVES

Technical and economic trends, together with applications requirements, will drive the future of mobile communications. For example, a forecast of the mobile communications market for Japan is shown in Figure 1.13.[27] In particular, the trend of mobile Internet that represents the main driver for multimedia applications is presented.

It is expected by the UMTS Forum that in Europe in 2010 more than 90 million subscribers will use mobile multimedia services and will generate about 60 percent

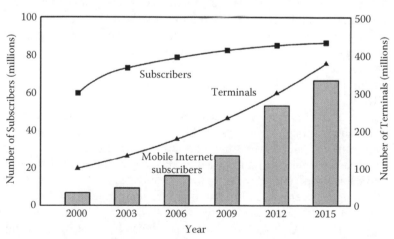

FIGURE 1.13 Mobile communications market forecast for Japan.

of the traffic in terms of transmitted bits. Additional frequency assignment will be
necessary for 3G to accommodate the growing demand. The bandwidth to be added
is assumed to be 160 MHz in 2010. However, the added bandwidth greatly depends
on the growth ratio of traffic per subscriber.

Therefore, the study of high-capacity cellular systems with improved spectrum
efficiency and new band width is necessary to accommodate growing traffic in 2010
and beyond. Higher data rates and wireless Internet access are key components of
the future mobile communications systems, but they are also actually key concepts
in 3G systems. Data rates up to 8 Mbps will be possible without making any drastic
change in the current standard. Future mobile communications systems should bring
something more to the table than faster data or wireless Internet access. Something
that we are missing today (even in 3G) is the flexible interoperability of various
existing networks like cellular, cordless, WLAN-type, short connectivity, and wired
systems. It will be a huge challenge to integrate the whole worldwide communication infrastructure into one transparent network allowing various ways to connect
into it depending on the user's needs, as well as the available access methods. The
heterogeneity of various access methods can be overcome either by using multimode
terminals and additional network services, or by creating a completely new network
system that will implement the envisioned integration. The first option implies only
further development of the existing networks and services and cannot be very flexible. The second option is more profound and can result in a more efficient utilization of networks and available spectrum. The creation of this new network requires
a completely new design approach. So far, most of the existing systems have been
designed in isolation without taking into account a possible interworking with other
access technologies. This method of system design is mainly based on the traditional
vertical approach to support a certain set of services with a particular technology.
The UTRA concept has already combined the FDD and TDD components to support
the different symmetrical and asymmetrical service needs in a spectrum-efficient
way. This is the first step to a more horizontal approach,[48] where different access
technologies will be combined into a common platform to complement each other
in an optimum way for different service requirements and environments. Due to the
dominant role of IP-based data traffic, these access systems will be connected to a
common, flexible, and seamless IP-based core network. This results in a lower infrastructure cost, faster provisioning of new features, and easy integration of new network elements, and could be supported by technologies like JAVA Virtual Machine
and CORBA.

The vision of the future network, including a variety of internetworking access
systems, is shown in Figure 1.14. The mobility management will be part of a new
media access system, and serve as the interface between the core network and the
particular access technology to connect a user via a single number for different access
systems to the network. Global roaming for all access technologies is required. The
internetworking between these different access systems in terms of horizontal and
vertical handover, and seamless services with service negotiation with respect to
mobility, security, and QoS will be a key requirement. The last will be handled in
the common media access system and the core network. Multimode terminals and
new appliances are also key components needed to support these different access

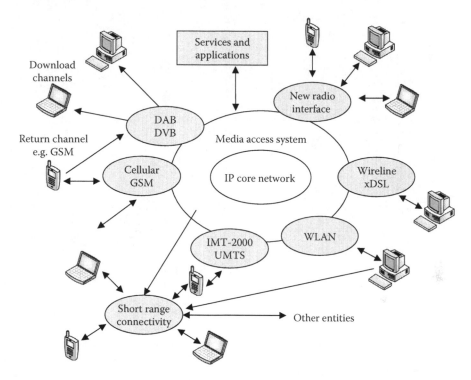

FIGURE 1.14 The network, including a variety of internetworking access systems.

technologies of the common platform seamlessly from the user perspective. These terminals may be adaptive, based on high signal processing power. Therefore, the concept of software-defined radio supported by software downloading can be a key technology in the future perspective.

To make the vision of a system beyond 3G happen, many technical challenges have to be solved by extensive research activities at different layers. In spite of the 2 Mbps data rates achievable by 3G systems, the overall economic capacity of these systems can still be only a small fraction of the actual need of the seamless information mobility.

2 Convergence Technologies

This chapter provides an overview of the key convergence technologies that offer many services from the network infrastructure point of view. After a short presentation of the next generation network (NGN) architecture, we deal with convergence technologies for third generation (3G) networks. This chapter also reviews technologies for 3G cellular wireless communication systems. Next, the third generation wideband code-division multiple access (WCDMA) standard has been enhanced to offer significantly increased performance for packet data and broadcast services through the introduction of high-speed downlink packet access (HSDPA), enhanced uplink, and multimedia broadcast multicast services (MBMS). Challenges in the migration to fourth generation (4G) mobile system conclude this chapter.

2.1 INTRODUCTION

As a term, convergence has been coined by both the telecom and datacom industries. From a telecom perspective, it is the expansion of the public switched telephone network (PSTN) to offer many services on the single network infrastructure. For Internet advocates, it is the death of the PSTN as its role is largely replaced by technologies such as voice over Internet Protocol (VoIP). In reality, the truth lies somewhere in the middle, and it is here that the cellular industry takes the best of both worlds to create an evolved network, where the goal is the delivery of effective services and applications to the end user, rather than focusing on a particular technology to drive this delivery. That said, the economy of scale and widespread acceptance of IP as a means of service delivery lead to it playing a central role in this process.[1]

The communications industry, particularly the cellular industry, is currently going through a state of enormous transition. Some years ago, many of the major cellular operators began deploying a network to support packet switched data services which will lead them to the 3G. This step to 3G involves a major change in the network infrastructure with the introduction of complex technologies such as asynchronous transfer mode (ATM), code division multiple access (CDMA), and the IP. For forward-looking operators, this transition also requires a clear, strategic transformation of their business model to grasp and maximize the benefits of the next generation's lucrative revenue streams. An operator requires both a highly motivated staff with a substantial skill set, as well as comprehensive, dynamic information systems. Also crucial is a clear understanding of the role the operator will play in this new model on the continuum from mere provision of a bit-pipe, to an organization offering full Internet service provider (ISP) capabilities and value-added services. This revised business model needs to incorporate integrated solutions for charging and billing, and provide a clear understanding of the new available revenue streams. Smooth convergence of network and telecommunications technologies and a proactive business strategy are fundamental to the success of the future mobile operator.

Many telecom engineers have little experience with the new packet and IP technologies. To remain competitive, it is essential that they learn the new packet-switched skills quickly. The older circuit-switched skills will be required for a long time, as circuit switching is not expected to disappear overnight and will probably be around for decades. However, new network components for telecom networks will be based around packet-switched technology.

Second generation (2G) cellular systems have been implemented commercially since the late 1980s. Since then, the systems have evolved dramatically in both size and reliability to achieve the level of quality subscribers expect of current networks. Mobile network operators have invested heavily in the technology and the infrastructure, and it is unreasonable to expect this to be simply discarded when a new 3G system is proposed.

Large bandwidth, guaranteed quality of service, and ease of deployment coupled with great advancement in semiconductor technologies make this converged wireless system a very attractive solution for broadband service delivery. The key applications evolved from the advancement of broadband wireless and the underlying technologies, including broadband wireless mobile (3G wireless and 4G mobile), broadband wireless access, broadband wireless networking, as well as broadband satellite solutions, will surely dominate the whole communications market and therefore will improve the business model in many aspects.

Convergence of broadband wireless mobile and access is the next topic in wireless communications. Fueled by many emerging technologies, including digital signal processing, software-definable radio, intelligent antennas, superconductor devices, as well as digital transceivers, the wireless system becomes more compact with limited hardware and more flexible and intelligent software elements. Reconfigurable and adaptive terminals and base stations help the system to be easily applied in wireless access applications. The compact hardware and very small portion of software (called the common air interface basic input-output system or CAIBIOS) will go the way the computer industry did in the past.[2]

Wireless mobile Internet will be the key application of the converged broadband wireless system. The terminal will be smart instead of dumb, compatible with mobile and access services including wireless multicasting as well as wireless trunking. This new wireless terminal will have the following features:

- 90 percent of the traffic will be data
- The security function will be enhanced (e.g., fingerprint chip embedded)
- A voice recognition function will be enhanced; keypad or keyboard attachments, as well as wireless versions will be options
- The terminal will support single and multiple users with various service options
- The terminal will be fully adaptive and software-reconfigurable

As wireless communications evolve to this convergence, 4G mobile wireless com-munications (4G mobile) will be an ideal mode to support high-data rate connection from 2 to 20 Mbps based on the spectrum requirement for IMT-2000, as well as the coexistence of the current spectrum for broadband wireless access. This 4G mobile system's vision aims at the following:

- Providing a technological response to accelerated growth in demand for broadband wireless connectivity
- Ensuring seamless services provisioning across a multitude of wireless sys-tems and networks, from private to public, from indoor to wide area
- Providing optimum delivery of the user's wanted service via the most appropriate network available
- Coping with the expected growth in Internet-based communications
- Opening new spectrum frontiers

Several important issues arise out of the convergence in mobile delivery of multimedia content. Among the regulatory issues are usage of spectrum, spectral coexistence of mobile phone and TV broadcasting services, and technical/opera-tional parameters. Another issue would be to define the players in the chain and the player(s) or the licensee(s) regulators should address.

Some interesting business issues are also emerging. Control of the revenue gath-ering system is an important matter. Most players would aim at gathering the revenue directly from the consumer but this is more easily said than done.

Content, the business driver, needs its own set of regulations, customized to the new types of products catering to the lifestyle of mobile consumers, size and resolu-tion of displays, battery consumption, and repurposing or redimensioning of archived content.

2.2   NEXT GENERATION NETWORK ARCHITECTURE

The future wireless network is an open platform supporting multicarrier, multiband-width, and multistandard air interfaces, with content-oriented bandwidth-on-demand (BoD) services dominant throughout the whole network. In this way, packetized transmission will go all the way from one wireless end terminal directly to another. Figure 2.1 shows this wireless network architecture. The major benefits of this archi-tecture are that the network design is simplified and the system cost greatly reduced. The base transceiver system (BTS) is now a smart open platform with a basic broad-band hardware pipe embedded with a CAIBIOS. Most functional modules of the sys-tem are software definable and reconfigurable. The packet switching is distributed in the broadband packet backbone (or core network, called packet-division multiplex, PDM). The wireless call processing, as well as other console processing, is handled in this network. The gateway (GW) acts as proxy for the core network and deals with any issues for the BTS, and the BTS is an open platform supporting various standards, optimized for full harmonization and convergence. The terminal (mobile station, MS) can be single user or multiuser oriented, supporting converged wireless applications.

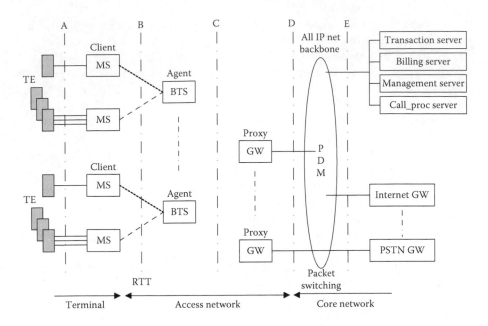

FIGURE 2.1　Wireless network architecture model. Reproduced with permission from W. Lee. "Compact multidimensional broadband wireless: The convergence of wireless mobile and access," *IEEE Communications Magazine* 38 (November 2000): 119–23.

Considering the signaling protocol, the client–server model is established between a wireless terminal and a core network. The BTS becomes the agent in both directions. This end-to-end direct signaling can ensure that the wireless terminal is smart and intelligent. Figure 2.2 shows the system protocol stack: (a) general protocol stack and (b) an example of support for wireless access applications.

Different services—ATM, IP, synchronous transfer mode (STM), Motion Picture Experts Group (MPEG)—can be supported through a service convergence layer. To guarantee wireless quality of service (QoS) and high spectrum utilization, dynamic bandwidth allocation (DBA) is required through the medium access control (MAC) DBA sublayer, which improves the conventional layer architecture. The DBA scheduler is the core of the MAC. To realize dynamic resource allocation, this scheduler is essential for the broadband wireless link, which in general helps

- Support class of service offerings
- Provide agnostic support for all network protocols
- Eliminate the need for traffic shaping and user parameter control
- Eliminate end-to-end packet and/or cell delay variation
- Increase spectrum utilization

The transmission convergence layer handles various transmission modulations, error corrections, segmentations, and interface mappings of wireless mobile and access in the physical layer.

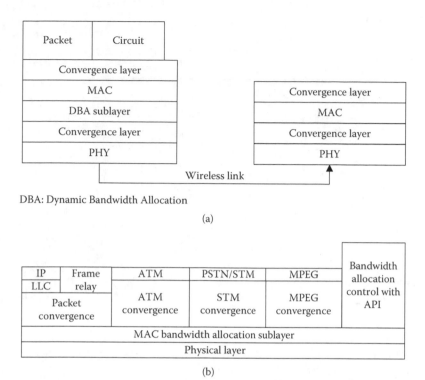

DBA: Dynamic Bandwidth Allocation

(a)

(b)

FIGURE 2.2 (*a*), General protocol stack; (*b*), protocol examples. Reproduced with permission from W. Lee. "Compact multidimensional broadband wireless: The convergence of wireless mobile and access," *IEEE Communications Magazine* 38 (November 2000): 119–23.

As telecommunications move into the twenty-first century, we are witnessing a move toward convergence of the traditional telephone networks such as the PSTN with the Internet. This convergence, along with changes in software technology (e.g., Java), will have a profound impact on the software deployed in the network. For example, with Internet telephony, switches are no longer large monolithic entities, but instead distributed pieces of hardware (e.g., routers and media gateways) with a softswitch core. One of the more prominent changes will be an opening up of the network. Application programming interfaces (APIs) will be available at many layers of the network, enabling more people to be involved in the service creation process. A demand will exist to make the converged network of the twenty-first century more like today's Internet in terms of application and service creation capabilities. This will create a large challenge for the software architects of the converged network: they must make the network as secure and reliable as today's PSTN, and provide the openness and programmability demanded by other networks.[3]

The NGN, as some call the future converged PSTN and Internet, will have a software architecture very different from that of today's PSTN. The most popular concept for supporting voice over IP and interworking with the PSTN includes the use of a softswitch or call agent.

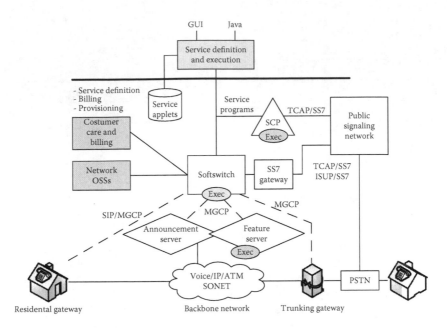

FIGURE 2.3   An NGN with a VoIP architecture. Reproduced with permission from S. Maye and A. Umar. "The impact of network convergence on telecommunications software," *IEEE Communications Magazine* 39 (January 2001): 78–84.

An NGN with a typical VoIP architecture is defined in Figure 2.3. In this architecture, we can see a centralized control entity, the softswitch, signals consumer end points via the Session Initiation Protocol (SIP) and/or Media Gateway Control Protocol (MGCP). MGCP is used to control trunking gateways that provide the physical interface between the packet-switched network and the PSTN. Finally, the softswitch communicates with the PSTN via the SS7 gateway, which converts transaction capability application part (TCAP) and integrated services digital network (ISDN) signaling user part (ISUP) over IP messages to TCAP and ISUP on the signaling system 7 (SS7) network. In addition, the softswitch also has interfaces to network operations systems (e.g., for provisioning and management), customer care and billing systems, service control points (SCPs) for real-time number translation and intelligent network service support, and feature servers that provide support for advanced intelligent services.

In the NGN architecture, we see that software performs all the switching logic, and also is responsible for play out of announcements (e.g., "the number you have dialed ..."), translation of phone numbers (e.g., for 800 calls and local number portability), and intelligent services (on the feature server). Service logic can be defined and executed by customers and/or third-party service creators/providers through APIs provided by the service provider through service applets (which can be constructed, e.g., through a visual programming environment). The services defined will execute in several places: on the SCP, on feature servers, in the softswitch, and in the customer/third-party service creator environment. These services will run in virtual partitioned execution environments (e.g., Java sandboxes) to protect the integrity of the network and its resources.

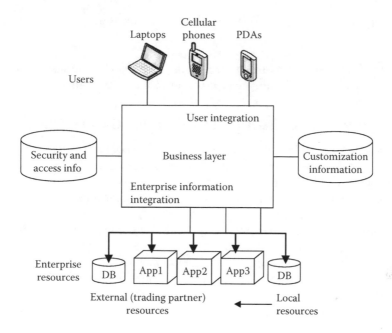

FIGURE 2.4 Architectural framework for next generation applications. Reproduced with permission from S. Maye and A. Umar. "The impact of network convergence on telecommunications software," *IEEE Communications Magazine* 39 (January 2001): 78–84.

Figure 2.4 shows an architectural view of next generation applications. This architectural framework consists of two integration layers that surround the business logic. The user integration layer takes into account the wide range of users with a diverse array of devices (laptops, Web browsers, personal digital assistants or PDAs, cellular phones) with which you communicate. The enterprise resource integration is used to connect to various local as well as remote (i.e., external trader) applications and databases. Note that both integration layers are triangular (i.e., the integration glue is thin in some cases but quite thick in others). For example, integration with Web-based applications requires less effort than a mainframe-based application. The integration effort also depends on whether you are interacting with local (i.e., within the same enterprise) or external applications. The two integration layers can greatly benefit from a converged network to minimize the development effort. This architecture must support applications that

- Operate on an Internet scale (i.e., tens of thousands of users instead of hundreds)
- Provide Internet connectivity (i.e., best-effort open Internet instead of a controlled local area network, LAN)
- Support multiple customers that require security and load balancing between multiple customers
- Allow multiple configurations (i.e., managing diverse user profiles and configurations) and provide high-volume infrastructure with scalable services to diverse populations

- Support commercial service provision (i.e., measuring and billing for services)

To satisfy these requirements, high telecommunications bandwidth and quick interconnection with multiple applications on the network are essential, among other things.

2.3   CONVERGENCE TECHNOLOGIES FOR 3G NETWORKS

More than ever before in other industrial sectors, the mobile communication industry has seen tremendous advances during the past years. Up to the beginning of the 1990s, mobile communication had been seen mainly as a tool for business people. That changed with the introduction of the digital communication system, especially when it came to the success of the market in the second half of the 1990s. Building on the success of Global System for Mobile (GSM) in Europe, and especially its leadership over the technology in the United States, the European Commission decided to put in place the foundations for a similar 3G mobile communications system by establishing the Universal Mobile Telecommunication System (UMTS) Task Force in 1994 through the RACE Project.[4-6] The main recommendations were published in the UMTS Task Force report (1996) and are summarized as follows:

- UMTS standards must be open to global network operators and manufacturers
- UMTS will offer a path from existing 2G digital systems GSM900, DCS1800, and DECT
- Basic UMTS, for broadband needs up to 2 Mbps, will be available from 2002
- Full UMTS services and systems for mass market services will be available from 2005

The acronym UMTS was defined by the Task Force as: "UMTS, the Universal Mobile Telecommunication System; it will take the personal communications user into the new information society. It will deliver information, pictures, and graphics direct to people and provide them with access to the next generation of information-based services. It moves mobile and personal communications forward from 2G systems that are delivering mass market low-cost digital telecommunication services."

This definition is important because it defines the essential differences between 2G and 3G systems. UMTS will be a mobile communication system that can offer significant user benefits including high-quality wireless multimedia services to a convergent network of fixed, cellular, and satellite components. It will deliver information directly to users and provide them with access to new and innovative services and applications. It will offer mobile personalized communications to the mass market regardless of location, network, or terminal used.

To better understand the 3G concept, the mobile communications space comprises four geographically distinct zones. These zones expand on the 2G principle of a cellular structure as shown in Figure 2.5.

In the early days, mobile multimedia was mainly seen as a technological platform and much less as a complete new system with new services. A question often arises:

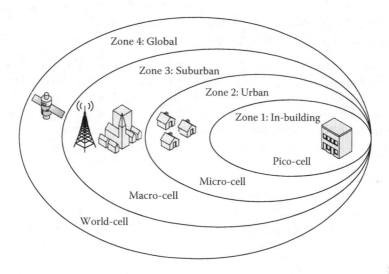

FIGURE 2.5   Four zone model of mobile communications.

What kind of new mobile services can we expect from UMTS? There are many opportunities which cannot yet be explained in detail because many of them are still in the development and trial stages. The opportunities of mobile multimedia broadband are presented in Figure 2.6. In terms of market penetration, the GSM/UMTS world is the most successful technology platform for 3G.

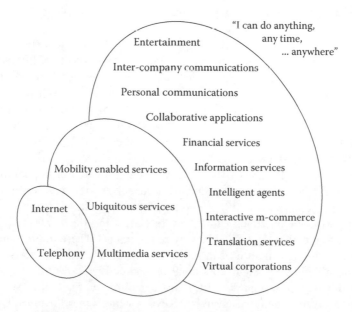

FIGURE 2.6   Opportunities of mobile multimedia broadband.

2.4   TECHNOLOGIES FOR 3G CELLULAR WIRELESS COMMUNICATION SYSTEMS

3G mobile communication systems based on the WCDMA and CDMA2000 radio access technologies have seen widespread deployment around the world. The applications supported by these commercial systems range from circuit-switched services, such as voice and video telephony, to packet-switched services, such as video streaming, e-mail, and file transfer. As more packet-based applications are invented and put into service, the need for better support for different QoS levels, higher spectral efficiency, and higher data rates for packet-switched services increases in order to further enhance the user experience while maintaining efficient use of the system resources. This need has resulted in the creation of CDMA2000 1x Evolution Data–Optimized (1xEV-DO) Revision 0 and Revision A by the 3GPP2 project, as well as the HSDPA and enhanced uplink (EUL) evolution of WCDMA in 3GPP.

CDMA2000 1xEV-DO Revision 0 more than doubled the forward link spectral efficiency of CDMA2000 1x for packet data applications. It was developed to provide efficient support for asymmetric best-effort packet data. Since its first commercial deployment in 2002, market feedback has revealed that some popular packet data applications actually result in symmetric traffic and some stringent latency requirements. This has led to the development of CDMA2000 1x EV-DO Revision A, which includes significant improvements on the 1x EV-DO reverse link, including increased total sector throughput for best-effort applications, and shortened delay for applications with low-latency requirements. Similarly, 3G developed HSDPA to address the need for further improved WCDMA forward-link (downlink) packet data access, followed by the development of EUL to improve the corresponding reverse (uplink) performance and capabilities. Both 3GPP and 3GPP2 have also developed techniques to more efficiently support broadcast and multicast services as an integrated part of the 3G networks.[7]

Many of the developed evolution steps are still under different stages of implementation and testing. The drive for the 3G Partnership Project (3GPP) and 3GPP2 to continue to develop new technologies with even better performance and capabilities continues. 3GPP is considering continuous evolution of WCDMA, as well as more substantial steps as part of the so-called 3GPP long-term evolution (LTE).

We now present key enhancements to CDMA2000 1x EV-DO systems. CDMA2000 is a registered trademark of the 3GPP2. CDMA2000 1x EV-DO Revision 0 (also referred to as DO Rev0) was driven by the design vision of a wide area–mobile wireless Ethernet.[8] The result was a high-rate wireless packet data system with substantial improvement in downlink capacity and coverage over traditional CDMA2000 systems such as IS-95 and IS-2000. In addition to high throughput, DO Rev0 provides QoS support to enable operators to offer a variety of applications with different throughput and latency requirements. These improvements were accomplished through the use of large packet sizes encoded with low-rate turbo-codes, transmitted using adaptive modulation and coding, downlink physical layer hybrid automatic repeat request (ARQ) and downlink multiuser diversity, together with antenna diversity at the receiver. DO Rev0 systems support per flow QoS on the downlink, and per terminal QoS on the uplink. A flow is a source with transmission requirements associated with an application such as video telephony, VoIP, gaming, Web browsing, and file transfer.

Increasing demand for high-speed wireless Internet access has resulted in rapid growth of the number of CDMA2000 1x EV-DO users worldwide. Operators have observed a strong demand for applications such as VoIP, video telephony, wireless gaming, and push-to-talk (PTT), along with demand for downlink-intensive applications such as Web browsing and file transfer. These applications demand a system that can support large numbers of simultaneous users while meeting their desired latency requirements. In order to meet this demand, 3GPP2 approved enhancements to CDMA2000 1x EV-DO Revision 0.

As in IS-95 and IS-2000 systems, the 1x EV-DO Revision 0 carriers are allocated the 1.25 MHz bandwidth and use a direct sequence (DS) spread waveform at 1.2288 Mchips/s. The fundamental timing unit for downlink transmissions is a 1.666-ms slot that contains the pilot and MAC channels, and a data portion that may contain the traffic or control channel. Unlike IS-2000, where a frame is 20 ms, a frame in 1x EV-DO Revision 0 is 26.66 ms.[9] 1x EV-DO Revision 0 uses a time-division multiplexed (TDM) downlink (transmitting to one user at a time). The traffic channel data rate used by the access network for transmission to an access terminal is determined by the data rate control (DRC) message previously sent by the access terminal on the uplink. The DRC indicates not only the data rate, but also the modulation, code rate, preamble length, and maximum number of slots required to achieve the desired physical layer error rate.

1x EV-DO Revision 0 introduced physical layer hybrid ARQ on the downlink. The access network transmits packets to an access terminal over multiple slots staggered in time. A three-slot separation between subpacket transmissions allows the access terminal (AT) to demodulate and decode the packet, and indicate to the access network whether or not the packet was successfully decoded.

The 1x EV-DO Revision 0 downlink traffic channel is a shared medium that provides high peak rate transmission to active access terminals. Addressing on the shared channel is achieved by a MAC index that is used to identify data transmission from a sector to a particular access terminal.

The CDMA2000 1x EV-DO Revision A system was standardized in March 2004 by 3GPP2 and the Telecommunication Industry Association (TIA) of North America. A 1x EV-DO Revision A network can provide downlink sector capacity of 1,500 kbps and uplink capacity of 500 kbps (two-way receive diversity), or 1,200 kbps (four-way receive diversity), with 16 active users per sector, using just 1.25 MHz of the spectrum.

The enhancements offered by CDMA2000 1x EV-DO Revision A are as follows:

- An uplink physical layer with hybrid ARQ higher-order modulation (quadrature phase shift keying, QPSK, and 8-PSK), higher peak rate (1.8 Mbps), and finer rate quantization
- An uplink multiflow MAC with QoS support, comprehensive network control of spectral efficiency and latency trade-off for each application flow, and a more robust interference control mechanism that permits system operations at higher load

- A downlink physical layer with higher peak rate (3.1 Mbps), finer rate quantization, and short packets for transmit delay reduction and improved link utilization
- A downlink MAC layer that permits the access network to serve multiple users with the same physical layer packet, improving not only transmission latency but also packing efficiency
- Rapid connection setup for applications that require instant connect use of shorter interpacket intervals and a higher rate access channel

1x EV-DO Revision A is designed to offer efficient support for both delay-sensitive and delay-tolerant applications. Features added to 1x EV-DO Revision A are short packets and multiuser packets on the downlink, and physical layer ARQ and multi-flow reverse link MAC layer on the uplink. The inclusion of these features provides substantial improvement in the performance of delay-sensitive applications such as VoIP, gaming, and videotelephony. 1x EV-DO Revision A is fully compatible with 1x EV-DO Revision 0 networks. 1x EV-DO Revision A systems deliver high spectral efficiency, support large numbers of mobile users, provide performance comparable to toll-quality applications, support end-to-end QoS that allows operators to maximize revenue through tiered services, and provide comprehensive network control over terminal behavior.

To conclude, the enhancements in 1x EV-DO Revision A provide significant gains in spectral efficiency and substantial improvement in QoS support relative to 1x EV-DO Revision 0. In particular, Revision A approximately doubles the reverse link spectral efficiency for best-effort packet applications requiring low latency (e.g., VoIP, videotelephony, wireless gaming, and PTT).

3GPP2 is taking a similar two-step approach by first developing CDMA2000 1x EV-DO Revision B to support even higher data rates by means of scalable bandwidth (multicarrier) techniques, followed by a longer-term evolution to be specified in Revision C. Air interface technologies that are considered for the long-term evolution and benchmarked against the performance of existing systems include multiple access schemes such as orthogonal frequency division multiple access (OFDMA) and space-division multiple access (SDMA); advanced multiple antenna technologies such as receiver-diversity, multiple-input multiple-output (MIMO), and beamforming antennas; higher-order modulation schemes such as 64-quadrature amplitude modulation (QAM); improved signal processing within the receiver such as interference cancellation; and powerful equalization techniques. The overall goal is to improve user experience by increasing the peak data rates and shortening the application delay, as well as reducing the cost for the operators by increasing the spectral efficiency and reducing the cost of network components. An evolutionary approach to system development and deployment enjoys the benefits of lower cost and faster time to market. Upgrading an existing network typically involves adding new channel cards and upgrading the system software, which is far less costly and time-consuming than building a new system from the ground up.

2.5   3G MOBILE COMMUNICATION SYSTEMS AND WCDMA

The 3G WCDMA standard has been enhanced to offer significantly increased per-formance for packet data and broadcast services through the introduction of HSDPA, enhanced uplink, and MBMS. The rapid widespread deployment of WCDMA and an increasing uptake of 3G services are raising expectations with regard to new ser-vices. Packet data services such as Web surfing and file transfer are already provided in the first release of WCDMA networks, release 99. Although this is a significant improvement compared to 2G networks, where such services have limited or no sup-port, WCDMA is continuously evolving to provide even better performance. Release 5 of WCDMA, finalized in early 2002 and with products starting to appear, intro-duced improved support for downlink packet data, often referred to as HSDPA. In release 6, finalized early 2005, the packet data capabilities in the uplink (enhanced uplink) were improved. Release 6 also brought support for broadcast services through MBMS, enabling applications such as mobile TV. The path from WCDMA to WCDMA Evolved is illustrated in Figure 2.7.

WCDMA has been evolving to meet the increasing demands for high-speed data access broadcast services. These two types of services have different characteristics, which influence the design of the enhancements.

For high-speed data access, data typically arrives in bursts, posing rapidly vary-ing requirements on the amount of radio resources required. The transmission is typically bidirectional and low delays are required for a good end-user experience. As the data is intended for a single user, feedback can be used to optimize the trans-mission parameters.

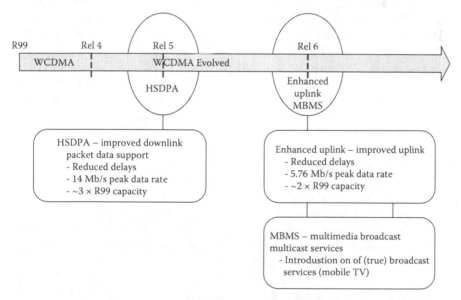

FIGURE 2.7   The path from WCDMA to WCDMA Evolved. Reproduced with permission from S. Pakval et al., "Evolving 3G mobile systems: Broadband and broadcast services in WCDMA," *IEEE Communications Magazine* 44 (February 2006): 68–74.

Broadcast/multicast services carry data intended for multiple users. Consequently, user-specific adaptation of the transmission parameters is cumbersome and diversity not requiring feedback is crucial. Due to the unidirectional nature of broadcast data, low delays for transmission are not as important as for high-speed data access.

High-Speed Data Access

To meet the requirement on low delays and rapid resource (re)allocation, the corresponding functionality must be located close to the air interface. In WCFMA this has been solved by locating the enhancement in the base station as part of additions to the MAC layer.

Traditional cellular systems have typically allocated resources in a relatively static way, where the data rate for a user is changed slowly or not at all. This approach is efficient for applications with a relatively constant data rate such as voice. For data with a bursty nature and rapidly varying resource requirements, fast allocation of shared resources is more efficient. In WCDMA, the shared downlink resource consists of transmission power and channelization codes in node B (the base station), while in the uplink the shared radio resource is the interference at the base station. Fast scheduling is used to control allocation of the shared resource among users on a rapid basis. Additionally, fast hybrid ARQ with soft combining enables fast retransmission of erroneous data packets. A short transmission time interval (TTI) is also employed to reduce the delays and allow the other features to adapt rapidly. Similar principles are used for both HSDPA and enhanced uplink, although the fundamental differences between downlink and uplink must be accounted for.

As an illustration, the architecture with HSDPA, enhanced uplink additions, and MBMS enhancements is shown in Figure 2.8. HSDPA, enhancement link, and MBMS can simultaneously be present in a single cell, although for illustrative purposes they are shown in different cells in the figure.

A number of radio network controllers (RNCs) are connected to the core network. Each RNC controls one or several node Bs, which in turn communicate with the user equipment (UE). The radio link control (RLC) entity in the RNC is unchanged compared to previous versions of WCDMA; it provides ciphering and also guarantees lossless data delivery if the hybrid ARQ protocol fails, for example, at an HSDPA cell change, where the node B buffers are flushed. Some functionality has also been added to the existing MAC functionality in the RNC to support flow control between the RNC and node B for HSDPA, and reordering and selection combining for enhanced uplink. Furthermore, the RNC handles mobility, for example, channel switching when a user is moving from a cell where a previous release of WCDMA is used. The RNC is also responsible for the overall radio resource management, for example, setting limits on the amount of resources to be used for HSDPA and enhanced uplink.

High-Speed Downlink Packet Access (HSDPA)

A key characteristic of HSDPA is the use of shared-channel transmission. This implies that a certain fraction of the total downlink radio resources available within

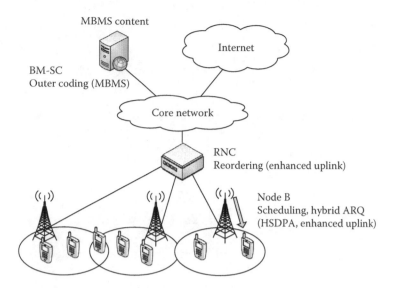

FIGURE 2.8 Architecture with HSDPA and enhanced uplink additions, and MBMS enhancements. Reproduced with permission from S. Pakval et al., "Evolving 3G mobile systems: Broadband and broadcast services in WCDMA," *IEEE Communications Magazine* 44 (February 2006): 68–74.

a cell can be seen as a common resource that is dynamically shared between users, primarily in the time domain. The use of shared-channel transmission, in WCDMA implemented through the high-speed downlink shared channel (HS-DSCH), enables the possibility to rapidly allocate a large amount of the downlink resources to a user when needed.

The basic HS-DSCH code and time structure are illustrated in Figure 2.9. The HS-DSCH code resource consists of a number of codes of spreading factor SF = 16, and the number codes is configurable between 1 and 15. Codes not reserved for HS-DSCH transmission are used for other purposes (e.g., related control signaling, MBMS, and circuit-switched services such as voice).

Allocation of the HS-DSCH code resource is done on a 2-ms TTI basis. The use of a short TTI reduces the overall delay and improves the tracking of fast channel variations exploited by the link adaptation and the channel-dependent scheduling as discussed below. Although the common code resource is shared primarily in the time domain, sharing in the code domain is also possible. The reasons are twofold: support of terminals not able to despread the full set of codes, and efficient support of small payloads (i.e., when the transmitted data does not require the full set of allocated HS-DSCH codes).

In addition to being allocated a part of the overall code resource, a certain part of the total available cell power should also be allocated for HS-DSCH transmission. Note that the HS-DSCH is not power controlled but rate controlled. This allows the remaining power (after serving other, power-controlled channels) to be used for HS-DSCH transmission and enables efficient exploitation of the shared power resource.

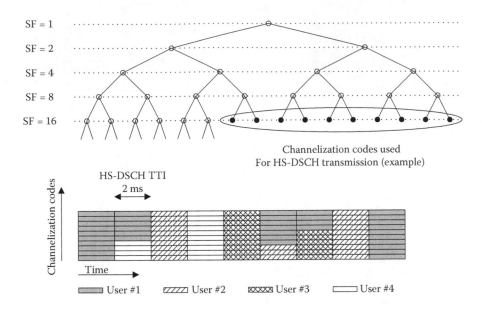

FIGURE 2.9   Code and time domain structure for HS-DSCH. Reproduced with permission from S. Pakval et al. "Evolving 3G mobile systems: Broadband and broadcast services in WCDMA," *IEEE Communications Magazine* 44 (February 2006): 68–74.

Control signaling necessary for successful reception of the HS-DSCH at the terminal is carried on shared control channels. There is also a need for transmitting power-control commands for the uplink in the downlink. These are carried either on a conventional dedicated channel, which can also carry non-HS-DSCH services, or on a new type of dedicated channel introduced in release 6, optimized to carry power control commands only and reducing the code space required by up to a factor of 10.

The scheduler is a key element and to a large extent determines the overall downlink performance, especially in a highly loaded network. In each TTI, the scheduler decides to which user(s) the HS-DSCH should be transmitted and, in close cooperation with the link-adaptation mechanism, at what data rate. A significant increase in capacity can be obtained if channel-dependent scheduling is used. Since the radio conditions for the users typically vary independently, at each point in time there is almost always a user whose channel quality is near its peak. The gain obtained by transmitting to users with favorable conditions is commonly known as multiuser diversity, and the gains are larger with larger channel variations and a larger number of users.[11]

A practical scheduler strategy exploits the short-term variations (e.g., due to multipath fading and fast interference variations) while maintaining some degree of long-term fairness between the users. In principle, the larger the long-term unfairness, the higher the cell capacity, and trade-off between the two is required. Additionally, traffic priorities should also be taken into account, for example, to prioritize streaming services before a file download. The scheduler algorithm is implementation specific.[12]

The next key feature of HSDPA is hybrid ARQ with soft combining, which allows the terminal to rapidly request retransmission of an erroneously received transport block, essentially fine-tuning the effective code rate and compensating for errors made by the link adaptation mechanism. Closed-loop power control has been used in CDMA systems to combat the fading variations in the radio channel and to maintain a constant signal-to-noise ratio (E_b/N_0). For services that can tolerate some jitter in the data rate, it is more efficient to control the signal-to-noise ratio by adjusting the data rate while keeping transmission power constant. This is known as link adaptation or rate adaptation.

Link adaptation is implemented by adjusting the channel-coding rate, and selecting between QPSK and 16-QAM. Higher-order modulation such as 16-QAM makes more efficient use of bandwidth than QPSK, but requires greater received E_b/N_0. Consequently, 16-QAM is mainly useful in advantageous channel conditions. In addition, the data rate also depends on the number of channelization codes assigned for HS-DSCH transmission in a TTI. The data rate is selected independently for each 2-ms TTI by node B, and the link-adaptation mechanism can therefore track rapid channel variations.

Soft combining implies that the terminal does not discard soft information in case it cannot decode a data block as in traditional hybrid ARQ protocols, but combines soft information from previous transmission attempts with the current retransmission to increase the probability of successful decoding. Incremental redundancy (IR) is used as the basis for soft combining, that is, the retransmissions may contain parity bits not included in the original transmission. It is well known that IR can provide significant gains when the code rate for the initial transmission attempts is high, as the additional parity bits in the retransmission result in a lower overall coding rate. Thus, IR is mainly useful in band-limited situations, for example, when the terminal is close to the base station and the amount of channelization codes (and not the transmission power) limits the achievable data rate.

Enhanced Uplink

The enhanced uplink relies on basic principles similar to those of the HSDPA downlink: scheduling and fast hybrid ARQ, implemented through an enhanced dedicated channel (E-DCH). The E-DCH is turbo-encoded and transmitted in a similar way as the DCH in previous releases. Simultaneous transmission on E-DCH and DCH is possible; the E-DCH is processed separately from the other channels.

In addition to the 10-ms TTI found in earlier releases, the E-DCH supports a TTI of 2 ms, thus reducing the delays and allowing for fast adaptation of the transmission parameters. One transport block of data can be transmitted in each TTI; the size of the transport block depends on the available power and the limitations set by the scheduling mechanism in the node B. Multiple data flows with different priorities can be multiplexed onto the E-DCH to support mixed services.

Unlike the downlink, the uplink is nonorthogonal and fast power control is therefore essential for the uplink to handle the near-far problem and to ensure coexistence with terminals and services not using the enhancements. The E-DCH is transmitted with a power offset relative to the power-controlled uplink control channel and, by adjusting the maximum allowed power offset, the scheduler can control the E-DCH data rate.

Soft handover is supported for the E-DCH for two reasons: first, receiving the transmitted data in multiple cells adds a macro-diversity gain, and second, power control from multiple cells is required in order to limit the amount of interference generated in neighboring cells.

For the E-DCH, the shared resource is the amount of tolerable interference, that is, the total received power at node B, and the purpose of the scheduler is to determine which terminals are allowed to transmit when and at what data rate. To efficiently support packet data services, the target is to allocate a large fraction of the shared resource to users momentarily requiring high data rates, while at the same time ensuring stable system operation by avoiding large interference peaks.

The scheduling framework is based on scheduling grants sent by the node B scheduler to control the transmission activity, and scheduling requests sent to request resources. The scheduling grants control the maximum allowed E-DCH-to-pilot power ratio the terminal may use; a large grant implies the terminal may use a higher data rate but also contribute more to the interference level (noise rise) in the cell. Based on measurements of the (instantaneous) interference level, the scheduler controls the scheduling grant in each terminal to maintain the interference level in the cell at a desired target. Unlike HSDPA, where typically only a single user is addressed in each TTI, the implementation-specific uplink scheduling strategy in most cases will schedule multiple users in parallel. The reason for this is the significantly smaller transmit power of a terminal compared to a node B; a single terminal typically cannot utilize the full cell capacity on its own.

Fast scheduling allows for a more relaxed connection admission strategy. A larger number of bursty high-rate packet-data users can be admitted to the system as the scheduling mechanisms can handle the situation when multiple users need to transmit in parallel. Without fast scheduling, the admission control would have to be more conservative and reverse a margin in the system in case of multiple users transmitting simultaneously.

Hybrid ARQ with soft combining can be exploited not only to provide robustness against unpredictable interference, but also to improve the link efficiency to increase capacity and/or coverage. One possibility to provide a data rate of x Mbps is to transmit at x Mbps and set the transmission power to target a low error probability (on the order of a few percent) on the first transmission attempt. Alternatively, the same resulting data rate can be provided by transmitting using an n times higher data rate at an unchanged transmission power and multiple hybrid ARQ retransmissions.

MULTIPLE BROADCAST MULTICAST SERVICES

In the past, cellular systems have mostly focused on the transmission of data intended for a single user and not on broadcast services. Broadcast networks, exemplified by radio and TV broadcasting networks, have on the other hand focused on covering very large areas and have offered limited or no possibilities for transmission of data intended for a single user. MBMS, introduced in Release 6, supports multicast/broadcast services in a cellular system, thereby combining multicast and unicast transmissions within a single network. With MBMS, the same content is transmitted to multiple users in a unidirectional fashion, typically by multiple cells in order to cover the large area in which the service is provided. Broadcast and multicast describe different (although closely related) scenarios:

In broadcast, a point-to-multipoint radio resource is set up in each cell as part of the MBMS broadcast area and all users subscribing to the broadcast service simultaneously receive the same transmitted signal. No tracking of users' movements in the radio access network is performed and users can receive the content without notifying the network. Mobile TV is an example of a service that can be provided through MBMS broadcast.

In multicast, users request to join a multicast group prior to receiving any data. Users' movements are tracked and the radio resources are configured appropriately. Each cell in the MBMS multicast area may be configured for point-to-point or point-to-multipoint transmission. In sparsely populated cells with only one or a few users subscribing to MBMS, point-to-point transmission may be appropriate, while in cells with a larger number of users, point-to-multipoint transmission is better suited.

Point-to-multipoint MBMS data transmission uses the forward access channel (FACH) with turbo-coding and QPSK modulation at a constant transmission power. Multiple services can be configured in a cell, either time multiplexed on one FACH or transmitted on separate channels.

2.6 3G PERSONAL COMMUNICATION SERVICES TECHNOLOGIES

3G personal communication services (PCS) are already offered in limited areas, and will be used more extensively in the future.[14] 3G PCS will convey multimedia traffic in mobile and wireless environments. UMTS and 3G PCS systems will gradually replace GSM. UMTS uses both WCDMA and hybrid time division multiple access (TDMA)/CDMA in the radio network.[15,16]

Since UMTS and most of the other PCS technologies are cellular architectures, they require a carefully designed and deployed infrastructure. Sometimes rapid deployment without extensive preplanning is needed. Tactical communication systems and networks used after disasters are examples of systems that require rapid deployment. In these systems, predeployment of a cellular infrastructure is often impossible. Therefore, infrastructureless routing algorithms and resource management schemes are needed to fulfill the rapid deployment requirement.

Ad hoc techniques have been developed to route data packets between mobile terminals through an infrastructureless network. Many ad hoc routing techniques have been proposed in the literature.[17] Available ad hoc routing algorithms, however, are not scalable enough to manage tens of thousands of nodes. Furthermore, they do not address the management of radio resources.

Virtual cell layout (VCL) leverages both cellular and ad hoc paradigms to handle a large number of mobile terminals in a rapidly deployable network. In VCL, the communication area is tessellated with fixed-size hexagons. Each hexagon represents a

VCL cell to which the available spectrum is assigned. Also, the CDMA codes are distributed among the fixed VCL cells. Hence, if a mobile access point can find out its geographic location, it can also determine the available set of carriers and codes without a need for a central topology database or a central resource manager.

2.7   CHALLENGES IN THE MIGRATION TO 4G MOBILE SYSTEM

The step to be taken in order to arrive at the goal of 4G is called beyond 3G (B3G). In other words, B3G is heterogeneous systems and networks working together, while 4G is a new air interface. Within the rapid development of wireless communications networks, it is expected that 4G mobile systems will be launched within decades. 4G mobile systems focus on seamlessly integrating the existing wireless technologies including GSM, wireless LAN, and Bluetooth. Also, 4G systems will support comprehensive and personalized services, providing stable system performance and quality of service.

Development of new wireless access technologies, services, and applications has been motivated by identifying future service needs, potentially available spectrum matching the capacity needs, and agreeing on the technical specifications to enable access to new services. This also has led to the emergence of a wide range of wireless digital transmission technologies and service platforms to comply with the new user needs requiring more capacity, support of multimedia traffic, extended support for mobility, and so on. This is the way, for instance, that GSM/GPRS/EDGE have been made available for wide-area mobile communications.

On the path toward 4G, maximum range of service platforms and access infrastructures have to be ensured. 4G has to be seen as the next generation communication systems technology, which may include new wireless access technologies, but in any case will be able to provide a unified framework to both ends of the communication system.[17]

Different research programs have their own visions regarding 4G features and implementations. Some key features, mainly from the user's point of view of 4G networks, are stated as follows:

- High usability (anytime, anywhere, and with any technology)
- Support for multimedia services at low transmission cost
- Personalization
- Integrated services

First, 4G networks are all-IP-based heterogeneous networks that will allow users to use any system, at any time, anywhere. Users carrying an integrated terminal will be able to use a wide range of applications provided by multiple wireless networks. Second, 4G systems provide not only telecommunication services, but also data and multimedia services. To support multimedia applications, high data rate services with good system reliability must be provided. At the same time, a low per-bit transmission cost will be maintained. Third, personalized service will be provided by this new generation network. It is expected that when 4G services are launched, users in widely different locations, occupations, and economic classes will use the services.

To meet the demands of these diverse users, service providers should design personal and customized services for them.[18] Finally, 4G systems will also provide facilities for integrated services. Users will be able to use multiple services from any service provider at the same time.

4G wireless communication systems will be made up of different radio networks providing access to an IP version 6 (IPv6)-based network layer.[19,20] Multimedia is expected to be a main application of 4G networks. However, multimedia streams can be sensitive to packet losses, which in turn can result in video artifacts. Such packet losses can often occur when there is an interruption to a connection as a user moves between autonomous networks.

Cooperation of heterogeneous access networks (cellular and broadcast in particular) is an area that has been being investigated for some time through a number of projects, with the aim of setting up technical foundations, developing specific services and architectures, and addressing network management aspects.[21]

Most of the technical barriers have been identified, but new approaches are needed to adapt to the current regulatory and business context characterized by openness of systems, diversification of system actors, and the search for productive investment. Network cooperation is probably one of the main clues for addressing the 4G technological landscape, but it needs to be driven by a number of requirements that are meaningful from the technological viewpoint, as well as from regulatory and business ones.

In recent years an excess of available technologies addressing killer applications has created probability issues for many companies, hence leading to the rethinking of requirements not only on the technical side, but also from the business and end-to-end perspectives. In parallel, there has been increasing interest in the push paradigm, in particular to groups of users supported by the spectrum efficiency of broadcast bearers, and the attractiveness of broadcast TV interactive services. The push toward an Information Society has motivated the development of new wireless access technologies, services, and applications.

Moving toward 4G, wireless ad hoc networks receive growing interest due to users' provisioning of mobility, usability of services, and seamless communications. In fading environments, ad hoc networks provide the opportunity to exploit variations in channel conditions and transmit to the user employing the current *best channel.*

Wireless ad hoc networking has recently attracted growing interest, and has emerged as a key technology for next generation wireless networking. Devices enabling the wireless ad hoc networking paradigm are becoming smaller and cheaper, with lots of embedded capabilities delivering services seamlessly to end users and paving the path toward 4G. In a wireless ad hoc network a node sends or forwards packets to its neighboring nodes by accessing the shared wireless channel. A significant characteristic of a wireless channel is time-varying fading due to the existence of multiple transmission paths between a source and a destination. In practice, the channel quality among surrounding nodes can vary significantly for both mobile and stationary nodes. Any change in the line-of-sight path or any reflected path will affect the channel quality and hence change the data rate that is feasible with multirate networks. Although traditionally viewed as a source of unreliability that needs to be mitigated, recent research suggests exploiting the channel fluctuations opportunistically when and where the channel is strong.

In wireless ad hoc networks there are two main classes of opportunistic transmission. The first is to exploit the time diversity of an individual link by adapting its transmit rate to the time-varying channel condition. The basic idea is to transmit more packets at higher rates when the channel condition is better. Exploiting multiuser diversity is another class of opportunistic transmission, which jointly leverages the time and spatial heterogeneity of channels to adjust rates. In wireless networks, a node may have packets destined to multiple neighboring nodes. Instantaneously selecting an *on-peak* receiver with the best channel condition improves system performance.[22]

However, most existing opportunistic transmission schemes do not consider the interaction among neighboring transmitters (i.e., a sender individually makes its local decision to maximize its own performance). It is hard to obtain the optimal overall system performance without leveraging node cooperation due to the following challenges. First, with a hidden terminal there is inequality in channel contention among nodes in wireless ad hoc networks, which can result in severe overall performance inefficiency. Second, with the shared wireless medium, co-channel interference has a deep impact on rate selection and flow scheduling in wireless ad hoc networks. Hence, neighboring transmitters should jointly determine the "on-peak" flows and their corresponding rate in a distributed way. Third, different QoS requirements of the system correspond to different optimization targets, for example, energy efficiency and throughput maximization, which call for different strategies. All these challenges require an efficient node cooperation mechanism to coordinate the transmissions among neighboring nodes.

Energy efficiency is one of the key issues in wireless ad hoc networks because most mobile devices are battery operated. An effective way to achieve energy efficiency is to reduce the transmission power whenever possible. However, in a multirate enabled network, reducing transmission power may result in reduced transmission rate. Moreover, in a wireless ad hoc network, the hidden terminal phenomenon will cause one node to have smaller contention probability than another node (say, a node in the *hidden* position); hence, different nodes will have different probabilities of winning the channel access (we call this phenomenon inequality of channel access). The inequality of channel access can result in severe overall energy inefficiency. Thus, node cooperation in rate adaptation to achieve high overall energy efficiency is called for.

One of the most important features of HSDPA is packet scheduling. The main goal of packet scheduling is to maximize system throughput while satisfying the QoS requirements of users. The packet scheduler determines to which user the shared channel transmission should be assigned at a given time. In HSDPA the packet scheduler can exploit short-term variations in the radio conditions of different users by selecting those with favorable instantaneous channel conditions for transmission, which is illustrated in Figure 2.10. This idea is based on the fact that good channel conditions allow for higher data rates (R) by using a higher-order modulation and coding schemes, thus resulting in increased system throughput.[23,24]

To quickly obtain up-to-date information on the channel conditions of different users, the functionality of the packet scheduler has been moved from the RNC

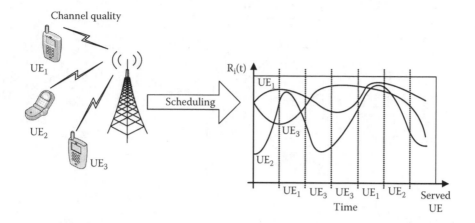

FIGURE 2.10 The user channel quality for scheduling decisions. Reproduced with permission from B. A. Manthari, H. Hassanien, and N. Nasser. "Packet scheduling in 3.5G high-speed downlink packet access networks: Breadth and depth," *IEEE Network* 21 (January/February): 41–46.

in UMTS to the MAC high-speed (MAC-hs) sublayer at the node B as shown in Figure 2.11.[25] The MAC-hs is a new sublayer added to the MAC layer at the node B in HSDPA in order to execute the packet scheduling algorithm. In addition, the minimum TTI (i.e., the time between two consecutive transmissions) has been reduced from 10 ms in UMTS Release 99 to 2 ms in Release 5 which includes HSDPA.

This is because HSDPA allows the packet scheduler to better exploit the varying channel conditions of different users in its scheduling decisions and increase the granularity of the scheduling process. It should be noted that favoring users with good channel conditions may prevent those with bad channel conditions from being served. A good design of a scheduling algorithm not only should take into account maximization of the system throughput, but also should be fair to users who use the

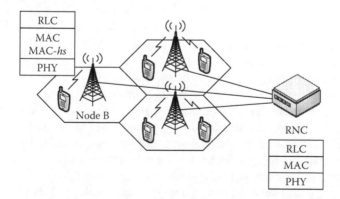

FIGURE 2.11 The MAC-hs at rate node B in HSDPA. Reproduced with permission from B. A. Manthari, H. Hassanien, and N. Nasser. "Packet scheduling in 3.5G high-speed downlink packet access networks: Breadth and depth," *IEEE Network* 21 (January/February): 41–46.

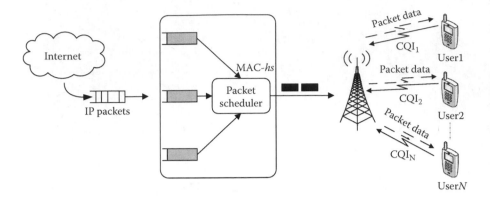

FIGURE 2.12 The packet scheduler model in HSDPA. Reproduced with permission from B. A. Manthari, H. Hassanien, H., and N. Nasser. "Packet scheduling in 3.5G high-speed downlink packet access networks: Breadth and depth," *IEEE Network* 21 (January/February): 41–46.

same service and pay the same amount of money. That is, scheduling algorithms should balance the trade-off between maximizing throughput and fairness.

We now briefly describe the packet scheduler model and how it works in HSDPA. As mentioned above, the packet scheduler for HSDPA is implemented at the MAC-hs layer of node B. Node B can serve N users simultaneously, $N \geq 1$, and selects one transmission user in a slot of fixed time duration. Also, and without loss of generality, it is assumed that each user has one connection request. Thus, a node B maintains one queue for every user. The packet scheduler model in HSDPA is shown in Figure 2.12.

Upon call arrival, the RLC layer receives traffic in the form of IP packets from higher layers. The packets are segmented into fixed-size protocol data units (PDUs). These PDUs are stored in the transmission queue of the corresponding user in a first-in/first-out fashion. Subsequently, the PDUs are transmitted to the appropriate mobile user according to the adopted scheduling discipline.

The packet scheduler works as follows. During every TTI, each user regularly informs the node B of its channel quality condition by sending a report known as a channel quality indicator (CQI) in the uplink to the node B. The CQI contains information about the instantaneous channel quality of user. This information includes the size of the transport block the node B should send to the user, the number of simultaneous channel codes, and the type of modulation and coding schemes the user can support. Node B then selects the appropriate mobile user according to the adopted scheduling discipline and sends data to the selected user at the specified rates. The user is able to measure its current channel conditions by measuring the power of the received signal from the node B and then using a set of models to determine its current supportable data rates (i.e., the rates at which it can receive data from the node B given its current channel condition). Therefore, users with good channel conditions will enjoy potentially higher supportable data rates by using higher modulation and coding rates, whereas users with bad channel conditions will experience lower data rates instead of adjusting their transmission power.[24,25]

HSDPA is designed to support non-real-time applications (interactive and background) and also to some extent real-time applications (streaming). Since real-time applications have different QoS constraints than non-real-time applications, the design of scheduling algorithms for real-time applications should be different from that for non-real-time applications. Therefore, scheduling algorithms can be classified into two groups: real-time (RT) and non-real-time (NRT) scheduling algorithms. In addition, scheduling algorithms within each group can be characterized by three factors[26]:

- Scheduling frequency—The rate at which users are scheduled. Scheduling algorithms that make use of the channel conditions of users need to make decisions every TTI to better exploit fast variation of channel conditions and are therefore called fast scheduling algorithms. Other scheduling algorithms that do not make a decision every TTI are called slow scheduling algorithms.
- Service order—The order in which users are served. For example, some scheduling algorithms schedule users based on their channel conditions, whereas others schedule them randomly.
- Allocation method—The method of allocating resources. For instance, some scheduling algorithms provide the same data amount for all users per allocation interval, while others give all users the same time, code, or power per allocation interval.

2.8 CONCLUDING REMARKS

The converged network will be a vital component of the next generation applications currently being developed for a wide range of business situations. The demand for making the converged network of the twenty-first century more like today's Internet will create a large challenge for the software architects of the converged network.

1x EV-DO Revision A is fully backward compatible with 1x EV-DO Revision 0 networks, and an upgrade involves only a change to the mobile station and base station. 1x EV-DO Revision A systems deliver high spectral efficiency, support large numbers of mobile users, provide performance comparable with toll-quality voice applications, and provide comprehensive network control over terminal behavior.

With the recent evolution to the WCDMA standard, support for packet data and broadcast services has been considerably improved to meet future demands. Fast adaptation to rapidly varying traffic and channel conditions has been applied to WCDMA through HSDPA and enhanced uplink, thereby providing high data rates to cellular users. Similarly, by combining the transmissions from multiple sites, true broadcast services are possible in WCDMA with the introduction of MBMS.

High-speed downlink packet access has been introduced in order to support high data rates beyond those that 3G/UMTS can offer. HSDPA promises a data rate of up to 10 Mbps, which allows support of new multimedia applications and improved QoS for already existing ones. HSDPA relies on new technologies to help achieve the high data rates it offers, among which is packet scheduling. The functionality of packet scheduling is crucial to the operation of HSDPA, since it controls the distribution of radio resources among mobile users.

3 Wireless Video

This chapter surveys wireless video that has been commercialized recently or is expected to go to market in third-generation (3G) and beyond mobile networks, mainly covering the corresponding technologies. We present a general framework that takes into account multiple factors, including source coding, channel resource allocation, and error concealment, for the design of energy-efficient wireless video communication systems. This framework can take various forms and be applied to achieve the optimal trade-off between energy consumption and video delivery quality during wireless video transmission. This chapter also reviews rate control in streaming video over wireless, which is an important issue. We continue with a short presentation of content delivery technologies. The emphasis is on layer ½ technologies, as well as technologies above layer 2. This chapter concludes with the H.264/AVC standard in the wireless video environment, together with a video coding and decoding algorithm, network integration, compression efficiency, error resilience, and bit error adaptivity. The applicability of all these encoding and network features depends on application, constraints such as the maximum tolerable delay, possibility of online encoding, and availability of feedback and cross-layer information.

3.1 INTRODUCTION

In the early 1990s, most of us could not have imagined the current popularity of multimedia communication over wireless networks. At the beginning of this century, we have experienced two mobile digital network generations: the second generation (2G), which brought us digital mobile communication, and the 3G, which is characterized by its ability to carry data at much higher rates, 2G–3G radio access networks include:

- General packet radio service (GPRS)
- Enhanced Data Global System for Mobile Communications (GSM) Environment (EDGE)
- Wideband code-division multiple access (W-CDMA), also known as Universal Mobile Telecommunications System (UMTS)
- High-speed downlink packet access (HSDPA)
- CDMA2000 1x Evolution, Data-Only (1x EV-DO)[1]

3G cellular network diffusion seems to be advancing steadily. Mobile multimedia communication has taken off with 3G networks.[2] Flat-rate pricing for unlimited access to data on 3G mobile networks is now becoming a common practice of operators. The combination of WiFi and 3G cellular networks will bring a realistic and comfortable solution beyond 3G, where 54 Mbps in hotspots and several hundred kilobits per second with wide coverage are available.

Video communication through wireless channels is still a challenging problem due to the limitations in bandwidth and the presence of channel errors. Because many video services are originally coded at a high rate and without considering the different channel conditions that may be encountered later, a means to repurpose this content for delivering over a dynamic wireless channel is needed.[3]

Transmitting video over wireless channels from mobile devices has gained increased popularity in a wide range of applications. A major obstacle to these types of applications is the limited energy supply in mobile device batteries. For this reason, efficiently using energy is a critical issue in designing wireless video communication systems.[4]

Rate control is important to multimedia streaming applications in both wired and wireless networks. First, it results in full utilization of bottleneck links by ensuring that sending rates are not too low. Second, it prevents congestion collapse by ensuring that sending rates are not too aggressive.[5]

Most emerging and future mobile client devices will significantly differ from those used for speech communications only: handheld devices will be equipped with a color display and a camera, and have sufficient processing power to allow presentation, recording, and encoding/decoding of video sequences. In addition, emerging and future wireless systems will provide sufficient bit rates to support video communication applications. Nevertheless, bit rates will always be scarce in wireless transmission environments due to physical bandwidth and power limitations; thus, efficient video compression is required. Nowadays H.263 and MPEG-4 Visual Simple Profile are commonly used in handheld products, but it is foreseen that H.264/AVC will be the video codec of choice for many video applications in the near future. The compression efficiency of the new standard outdoes prior standards roughly by at least a factor of two. Although compression efficiency is the major feature for a video codec to be successful in wireless transmission environments, it is also necessary that a standard provides means to be integrated easily into existing and future networks, as well as address the needs of different applications.[7]

This chapter starts with a short overview of video over wireless. Then, it seeks to describe an energy-efficient wireless communication system. A brief description of rate control in streaming video over wireless is also presented. Content delivery technologies including layer ½ technologies and techniques above layer 2 are also emphasized. An outline of the H.264/AVC standard in the wireless video environment concludes this chapter.

3.2   VIDEO OVER WIRELESS

A wireless transmission system might delay, lose, or corrupt individual data units. The unavailability of a single data unit usually has significant impact on perceived quality due to spatiotemporal error propagation. In modern wireless system designs, data transmission is usually supplemented by additional information between the sender and the receivers, and within the respective entities. Abstract versions of available messages are included.

The video encoder generates data units containing the compressed video stream, possibly stored in an encoder buffer before transmission. Abstract versions of available messages are included in Figure 3.1. In fact, this is an abstraction of an

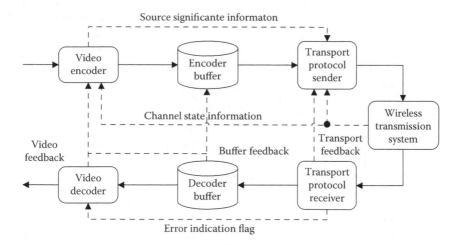

FIGURE 3.1 An end-to-end video transmission system. Reproduced with permission from T. Stockhammer and M.M. Hannukesela, "H.264/Ave Video for wireless transmission." *IEEE Wavelength Communications* 12 (August 2005): 6–13.

end-to-end video transmission system. Each processing and transmission step adds some delay which can be fixed, deterministic, or random. The encoder and decoder buffers allow compensating for variable bit rates produced by the encoder, as well as channel delay variations to keep the end-to-end delay constant and maintain the timeline at the decoder. If the initial play-out delay is not or cannot be too extensive, late data units are commonly treated as lost.

In a typical video distribution scenario as shown in Figure 3.2, video content is captured, then immediately compressed and stored on a local networks. At this stage, compression efficiency of the video signal is most important as the content is usually encoded with relatively high quality and independently of any actual channel characteristics. We note that the heterogeneity of client networks makes it difficult for the encoder to adaptively encode the video contents for a wide degree of different channel conditions; this is especially true for wireless clients. Subsequently, for transmission over wireless or highly congested networks, the video bit stream first passes through a network node, such as mobile switch/base station or proxy

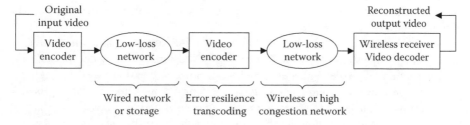

FIGURE 3.2 Video transmission scenario with error resilience transcoding. Reproduced with permission from A-Vetro, J. Xin and H. Sun, "Error resilience video transcoding for wireless communications." *IEEE Wireless Communications* 12 (August 2005): 14–21.

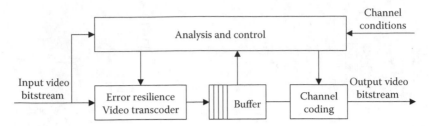

FIGURE 3.3 Error resilience video transcoding. Reproduced with permission from A Vetro, J.Xin and H. Sun. "Error resilience video transcoding for wireless communications," *IEEE Wireless Communications* 12 (August 2005): 14–21.

server, which performs error resilience transcoding. In addition to satisfying rate constraints of the network and display or computational requirements of a terminal, the bit stream is transcoded so that an appropriate level of error resilience is injected in the bit stream. The optimal solution in the transcoder is one that yields the highest reconstructed video quality at the receiver.[8]

The process of error resilience video transcoding is not achieved by the addition of bits into the input bit stream to make the output bit stream more robust to errors. Such an approach is closer to conventional channel coding approaches in which some overhead channel bits are added to the source payload for protection and possible recovery. Rather, for the video source, a variety of strategies exist that affect the bit stream structure at different levels of the stream (e.g., slice vs. block level). Among the different techniques to localize data segments to reduce error propagation, are partitioning of the stream so that unequal error protection can be applied, or redundancy added to the stream to enable a more robust means of decoding.

Error resilience transcoding of video based on analysis of the video bit stream, channel measurement, and buffer analysis is illustrated in Figure 3.3. From the source side, characteristics of the video bit stream are usually extracted to understand the structure of the encoded bit stream and begin building the end-to-end rate distortion model of the source, while from the network side, characteristics of the channel are obtained. Both the content and channel characteristics, as well as the current state of the buffer, are used to control the operation of the error resilience transcoder. It is also possible to jointly optimize the source and channel coding. The transcoding of stored video is not necessarily the same as that for live video. For example, preanalysis may be performed on stored video to gather useful information that may be used during the transcoding process.

We focus now on some of the major conceptual components in a wireless video communication system shown in Figure 3.4. At the sender side, video packets are first generated by a video encoder, which performs compression by exploiting both temporal and spatial redundancies. After passing through the network protocol stack, transport packets are generated and then transmitted over a wireless channel that is lossy in nature. Therefore, the video sequence must be encoded in an error-resilient way that minimizes the effects of losses on the decoded video quality. The set of source coding parameters directly control video delivery quality; this includes prediction mode and quantization step size.

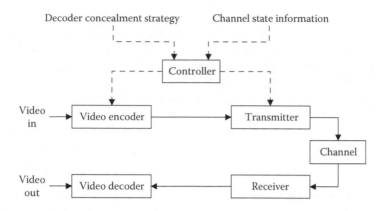

FIGURE 3.4   Conceptual components in wireless communication system.

To combat channel errors, forward error correction (FEC) may be applied at the lower layers such as the link and physical layers. In addition, at the physical layer, modulation modes and transmitter power may be able to be adjusted according to the changing channel conditions. Scheduling the transmission of each packet may also be an adaptable parameter.

The functionality of the lower layer adaptations is indicated by the transmitter block. The set of channel parameters can be controlled at the transmitter. At the receiver, the demodulated bit stream is processed by the channel decoder, which performs error detection and/or correction. This functionality is represented by the receiver block. Corrupt packets are usually discarded by the receiver, and are therefore considered lost. In addition, packets that arrive at the receiver beyond their display deadlines are also treated as lost. This strict delay constraint is another important difference between video communications and many other data transmission applications. The video decoder then decompresses video packets and displays the resulting video frames in real time.

The video decoder employs concealment techniques to mitigate the effects of packet losses. Here, the goal is to achieve the best video delivery while using a minimum amount of transmission energy. Wireless channels typically exhibit high variability in throughput, delay, and packet loss. Providing acceptable video quality in such an environment is a demanding task for the video encoder and decoder, as well as the communication and networking infrastructure. In each of these components, a number of coding and transmission parameters may be adapted based on source content and available channel state information (CSI).

In addition, factors affecting transmission energy consumption include the power used for transmitting each bit, the modulation mode, and the channel coding rate at the link layer or physical layer. To save energy, those parameters should also be adapted to the video content and the CSI.

3.3   ENERGY-EFFICIENT WIRELESS COMMUNICATION SYSTEM

In an increasing number of applications, video is transmitted to and from portable wireless devices such as cellular phones, laptop computers connected to wireless local area networks (WLANs), and cameras in surveillance and environmental tracking systems. For example, the dramatic increase in bandwidth brought by new technologies, such as the present 3G and emerging fourth generation (4G) wireless systems, and the IEEE802.11 WLAN standards, is beginning to enable video streaming capability in personal communications. Although wireless video communications is highly desirable in many applications, a major limitation in any wireless system is the fact that mobile devices typically depend on a battery with a limited energy supply. Such a limitation is especially a concern because of the high energy consumption rate in encoding and transmitting video bit streams. Thus, efficient use of energy becomes highly important, and sometimes the most critical part in the deployment of wireless video applications.

To design an energy-efficient communication system, the first issue is to understand how energy is consumed in mobile devices. Generally speaking, energy in mobile devices is mainly used for computation, transmission, display, and driving the speakers. Among those, computation and transmission are the two largest energy consumers. During computation, energy is used to run the operating system software, and encode and decode the audio and video signals. During transmission, energy is used to transmit and receive the radio frequency (RF) audio and video signals. It should be acknowledged that computation has always been a critical concern in wireless communications. For example, energy-aware operating systems have been studied to efficiently manage energy consumption by adapting the system behavior and workload based on the available energy, job priority, and constraints.

Computational energy consumption is especially a concern for video transmission, because motion estimation and compensation, forward and inverse discrete cosine transforms (DCTs), quantization, and other components in a video encoder all require a significant number of calculations.[9]

In energy consumption in computation, a power rate distortion model is proposed to study the optional trade-off among computational power, transmission rate, and video distortion.[4] Advances in very large scale integration (VLSI) design and integrated circuit (IC) manufacturing technologies have led to ICs with higher and higher integration densities using less and less power. According to Moor's law, the number of transistors on an IC doubles every 1.5 years. As a consequence, the energy consumed in computation is expected to become a less significant fraction of the total energy consumption. Therefore, focusing on the problem of how to encode a video source and send it to the base station in an energy-efficient way is very important. The goal is to minimize the amount of distortion at the receiver given a limited amount of transmission energy, or vice versa, to minimize the energy consumption while achieving a targeted video delivery quality.

One difference between video transmission and more traditional data communications is that video packets are of different importance. To efficiently utilize energy, unequal error protection (UEP) is usually preferred (e.g., it is more efficient to use more power to provide more protection when transmitting the more important packets). This requires a *cross-layer* perspective,[10] where the source and network layers are jointly considered. Specifically, the lower layers in a protocol stack, which directly control transmitter power, need to obtain knowledge of the importance level of each video packet from the video encoder, which is located at the application layer. On the other hand, it can also be beneficial if the source encoder is aware of the estimated CSI passed from the lower layers, and which channel parameters at the lower layers can be controlled, so that it can make smart decisions when selecting the source coding parameters to achieve the best video delivery quality. For this reason, joint consideration of video encoding and power control is a natural way to achieve the highest efficiency in transmission energy consumption.

3.4 STREAMING VIDEO OVER WIRELESS: RATE CONTROL

A widely popular rate control scheme for streaming in wired networks is equation-based rate control, known as Transmission Control Protocol (TCP)-friendly rate control (TFRC). In TFRC the TCP-friendly rate is determined as a function of packet loss rate, round-trip time (RTT), and packet size to mimic the long-term steady performance of TCP.[11] There are basically three advantages to rate control using TFRC. First, it can fully utilize bottleneck capacities while preventing congestion collapse. Second, it is fair to TCP flows, which are the dominant source of traffic on the Internet. Third, the TFRC results in small rate fluctuation, making it attractive for streaming applications that require constant video quality. The key assumption behind TCP and TFRC is that packet loss is a sign of congestion. In wireless networks, however, packet loss is dominated by physical channel errors, violating this key assumption. Neither TFRC nor TCP can distinguish between packet loss due to buffer overflow and that due to physical layer errors.

There have been a number of efforts to improve the performance of TCP or TFRC over wireless.[5] These approaches either hide end hosts from packet loss caused by wireless channel error, or provide end hosts the ability to distinguish between packet loss caused by congestion and that caused by wireless channel error.

Snoop is a TCP-aware local retransmission link layer approach. A Snoop model resides on a router or base station on the last hop (i.e., the wireless link) and records a copy of every forwarded packet. Assuming a Snoop module can access TCP acknowledgment (ACK) packets from the TCP receiver, it looks into the ACK packets and carries out local retransmissions when a packet is corrupted by wireless channel errors. While doing the local retransmission, the ACK packet is suppressed and not forwarded to the TCP sender. These schemes can potentially be extended to TFRC in order to improve the performance by using a more complicated treatment of the ACK packets from the TRFC receiver.[12–14]

End-to-end statistics can be used to help detect congestion when a packet is lost.[13,14] One-way delay can be associated with congestion in the sense that it monotonically increases if congestion occurs as a result of increased queuing delay, and remains constant otherwise.

An end-to-end based approach can be used to facilitate streaming over wireless.[5] Packet interarrival times and relative one-way delay are combined to differentiate between packet loss caused by congestion and that due to wireless channel errors. There are two key observations behind this approach: first, relative one-way delay increases monotonically if there is congestion; second, interarrival time is expected to increase if there is packet loss caused by wireless channel errors. Therefore, these two statistics can help differentiate between congestion and wireless errors. However, the high wireless error misclassification rate may result in underutilization of the wireless bandwidth. It is possible to use a similar approach to improve video streaming performance in the presence of wireless error, under the assumption that the wireless link is the bottleneck.

Other schemes such as those in References 13 and 14 that use end-to-end statistics to detect congestion can also be combined with TFRC for rate control. The congestion detection scheme can be used to determine whether or not an observed packet loss is caused by congestion; TFRC can then take into account only those packet losses caused by congestion when adjusting the streaming rate.

The disadvantage of end-to-end statistics-based approaches is that congestion detection schemes based on statistics are not sufficiently accurate, and require either cross-layer information or modifications to the transport protocol stack.

Another alternative is to use non-loss-based rate control schemes. For instance, TCP Vegas,[15] in its congestion avoidance stage, uses queuing delay as a measure of congestion, and hence could be designed not to be sensitive to any kind of packet loss, including that due to wireless channel error. It is also possible to enable routers with explicit congestion notification (ECN). As packet loss no longer corresponds to congestion, ECN-based rate control does not adjust sending rate upon observing a packet loss.

3.5   CONTENT DELIVERY TECHNOLOGIES

Supporting technologies in light of wireless video communication, from physical layers to application layers, are discussed in this section. They are currently in use or planned in the near future.

LAYER ½ TECHNOLOGIES

Adaptive modulation and coding (AMC) adaptively changes the level of modulation—binary phase shift keying (BPSK), quadrature PSK (QPSK), 8-PSK, 16-quadrature amplitude modulation (QAM), and so on, as well as the amount of redundancy for an error correction code. A higher level of modulation (e.g., 16-QAM) with no error correction code can be used by users with good signal quality (close to base station) to achieve higher bandwidth. A lower level of modulation (e.g., BPSK) with more redundancy for error correction is used by users with bad signal quality (in the cell edge) to keep the channel condition, but results in lower bandwidth. The idea is to limit the number of link errors by adjusting the dedicated bandwidth through AMC in general. W-CDMA also adjusts the spreading factor and number of multiplexing spreading codes.

Regarding adaptive modulation, 1xEV-DO and WLAN have a similar strategy. Typically, the average bit error rate (BER) requirement is set beforehand, depending on the class of application. The adaptive modulation is applied to ensure the quality of service (QoS) that is evaluated directly by the signal-to-interference ratio (SIR), or indirectly measured by BER. Layer ½ transport control generally provides two distinct states, quasi-error-free and burst errors, during fading periods, when there is a large variation of bandwidth and delay.

Possible wireless video delivery technologies are presented in Table 3.1. The bottom row shows the practical operational requirements recommended by 3G Partnership Project (3GPP), where PLR indicates the packet loss rate at the transport layer.

TABLE 3.1
Possible Wireless Video Delivery Technologies

Layer	Video Telephony	Packet Streaming	Messaging + Progressive Download	MBMS Data
Source coder (application layer)	Error concealment, feedback-based error control (Adaptive Intra Fresh) reversible VLC	Error concealment, feedback-based error control	Compression is essential	Point-to-point data repair
Error resilience tools Network adaptation	Resync marker Data partitioning	Slice interleaving, data partitioning, redundant pictures, packet scheduling, selective ARQ	Interleaving for progressive download	Interleaving for FEC above layer 2
End-to-end transport	Selective ARQ (AL3)	RTP/UDP+RTCP	Wireless-TCP/IP	FLUTE
Layer-½ transport	FEC (BER: 1e-4)	FEC+ARQ (PLR: 1e-4)	FEC+ARQ (PLR: <1e-4)	FEC

For multimedia broadcast multicast services (MBMS), the error requirement is defined in terms of the block error rate (BLER), where block is defined as a data block passed by the physical layer to the medium access control (MAC) layer for a given transport channel (e.g., physical layer error rate). The block size varies with bit rate and transmission time interval (TTI), and is defined as bit rate × TTI.

Selective combining (SC) is an enhancement in the recent 3GPP standard by which the network simulcasts the MBMS content, and the terminal simultaneously receives and decodes the MBMS data from multiple radio links. Selection of the radio link is performed on a layer ½ transport block basis at the radio link controller (RLC), based on cyclic redundancy check (CRC) results and sequence numbers. The above conditions are optimistic, where three radio links (RLs) are used in a pedestrian condition. To provide 90 percent coverage and 1 percent BLER in a cell, it is calculated that the fraction of cell transmission power must be about 10.7 percent.[2] This means an MBMS service will occupy more than 10 percent of a cell radio resource even in every optimistic conditions.

TECHNOLOGIES ABOVE LAYER 2

Major technical challenges exist when pursuing optimal media delivery in conjunction with application layer characteristics. When applying taxonomy to wireless video technologies in 3G and beyond, it is hard to provide a neat classification because the combination of real time/non-real time, streaming/file download, and bi-/unidirectional communication complicates the abstraction of technical elements.[16]

Table 3.1 provides a summary of our practical review of recent wireless video technologies. As for source coders, we assume that MPEG-4 will be selected for videotelephony and H.264 for packet streaming. MBMS is not limited to video application delivery, although we assume that MBMS is used for MMS broadcast and video clip distribution applications.

In MPEG-4 video, application layer error resilience tools were developed on the assumption that there is benefit in having damaged data delivery from H.323. At the source coder layer, these tools provide synchronization and error recovery functionalities. Efficient tools are Resynchronization Marker and Adaptive Intra-Refresh (AIR). The marker localizes transmission errors by inserting a code to mitigate errors. AIR prevents error propagation by frequently performing intraframe coding on motion domains. These application layer technologies are comprehensively described in References 17 and 18.

As for H.264/AVC, there is considerable literature on the error resilience structure. H.264 is based on hybrid video coding, and is similar in spirit to other standards, such as MPEG-4, but with new sophisticated coding technologies, such as its prediction scheme. It said that H.264 attains substantial bit rate savings (up to 50 percent) relative to other standards, such as MPEG-4, at the same subjective visual quality. The decoding complexity at least doubles that of MPEG-4, while the encoding complexity may triple (or more) that of MPEG-4, depending on the degree of rate–distortion (R-D) optimization and complexity of motion estimation. Complexity is most important in the messaging and progressive download applications.

In contrast to the MPEG-4 error resilience structure, the H.264 error resilience structure is based on a flexible network adaptation structure called network abstraction layer (NAL). Unlike MPEG-4, the H.264 error resilience structure is based on the assumption that bit-erroneous packets have been discarded by the receiver. The error resilience structure has been designed mainly for packet loss environments, and this design concept is valid, since Internet protocols are currently widely used for content delivery. The NAL unit syntax structure allows greater customization of the method of carving the video content to the transport layer. A typical example is packetization to the Real-Time Transport Protocol (RTP) payload. For example, the flexible syntax enables robust packet scheduling, where we send important packets earlier and retransmit lost packets so that we can improve the transmission reliability for important pictures.

Let us move to end-to-end transport technologies. Error concealment has a long history; it has been available since H.261 and MPEG-2. The easiest and most practical approach is to hold the last frame that was successfully decoded. The best-known approach is to use motion vectors that can adjust the image more naturally when holding the previous frame. More sophisticated error concealment methods consist of a combination of spatial/spectral and temporal interpolations with motion estimation.

One can add optimization techniques to feedback-based error control, especially for streaming services. A feedback channel indicates which parts of the bit stream were received intact, and which parts of the video signal could not be decoded and had to be concealed. Having this feedback, the source coder typically uses the

INTRA mode for some macroblocks (MBs) to stop interframe error propagation. This is also called a selective automatic repeat request (ARQ) method. Reference picture selection is another well-known feedback-based control technique performed at the source coder layer.

Modern wireless networks provide many different means to adapt quality of service, such as forward error correction methods on different layers, and end-to-end or link layer retransmission protocols. In what follows, we introduce the features of the H.264/AVC coding standard[19–27] that make it suitable for wireless video applications, including features for error resilience, bit rate adaptation, integration into packet networks, interoperability, and buffering considerations.

3.6 H.263/AVC STANDARD IN WIRELESS VIDEO ENVIRONMENT

Emerging and future wireless systems will provide sufficient bit rate to support video communication applications. Nevertheless, bit rates will always be scarce in wireless transmission environments due to physical bandwidth and power limitations; thus, efficient video compression is required. Today, H.263 and MPEG-4 Visual Simple Profile are commonly used in handheld products, but it is foreseen that H.264/AVC will be the video codec of choice for many video applications in the near future.[28] The compression efficiency of the new standard outdoes prior standards roughly by at least a factor of two.

VIDEO CODING AND DECODING ALGORITHMS

In common with earlier coding standards, H.264 does not explicitly define a CODEC (enCOder/DECoder pair) but rather defines the syntax of an encoded video bit stream together with the method of decoding this bit stream. With the exception of the deblocking filter, most of the basic functional elements (prediction, transform, quantization, entropy coding) are present in previous standards (MPEG-1, MPEG-2, H.261, H.263), but the important changes in H.264 occur in the details of each functional block. The H.264 encoder and decoder are shown in Figure 3.5 and Figure 3.6, respectively.[31]

The encoder includes two dataflow paths, a forward path (left to right) and a reconstruction path (right to left). The dataflow path in the decoder is shown from right to left to illustrate the similarities between encoder and decoder.

An input frame or field F_n is processed in units of macroblocks. Each macroblock is encoded in intra- or intermode and, for each block in macroblock, a prediction, PRED (marked P in Figure 3.5), is formed based on reconstructed picture samples. In intramode, PRED is formed from samples in the current slice that have been previously encoded, decoded, and reconstructed (uF'_n in the figure; u indicates that is unfiltered). In intermode, PRED is formed by motion-compensated prediction (MC) from one or two reference picture(s) selected from the set of list 0 and/or list 1 reference pictures. In the figures, the reference picture is shown as the previous encoded picture F'_{n-1}, but the prediction reference for each macroblock partition (in intermode) may be chosen from a selection of past or future pictures (in display order) that have already been encoded, reconstructed, and filtered (identical to the D'_n shown in the encoder). Using the header information decoded from the bit stream, the decoder

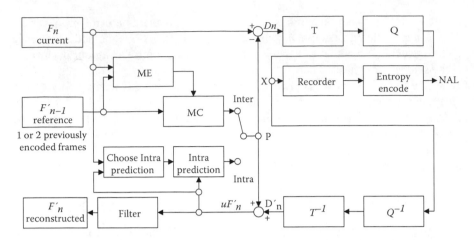

FIGURE 3.5   H.264 encoder. Reproduced with permission from K. R. Rao, Z. S. Bojkovic, and D. A. Milovanovic. *Introduction to Multimedia Communications: Applications, Middleware, Networking.* New York: John Wiley & Sons, 2006.

creates a prediction block PRED, identical to the original prediction PRED formed in the encoder. PRED is added to D'_n to produce uF'_n, which is filtered to create each decoded block F'_n.

The prediction PRED is subtracted from the current block to produce a residual (difference) block D_n that is transformed (using a block transform) and quantized to give X, a set of quantized transform coefficients which are reordered and entropy encoded. The entropy-encoded coefficients, together with side information required to decode each block within the macroblock (prediction modes, quantizer parameter, motion vector information, etc.) form the compressed bit stream which is passed to the NAL for transmission or storage.[32]

To encode and transmit each block in a macroblock, the encoder decodes (reconstructs) it to provide a reference for further predictions. The coefficients X are scaled

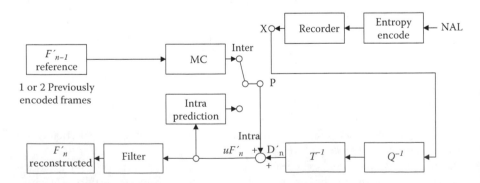

FIGURE 3.6   H.264 decoder. Reproduced with permission from K. R. Rao, Z. S. Bojkovic, and D. A. Milovanovic. *Introduction to Multimedia Communications: Applications, Middleware, Networking.* New York: John Wiley & Sons, 2006.

(Q^{-1}) and inverse transformed (T^{-1}) to produce a difference block D'_n. The prediction block PRED is added to D'_n to create a reconstructed block uF'_n (a decoded version of the original block). A loop filter is applied to reduce the effects of blocking distortion and the reconstructed reference picture is created from a series of blocks F'_n. The decoder receives a compressed bit stream from the NAL, and entropy decodes the data elements to produce a set of quantized coefficients X. These are scaled and inverse transformed to give D'_n.

Although compression efficiency is the major feature for a video codec to be successful in wireless transmission environments, it is also necessary that a standard provides means to be integrated easily into existing and future networks, as well as address the needs of different applications. In the following subsections we introduce the essential features of H.264/AVC for wireless systems, categorized as network integration, compression efficiency, error resilience, and bit rate adaptivity. It is important to understand that most features are general enough to be used for multiple purposes rather than assigned to a specific application.

Network Integration

The elementary unit processed by an H.264/AVC codec is called the NAL unit, which can easily be encapsulated into different transport protocols and file formats, such as the MPEG-2 transport stream, RTP, and MPEG-4 file format. There are two types of NAL units, video coding layer (VCL) NAL units and non-VCL NAL units. VCL NAL units contain data that represent the values and samples of video pictures in the form of slices or slice data partitions. One VCL NAL unit type is dedicated for a slice in an instantaneous decoding refresh (IDR) picture. A non-VCL NAL unit contains supplemental enhancement information (SEI), parameter sets, picture delimiter, or filler data.[29] Each NAL unit consists of a 1-byte header and the payload byte string. The header indicates the type of NAL unit and whether a VCL NAL unit is part of a reference or nonreference picture. Furthermore, syntax violations in the NAL unit and the relative importance of the NAL unit for the decoding process can be signaled in the NAL unit header.

H.264/AVC allows sending of sequence and picture level information reliably, asynchronously, and in advance of the media stream that contains the VCL NAL units by the use of parameter sets. Sequence and picture level data are organized into sequence parameter sets (SPS) and picture parameter sets (PPS), respectively. An active SPS remains unchanged throughout a coded video sequence (i.e., until the next IDR picture), and an active PPS remains unchanged within a coded picture. The parameter set structures contain information such as picture size, optional coding modes employed, and macroblocks for the slice group map. In order to be able to change picture parameters such as picture size without the need to transmit parameter set updates synchronously to the slice packet stream, the encoder and decoder can maintain a list of more than one SPS and PPS. Each slice header contains a coded word that indicates the SPS and PPS use.

COMPRESSION EFFICIENCY

Compression efficiency is the major attribute necessary for a video codec to be successful in wireless environments. Although the design of the VCL of H.264/AVC basically follows the design of prior video coding standards, it contains many new details that enable significant improvement in terms of compression efficiency. The gains do not come from a single new technique, but from an ensemble of advanced prediction, quantization, and entropy coding schemes.

The encoder implementation is responsible for appropriately selecting a combination of different encoding parameters, the so-called operational coder control. When using a standard with a complete specified decoder, parameters in the encoder should be selected such that good rate-distortion performance is achieved. For a video coder like H.264/AVC, the encoder must select parameters, such as motion vectors, macroblock modes, quantization parameters (QPs), reference frames, and spatial and temporal resolutions, to provide good quality under given rate and delay constraints. To simplify matters in deciding on good selection of the decoding parameters, this task is commonly divided into three levels:[30]

- Encoder control performs local decisions, such as the selection of macroblock modes, reference frames, or motion vectors, at the macroblock level and below, most appropriately based on a rate-distortion optimized mode selection applying Lagrangian techniques.
- Rate control mainly controls the timing and bit rate constraints of the application by adjusting the QP or Lagrange parameter and is usually applied to achieve a constant bit rate (CBR) encoded video suitable for transmission over CBR channels. The aggressiveness of the quantization/Lagrangian parameter change allows a trade-off between quality and instantaneous bit rate characteristics of the video stream.
- Global parameter selection selects the appropriate temporal and spatial resolutions of the video based on application, profile, and level constraints. Also, packetization modes, like slice sizes, are usually fixed for an entire session. The parameters are mainly determined by general application constraints.

ERROR RESILIENCE

In the following, we introduce different error resilience features included in H.264/AVC VCL with respect to their functionalities.

Slice Structured Coding. Slices provide spatially distinct resynchronization points within the video data for a single frame. This is accomplished by introducing a slice header, which contains syntactical and semantical resynchronization information. In addition, intraprediction and motion vector prediction are not allowed over slice boundaries. The encoder can select the location of the synchronization points at any macroblock boundary.

Flexible Macroblock Ordering. FMO allows mapping of macroblocks to slice groups, where a slice group itself may contain several slices. Therefore, macroblocks

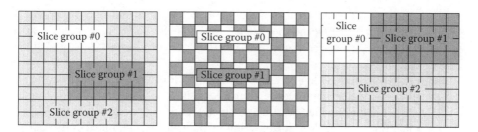

FIGURE 3.7 Macroblock allocation maps: foreground slice groups with one-left over background slice group; checker board-like pattern with two slice groups; and sub-pictures within a picture. Reproduced with permission from 2005 IEEE.

might be transmitted out of raster scan order in a flexible and efficient way. Some examples of macroblock allocation maps for different applications are shown in Figure 3.7. Dispersed macroblock allocations are especially powerful in conjunction with appropriate error concealment (i.e., when the samples of a missing slice are surrounded by many samples of correctly decoded slices).

Arbitrary Slice Ordering. With ASO, the decoding order of slices within a picture may not follow the constraint that requires the address of the first macroblock within a slice to monotonically increase within the NAL unit stream for a picture. ASO permits, for example, reduction of decoding delay in case of out-of-order delivery of NAL units.

Slice Data Partitioning. In data partitioning mode, each slice can be separated into a header and motion information, intra information, and intertexture information by simply distributing the syntax elements to individual NAL units. Due to this reordering at the syntax level, coding efficiency is not reduced, but obviously the loss of individual partitions still results in error propagation.

Intra-coding. H.264/AVC distinguishes IDR pictures and regular intrapictures, whereby the latter do not necessarily provide the random access property. Pictures before the intrapictures may be used as reference for succeeding predictively coded pictures. H.264/AVC also allows intracoding of single macroblocks for regions that cannot be predicted efficiently, or due to any other case where the encoder decides for nonpredictive mode. The intramode can be modified such that intraprediction from predictively coded macroblocks is disallowed.

Redundant Slices. A redundant coded slice is a coded slice that is a part of a redundant coded picture, which itself is a coded representation of a picture that is not used in the decoding process if the corresponding primary coded picture is correctly decoded. The redundant slice should be coded such that there is no noticeable difference between any area of the decoded primary picture and a decoded redundant picture.

Flexible Reference Frame Concept. H.264/AVC allows reference frames to be selected in a flexible way on a per macroblock basis, which provides the possibility to use two weighted reference signals for macroblock interprediction, allows frames to be kept in short-term and long-term memory buffers for future reference, and finally provides temporal scalability. The classical I, P, B frame concept is replaced by a highly flexible and general concept that can be exploited by the encoder for different purposes. However, this concept also requires that not only is the HRD (Hypothetical Reference Decoder) specified in the bit stream domain, but it is also necessary that the encoder be constrained in the amount of frames to be stored in the decoded picture buffer.

Switching Pictures. H264/AVC allows applying mismatch-free predictive coding even where there are different reference signals. So-called primary SP-frames are introduced in the encoded bit stream, which are in general slightly less efficient than regular P-frames but significantly more efficient than regular I-frames. The major benefit results from the fact that this quantized reference signal can be generated mismatch-free using any other prediction signal. If this prediction signal is generated by predictive coding, the frame is referred to as secondary SP-pictures, which are usually significantly less efficient than P-frames as an exact reconstruction is necessary. To also generate this reference signal without any predictive signal, so-called switching-intra (SI) pictures can be used. SI pictures are only slightly less inefficient than common I pictures and can also be used for adaptive error resilience purposes.

BIT RATE ADAPTIVITY

Bit rate adaptivity is one of the most important features necessary for applications in wireless systems to react to the dynamics due to statistical traffic and variable receiving conditions, as well as handovers and random user activity. Due to the applied error control features, these variations mainly result in varying bit rates in different timescales. For applications where online encoding is performed and the encoder has sufficient feedback on the expected bit rate on the channel by some channel state or decoder buffer fullness information, rate control for VBR channels can be applied. H.264/AVC obviously supports these features, mainly by the possibility of changing QPs dynamically, but also by the changing temporal resolution.

When channel bit rate fluctuations are not *a priori* known at the transmitter, or there is no sufficient means or necessity to change the bit rate frequently, play-out buffering at the receiver can compensate for bit rate fluctuations to some extent. In addition, for anticipated buffer under run, techniques such as adaptive media play out allow a streaming media client, without involvement of the server, to control the rate at which data is consumed by the play-out process.

However, these techniques might not be sufficient to compensate for bit rate variations in wireless applications. In this case, rate adaptation has to be performed by modifying the encoded bit stream. In today's systems rate adaptation is typically carried out by streaming servers. It is well known that intelligent decisions to drop less important packets rather than dropping random packets—this is treated under the framework of error resilience—can significantly enhance the overall quality. A formalized framework called rate-distortion optimized packet scheduling has been introduced and serves as the basis for several publications. Applying this framework is easiest when important and less important packets are identified in the encoding process. H.264/AVC provides several approaches to support packets with different importance levels for bit rate adaptivity. First, the temporal scalability features of H.264/AVC which rely on the reference frame concept can be used. Second, if frame dropping is not sufficient, one might apply data partitioning, which can be viewed as a very coarse but efficient method for SNR scalability. Third, flexible macroblock ordering may also be used for prioritization of regions of interest. For example, a background slice group can be dropped in favor of a more important foreground slice group.

For many user cases it is necessary to adapt the bit rate dynamically in the application to larger bit rates, and timescales larger than the initial play-out delay allows. In wireless streaming environments, bit stream switching provides a simple but powerful means to support bit rate adaptivity. In this case, the streaming server stores the same content encoded with different versions in terms of rate and quality. In addition, each version provides a means to randomly switch into it. IDR pictures provide this feature, but they are generally costly in terms of compression efficiency.

3.7   CONCLUDING REMARKS

Due to the limitations in bandwidth and the presence of channel errors, video communication through wireless channels is a challenging problem. Because many video sources are coded at a high rate and without considering the different channel conditions, a means to repurpose this content for delivery over a dynamic wireless channel is needed. To reduce the rate and change the format of the originally encoded video service to match network conditions and terminal capabilities, transcoding is used. Given the existence of channel errors that can easily corrupt video quality, there is also the need to make the bit stream more resilient to transmission errors. A key factor determining the effectiveness of a mobile device for wireless video transmission is its energy management strategy. A general framework for studying this problem is to achieve the best video delivery quality with the minimum energy consumption.

Rate control is an important issue in video streaming applications for wireless networks. A widely accepted rate control method in wired networks is TFRC. Here, the TCP-friendly rate is determined as a function of packet loss rate, round-trip time, and packet size. TFRC assumes that packet loss is primarily due to congestion. TFRC connections as an end-to-end rate solution for wireless video streaming can be proposed. This approach not only avoids modifications to the network infrastructure or network protocol, but also results in full utilization of the wireless channel.

To highlight wireless video it must be pointed out that layer ½ transport tends to provide two different conditions: quasi error free and burst errors during fading periods. In the former condition, upper layer error control technologies have a limited role. When considering this role, extraordinary adaptability of error control to the latter condition is essential. Rate control can be identified as the essential technology that provides extraordinary adaptability to varying bandwidth.

The H.264/AVC standard in wireless environments provides features that can be used in one or several application scenarios. It also allows easy integration in most networks. The selection and combination of different features strongly depends on the system and application constraints, namely, bit rates, maximum tolerable payout delays, error characteristics, and on-line encoding possibility, as well as availability of feedback and cross-layer information. Although the standardization process is finalized, the freedom at the encoder, as well as the combination with transport models such as FEC and retransmission strategies, promise optimization potentials. Therefore, further research in the areas of optimization, cross-layer design, feedback adaptation, and error concealment is necessary to fully understand the potential of H.264/AVC in wireless environments.

4 Wireless Multimedia Services and Applications

This chapter seeks to contribute to a better understanding of the current issues and challenges in the field of wireless multimedia services and applications. The chapter begins with real-time Internet Protocol (IP) multimedia services including the evolution from short to multimedia messaging services. It also covers information technology (IT)-based service delivery platform (SDP) deployment. Extended IP Multimedia System (IMS) architecture is also discussed. After that, we examine the current IMS policy control to update the significant changes in the core network. A number of service delivery platforms have already been developed and commercially deployed. The existing implementations focus on the IT systems and domain for service design and delivery, and accommodate the existing network implementations in their architectures.

4.1 INTRODUCTION

It is well known that the widely supported evolving path of wireless networks today is the path toward IP-based networks, also known as all-IP networks.[1] With the rapid growth of wireless networks and the great success of Internet video, wireless multimedia services are expected to be widely deployed in the near future. Namely, different types of wireless networks are converging into all-IP networks. The current trends in the development of real-time Internet applications, and the rapid growth of mobile systems, indicate that the future Internet architecture will need to support various applications with different quality of service (QoS) requirements.[2] Enabling QoS becomes more challenging when introducing it in an environment involving mobile hosts under different wireless access technologies, since available resources (bandwidth, battery life, etc.) in wireless networks are scarce and dynamically change over time. To support end-to-end QoS for video delivery over wireless Internet, there are several fundamental challenges[3]:

- QoS support encompasses a wide range of technological aspects.
- Different applications have very diverse QoS requirements in terms of data rates, delay bounds, and packet loss probabilities.
- Different types of networks inherently have different characteristics.
- There is dramatic heterogeneity among end users.

To address the above challenges, one should support QoS requirement in all components of the video delivery system from end to end, which include QoS provisioning from networks, scalable video presentation from applications, and network

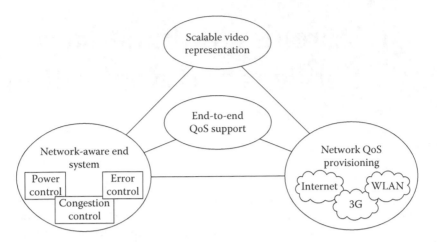

FIGURE 4.1   Components for end-to-end QoS.

adaptive congestion/error power control in end systems. Figure 4.1 illustrates key
components for end-to-end QoS support. They are network QoS provisioning, scalable video representation, and network-aware end system.

Network QoS provisioning. The best-effort nature of the Internet has promoted
the Internet Engineering Task Force (IETF) community to seek QoS support through
network layer mechanisms. The best-known mechanisms are the integrated services
(IntServ)[4] and differentiated services (DiffServ).[5] The approaches to providing QoS
in wireless networks are quite different from their Internet counterparts. General
Packet Radio Service (GPRS)/Universal Mobile Telecommunications System
(UMTS) and IEEE 802.11 have totally different mechanisms for QoS support.

Scalable video representation. In scalable coding, the signal is separated into
multiple layers of different visual importance. The base layer can be independently
decoded and it provides basic video quality. The enhancement layers can only be
decoded together with the base layer, and these further refine the video quality.
Enhancements of layered scalable coding have been proposed to provide further fine
granularity scalability.[6–8] Scalable video representation provides fast adaptation to
bandwidth variations, as well as inherent error resilience and complexity scalability properties that are essential for efficient transmission over error prone wireless
networks.

Network-aware end system. When network conditions change, the end systems
can employ adaptive control mechanisms to minimize the impact on user perceived
quality. Power control, congestion control, and error control are three main mechanisms to support QoS for robust video delivery over wireless Internet. Power control
is performed collectively from the group point of view by controlling transmission power and spreading gain for a group of users so as to reduce interference.
Congestion control and error control are conducted from the individual user's point of
view to effectively combat the congestions and errors occurring during transmission
by adjusting the transmission rate and allocating bits between source and channel
coding.

The demand for multimedia services in mobile networks has raised several technical challenges, such as the minimization of handover latency. In this context, soft and softer handover techniques have played a key role and provided the means for eliminating the handover latency, thus enabling the provision of mobile multimedia services. However, not all radio access networks support soft handover techniques. For example, the notorious IEEE 802.11 wireless local area networks (WLANs) support only hard handovers; consequently the use of multimedia services over such WLANs raises considerable concerns. Another major challenge in both fixed and wireless networks today is QoS provision. For wireless access networks, such as WLANs, the IntServ framework, standardized by the Internet Engineering Task Force (IETF), provides the necessary means for requesting and obtaining QoS per traffic flow. IntServ uses the Resource Reservation Protocol (RSVP) for implementing the required QoS signaling. RSVP is problematic in wireless networks, basically due to the need for reestablishing resource reservations every time a mobile node (MN) changes its point of attachment and the IP route with its corresponding node (CN) has to be updated end-to-end, resulting in increased handover latency.

RSVP was designed to enable hosts and routers to communicate with each other to set up the necessary states for IntServ support. It defines a communication session to be a data flow with a particular destination and transport layer protocol, identified by the triplet (destination address, transport-layer protocol type, and destination port number). Its operation applies only to packets of a particular session, and therefore every RSVP message must include details of the session to which it applies. Usually, the term flow is used instead of RSVP session. The RSVP defines seven types of messages, of which the PATH and RESV messages carry out the basic operation of resource allocation. The PATH message is initiated by the sender and travels toward the receiver to provide characteristics of the anticipated traffic, as well as measurements for the end-to-end path properties. The RESV message is initiated by the receiver, upon receipt of the PATH message, and carries reservation requests to the routers along the communication path between the receiver and the sender. After path establishment, PATH and RESV messages should be issued periodically to maintain the so-called soft states that describe the reservations along the path.

RSVP assumes fixed end points and for that reason its performance is problematic in mobile networks. When an active MN changes its point of attachment with the network (e.g., upon handover), it has to reestablish reservations with all its CNs along the new paths. For an outgoing flow, the MN has to issue a PATH message immediately after the routing change, and wait for the corresponding RESV message before starting data transmission through the new attachment point. Depending on the hops between the sender and the receiver, this can cause considerable delays, resulting in temporary service disruption. The effects of handover are even more annoying in an incoming flow because the MN has no power to immediately invoke the path reestablishment procedure. Instead, it has to wait for a new PATH message, issued by the sender, before responding with a RESV message in order to complete the path reestablishment. Simply decreasing the period of the soft state timers is not an efficient solution because this could significantly increase signaling overhead. A number of proposals can be found in the literature, extending RSVP for either intersubnet or intrasubnet scenarios. For intrasubnet scenarios, proposals that combine RSVP with micromobility solutions, such as cellular IP, can reduce the effects of handover on RSVP, as only the last part of the virtual circuit has to be reestablished. For inter-subnet scenarios, the existing proposals include advance reservations, multicasting, RSVP tunneling, and so on.

4.2   REAL-TIME IP MULTIMEDIA SERVICES IN UMTS

As communications technology develops, user demand for multimedia services is rising. Meanwhile, the Internet has enjoyed tremendous growth in recent years. Consequently, there is great interest in using IP-based networks to provide multimedia services. One of the most important arenas in which the issues are being debated is in standards development for the Universal Mobile Telecommunications System (UMTS). UMTS is the third generation (3G) mobile wireless cellular system that is evolving from the Global System for Mobile Communications (GSM). Given the large user base of GSM and the expected large user base of UMTS, the way IP multimedia services are provided in UMTS will significantly influence the success of IP multimedia services and the development of related technologies and protocols, such as IP version 6 (IPv6) and Session Initiation Protocol (SIP).

In the past, people's impression of mobile wireless cellular systems was that they mostly provided voice telephony service with a few frills like short message service (SMS). The incorporation of IP multimedia services into UMTS is one of the major features that will transform this image, as many new services like videoconferencing are now available. One of the main reasons multimedia services can be provided in UMTS is the increased data rates possible in 3G mobile wireless networks. Conversely, the potential of multimedia applications helps justify investments in 3G, as well as in 3G and beyond, and fourth generation (4G) systems.

IP MULTIMEDIA SERVICES

As the name multimedia implies, different types of media—video, audio, voice, and text—are involved. When multiple media types are involved in a session (e.g., video and audio), they need to be synchronized for simultaneous presentation to the user. Furthermore, multimedia services may be real-time or non-real-time. Videoconferencing is an example of real-time multimedia service. For real-time services, the time relation between successive data packets must be preserved, with little tolerance for jitter in packet arrival times. An example of a non-real-time multimedia service is multimedia messaging service (MMS), an enhancement of the popular SMS that allows multimedia messages, not just text messages. We focus on real-time multimedia services and the UMTS IP multimedia subsystem (IMS) that enables these services in UMTS.

The voice-centric circuit-switched telephone network and conventional cellular networks are not optimized to offer multimedia services. New network elements and protocols are necessary to provide multimedia services. Although circuit-switched connections can be used for data services, they are inefficient for transporting intermittent or bursty data. Meanwhile, packet-switching technology has been developing rapidly. It allows statistical multiplexing of traffic and can efficiently handle both

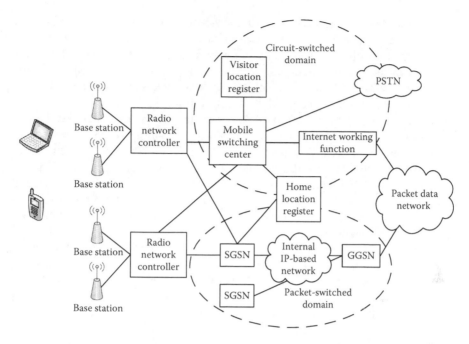

FIGURE 4.2 Generalized GPRS service architecture. Reproduced with permission from K. D. Wong and V. K. Varma. "Supporting real-time IP multimedia services in UMTS," *IEEE Commun. Magazine* 41 (November 2003): 148–55.

bursty and nonbursty traffic. Moreover, data traffic volume continues to grow more rapidly than voice traffic volume. For cost and efficiency reasons, and to achieve better service integration, it is more appropriate to use a common IP-based technology capable of supporting both multimedia and other kinds of traffic.[9]

The IMS will enable two levels of services.[10] The first level includes basic services like initiation, modification, and termination of multimedia sessions, as well as advanced services like multiparty sessions, session forwarding, blocking, caller ID, and so on. The second level of services includes services that are not being standardized but can be provided by third-party providers using capabilities provided by the network. For example, location-based advertising can be enabled using location information from the UMTS network.

The introduction of Gateway Packet Radio System (GPRS) was a major step in the evolution of GSM to UMTS. GSM Phase 1 and Phase 2 provided only circuit-mode connections. GPRS architecture is shown in Figure 4.2.[11] GPRS provides an end-to-end connectionless packet service, and includes a packet-mode transfer over the air and a packet-switched backbone. Subscribers benefit from higher peak data rates and charging based on traffic volume rather than hold time, while operators benefit from the efficient use of spectrum by allowing multiple users to dynamically share resources.

The GPRS network was initially developed to provide packet routing to external packet networks for the GSM radio access network. For UMTS Release 5 and beyond, GPRS takes on added significance, as it also provides transport for IP multimedia traffic within the mobile network. Resources over the air, as well as the radio access network (the base stations and radio network controllers), are dynamically shared between circuit-mode and packet-mode traffic. Circuit-mode and packet-mode traffic take different paths only in the core network behind the radio access network. On the one hand, circuit-mode traffic uses the circuit-switched domain, comprising traditional entities like mobile switching centers and visitor location registers that were present in the GSM architecture from the beginning. On the other hand, packet-mode traffic uses the new packet-switched domain. The packet domain functionality in the core network is implemented in gateway GPRS support nodes (GGSNs) and serving GPRS support nodes (SGSNs). SGSNs and GGSNs are IP routers with additional capabilities to control access and track location of mobile stations. The SGSN provides security functions, packet switching, and routing, and keeps track of the location of individual mobile stations. The GGSN provides interworking with external packet-switched networks and communicates with SGSNs via a packet-switched network.

A mobile station desiring packet access must be attached to the GPRS network. To use a specific packet protocol like IP with specified QoS, security, and other requirements, it activates one or more Packet Data Protocol (PDP) contexts. A PDP context is a set of configuration and usage settings related to the transport of data packets over a GPRS network. GPRS can support PDP contexts for various data network protocols, including X.25 and IP. After the establishment of an IP PDP context, for example, when an IP packet destined for the mobile station arrives at the GGSN, the GGSN tunnels it to the SGSN that currently serves the mobile station.

UMTS RELEASES

There are multiple releases of UMTS because work on UMTS continues as the specifications are continually enhanced. Each release provides a stable version of specifications for vendors and operators to use in implementation and deployment. There used to be an annual release beginning in 1996, where changes to GSM were incorporated (Release 96, Release 97, etc., by year). However, what was to become Release 2000 was split to become Release 4 and Release 5. Release4 was frozen in early 2001 and contains the groundwork for the future all-IP wireless network. Many features, including the IP multimedia session control features, were postponed and first specified in Release 5, which was frozen in mid-2002. The Third Generation Partnership Project (3GPP) worked on Release 6.

UMTS Release 4 provides IP connectivity, but neither IP multimedia services nor IP multimedia session control. In Release 4, a common IP-based packet-switched transport for voice and data may be used within the mobile network, unlike in previous releases where voice and data used separate circuit-switched and packet-switched transports, respectively. However, this is an innovation only in the circuit-switched domain in the core network, allowing voice to be packetized at media gateways controlled by mobile switching center servers, and packet-switched internal to the provider network. Packet-switched voice in the circuit-switched domain is not end-to-end, and the control signaling is still based on circuit-centric GSM-evolved control protocols. For data traffic, however, the network does not provide capabilities for IP multimedia session control (setting up sessions, tearing down sessions, etc.). It only provides IP transport, not end-to-end IP multimedia services. In contrast, from Release 5 onward, end-to-end IP multimedia sessions can be managed.

EVOLUTION FROM SHORT TO MULTIMEDIA MESSAGING SERVICES

Owing to the enormous success of SMS, new messaging types have been defined to enrich the enhanced contents. From the year 2000, it was evident that the interest for messaging was high with an exponential increase over time. Therefore, the principal operators worked on the possibility of sending messages with multimedia content (i.e., images, videos, etc.) using 3G mobile communication networks. We focus on the evolution from SMS, to enhanced messaging services (EMS), to MMS.

MMS is an innovative messaging system, standardized by both the Wireless Application Protocol (WAP) Forum (now Open Mobile Alliance, OMA) and 3GPP. It is similar to SMS and EMS, but allows one to insert multimedia contents in a message such as music, pictures, video clips, text, and so forth. MMS is independent of the underlying mobile network (e.g., GSM, wideband code-division multiple access or WCDMA, etc.) and allows content delivery to either mobile phones or e-mail addresses. On the basis of these features, it is estimated that this messaging service will increase telecommunication profits, addressing both professionals and personal user markets. Due to its high efficiency, the delivery of MMS to mobile phones through the air interface is based on the WAP protocol stack. Note that here we use MMS to denote both the service and the message itself; the distinction between these two meanings will be evident from the context.

It is possible to distinguish two scenarios for MMS messaging:

- Person-to-person MMS, that is, messaging between people. This scenario is associated with the availability of multimedia accessories (i.e., cameras) that may be connected to the mobile phone. In this case, the user has the possibility of taking a snapshot of a scene and sending it to one or more recipients (i.e., Internet users, users of handsets, etc.).
- Content (machine)-to-person MMS, a value-added service (VAS) may provide information about traffic, entertainment, and museums through multimedia messages. In this context, the user can activate several services and then receive information from content providers through MMS.

There are four key elements in the MMS architecture, defined in the MMS Environment (MMSE), and shown in Figure 4.3:

- MMS relay provides access to different architectural elements for MMS provision and supports interactions with other messaging systems. It could be implemented together with the MMS server.
- MMS server stores messages (this could include an e-mail server, short message service center, or fax).
- MMS user database has databases with the subscriber profile and information about user mobility.
- MMS user agent is the MMS-enabled mobile phone.

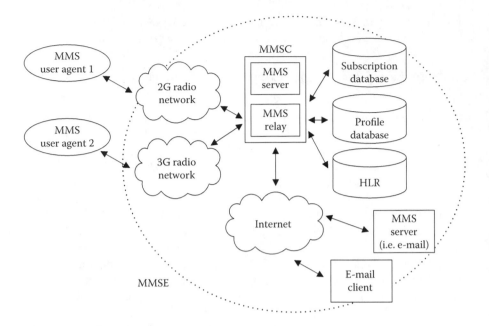

FIGURE 4.3 Key elements of the multimedia messaging services.

These different elements interact in MMSE in order to provide functionalities across different systems. Note that MMS relay and MMS server can be integrated into a single platform, called the MMS center (MMSC).

MMS has been defined through a standardization process aimed at integrating different media (i.e., text, audio, and video). Several standardization bodies have been involved in this process. In particular, 3GPP has defined high-level requirements: the MMS architecture, codecs, and streaming protocols. The WAP Forum (now OMA) has defined lower-layer aspects in order to connect the MMS-enabled phone to the WAP environment. The MMS Interoperability Group (MMS-IOP), belonging to OMA, has focused on end-to-end operability and interoperability of MMS. Further standardization efforts are currently in progress in order to enrich the MMS specification. The first 3GPP standard for MMS was MMS Release'99 (for OMA, the standard is MMS 1.0), followed by MMS Release 4 (MMS 1.1), and then, by MMS Release 5 (MMS 1.2).

MMS network representation is shown in Figure 4.4. In an MMS network it is possible to identify various system elements and interfaces. In particular, one or more MMS servers can be present, depending on the specific service they provide. The different elements are as follows:

- MMS client, which is implemented as an application that interacts with the users on their mobile terminal
- MMS proxy-relay and MMS server
- E-mail server, for Internet e-mail services, supporting the Simple Mail Transfer Protocol (SMTP)

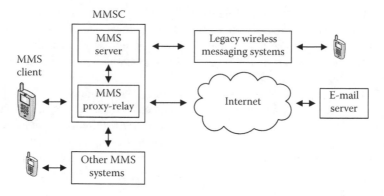

FIGURE 4.4   Multimedia message services network representation.

- Legacy wireless messaging systems, representing various existing support systems for wireless messaging (i.e., paging or SMS systems)

4.3 INTERNET PROTOCOL MULTIMEDIA SUBSYSTEM

The two most important aspects of 3G wireless networks with respect to IP-based multimedia services are the IMS and the multimedia broadcast/multicast services.[12] IMS is the next generation networking (NGN) architecture for mobile and fixed multimedia services standardized by the 3GPP. IMS utilizes protocol standardized by the IETF. The foundation of IMS is a version of the IETF SIP. SIP can create and terminate sessions with one or more participants. Also, SIP supports user mobility by proxying and redirecting requests to the user's current location.[14]

IMS enables complex IP-based multimedia sessions to be created with guaranteed QoS for each media component. Example applications include voice telephony and video conferencing. The IMS interoperates with both traditional telephony services and external IP-based multimedia services. It will enable new services, but also is an integration platform for both current and future telecommunications and Internet services. IMS truly merges the Internet with the mobile world by allowing the use of SIP and other IP-based technologies for application development. Modem service-oriented architectures (SOAs) allow the creation and operation of complex systems built on the actual implementation and location of various services. A service in this context is defined as a contractually defined behavior that can be implemented as a component for use by another component. One of the best-known examples of service-oriented technology is Web services, defined as a networked application that is able to interact using standard application-to-application Web protocols over well-defined interfaces, and that is described using a standard functional description language. Besides combining multiple media components, multimedia services include at least one continuous, that is, time sensitive, media component, such as audio or video, and all media components are required to be synchronized with each other. Therefore, while simple IP connectivity *allows*, real multimedia services *require* additional support from the network, at least in the area of QoS provisioning for the continuous media components. Despite many efforts to provide such support, the Internet remains a best-effort network, providing no guarantees about end-to-end packet transmission delay or reliability. Universal Mobile Telecommunications System (UMTS) networks have made significant progress in this direction. One aspect of this progress is the addition of the IMS components. Another aspect is the provisioning of guaranteed QoS.

SIP is the cornerstone of the IMS specification. Major functions of SIP are session control, addressing, and mobility management on the service level. SIP is a text-based protocol for establishment, control, and finalization of sessions between end points and also for control of the media channels between them. After a session is established, other protocols can be used for communication between applications, for example, Real-Time Protocol (RTP)/Real-Time Control Protocol (RTCP) for streaming of voice or video.

SIP is a session control protocol. It can create and terminate sessions with one or more participations. SIP supports user mobility by proxying and redirecting requests to the user's current location. SIP allows the specification of session properties during session initiation via the Session Description Protocol (SDP). SDP is used for describing multimedia sessions for the purposes of session announcement, session invitation, and other forms of multimedia session initiation. For example, an SDP description can include the session name and purpose, the time the session is active, the media compressing the session, information on how to receive such media (addresses, ports, formats, etc.), as well as many other session attributes. For new types of sessions, new SDP description formats can be introduced and standardized.

To summarize, SIP is designed to initiate and direct delivery of streaming media data from media servers. Video streaming is an important component of many Internet multimedia applications, such as distance learning, digital libraries, home shopping, and video-on-demand.[16] The best-effort nature of the current Internet poses many challenges to the design of streaming video systems.

IMS CONCEPTION

A simplified version of the UMTS Release 5 network architecture is shown in Figure 4.5. It shows the network elements used in providing real-time IP multimedia services. Because the IMS uses only the packet-switched domain for transport and local mobility management, it can be deployed without a circuit-switched domain. Hence, circuit-switched domain elements like mobile switching centers are not shown in the figure. Other network elements pertinent to the provision of services other than IP multimedia services (e.g., SMS) are also omitted.

The IMS includes one or more call session control functions (CSCFs), media gateway control functions (MGCFs), IMS media gateways, multimedia resource function controllers (MRFCs), multimedia resource function processors (MRFPs), subscription locator functions, breakout gateway control functions (BGCFs), and application servers. The roles of these elements are explained as we now address how the IMS performs major functions and adds multimedia support features to UMTS.

IMS signaling and session traffic are carried over IP. Moreover, not just any version of IP can be used; IMS boldly requires that only IPv6 must he used in the IMS

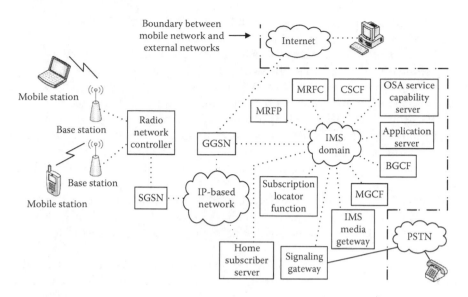

FIGURE 4.5 UMTS network architecture. Reproduced with permission from K. D. Wong and V. K. Varma. "Supporting real-time IP multimedia services in UMTS," *IEEE Commun. Magazine* 41 (November 2003): 148–55.

domain.[17] Although it would make UMTS the first global commercial system with widespread deployment of IPv6, it is a bold requirement because

- The IMS may be (but does not have to be) a different IP addressing domain than that used for the packet-switched domain backbone (which may be IPv4) and the circuit-switched domain backbone, so interworking issues between IPv4 and IPv6 need resolution.
- There is a lack of experience in IPv6.

The IP address used by the mobile station to access IP multimedia services must be from the IMS addressing domain, which can be arranged in PDP context activation when setting up IP connectivity. It is considered a possible benefit in terms of routing efficiency that this address be obtained from a GGSN in the serving network rather than the home network. Early in the standards process, it was required that this address be obtained from a gateway GPRS support node in the serving network, but the requirement has since been removed, allowing more flexibility in the choice of GGSN.

The CSCF plays a central role in call control and uses SIP. The CSCF is analogous to the signaling and control part of mobile switching centers for circuit-switched voice calls. However, it is capable of supporting multimedia sessions, not just voice calls. SIP is used for signaling between mobile stations and CSCF, between CSCFs, between CSCF and MGCF, and between CSCF and application servers. The CSCF plays different roles: as a proxy CSCF (P-CSCF), it is the first point of contact between a mobile station and the IMS, as a serving CSCF (S-CSCF) for session control and as an interrogating CSCF (I-CSCF), it is the main contact point in a network for all IMS-related signaling for its mobile stations. Call flows in the next section illustrate these CSCF roles.

While at least one end of the session is an IMS subscriber, the other end for voice-only sessions may often be a public switched telephone network (PSTN) subscriber. Four new functional elements—the IMS media gateway, the MGCF, the signaling gateway, and the BGCF—support scenarios where calls go between the PSTN and the PS domain. The IMS media gateway performs media translation: translation between media signals encoded in one format on one side and signals encoded in another format on the other side. The MGCF controls the IMS media gateway, provides application-level signaling translation between SIP-based and ISUP-based (ISDN User Part, which is used on the PSTN side) signaling, and communicates with the S-CSCF. Transport-level signaling translation between IP-based and SS7-based transport is performed at the signaling gateway (SGW). The BGCF identifies the network and the MGCF within that network, where the breakout to the PSTN should occur.

The home subscriber server (HSS) is a master database that stores information like subscription and location information. It can be thought of as an enhanced version of the home location register found in GSM. Since there may be multiple HSSs in a network, the subscription location function is queried by CSCFs during registration or session setup to find the HSS that has the desired subscriber information. The subscription location function is not needed in a single HSS environment because the CSCFs would know which HSS to use.

Various application servers arc supported, including SIP-based application servers and US application servers. Application servers are servers hosting applications that can implement all kinds of services, such as voice announcement services and prepaid billing enforcement. The support of a variety of application servers facilitates the addition of innovative new services. In both cases, SIP is used for signaling with S-CSCFs. However, in the case of an OSA application server, an OSA service capability server is inserted between the OSA application server and the S-CSCF.

The MRFP and media resource function controller (MRFC) support multiparty multimedia conferencing and media resource (e.g., for playing announcements) capabilities. The MRPC controls the media stream resources of the MRFP, which processes and mixes the actual media streams.

IP MULTIMEDIA SUBSYSTEM ARCHITECTURE

The IMS enhances the basic IP connectivity of UMTS by adding network entities that handle multimedia session setup as well as control and QoS provisioning. A session includes the senders, receivers, and data streams participating in an application. The new IMS entities ensure that multimedia sessions will be able to reserve the resources they need in order to perform satisfactorily. A session is able to request different QoS levels for each of its media components and modify these levels during its time. The IMS does not standardize any applications, only the service capabilities required to build various services. As a result, real-time and non real-time multimedia services can easily be integrated over a common IP-based transport. These services can directly interwork with all the services available over the Internet. Some of the services that can be provided over IMS are voice and telephony, presence services, instant messaging, chat rooms, voice and video conferencing, and multiparty gaming. The possible services are based on a small set of capabilities.[18,19]

- End-point identities including telephone numbers and Internet names
- Media description capabilities, including coding formats and data rates
- Person-to-person real-time multimedia services, including voice telephony
- Machine-to-person streaming multimedia services, including TV channels
- Generic group management, enabling chat rooms and messaging
- Generic group communication, enabling voice and video conferencing

The basic services can be controlled by an external application server (AS) in order to provide actual applications. Depending on the functionality of the AS, the application built on the top of these capabilities may be a video conference, a chat room, or other multimedia applications.

To maximize flexibility, the IMS organizes its functionality in three layers, as shown in Figure 4.6. The transport and end-point layer initiates and terminates the signaling needed to set up and control sessions, and provides bearer services and support for media conversions. The session control layer provides functionality that allows end points to be registered with the network, sessions to be set up between them, and media conversions to be controlled. In this layer multiple transport services may be combined in a single session. The AS layer allows sessions to interact with various AS entities. In this layer multiple sessions may be combined in a single application.

The relationship of the IMS to the packet-switched (PS) and circuit-switched (CS) domains of a UMTS network is shown Figure 4.7. In fact, this is IMS within a UMTS

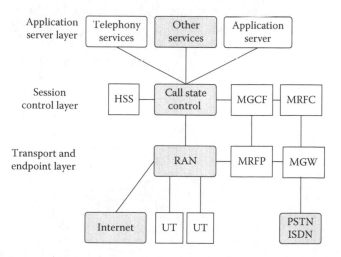

FIGURE 4.6 IMS layered service architecture. HSS, home subscriber system; MRFC, media resource function controller; MGCF, media gateway control function; RAN, radio access network; MRFP, media resource function processor; MGW, media gateway; UT, user terminal; PSTN, public switched telephone network; ISDN, integrated services digital network.

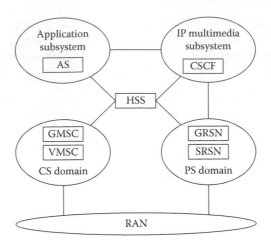

FIGURE 4.7 IMS within a UMTS core. CSCF, call session control functions; GMSC, gateway mobile services switching centers; VMSC, visitor mobile switching centers; RAN, radio access network; GGSN, gateway GPRS support nodes; SGSN, serving GPRS support nodes.

core network (CN). It can be seen that the IMS is indeed a subsystem of the UMTS core network that depends on the PS domain.

The actual data transfer services are provided by the existing IP-based mechanisms offered by GPRS. IMS provides flexible multimedia session management using IP bearer services. For complete application functionality to be provided, the IMS may have to rely on services provided by an external AS. The IMS itself provides only session setup and control functions, media processing functions, and media and signaling interworking functions.[20]

All media should be transported using the RTP over User Datagram Protocol (UDP), and, most importantly IMS should use IPv6. The general architecture of the IMS is given in Figure 4.8.[21]

Multimedia sessions are set up and controlled via various types of CSCF. The P-CSCF is the local contact point of the UT in the network it is visiting, analogous to SGSN in GPRS. Because a network may contain many serving CSCFs (S-CSCFs)

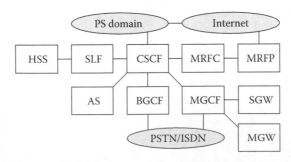

FIGURE 4.8 IMS architecture for mobile services. SLF, subscription local function; CSCF, call session control functions; BGCF, breakout gateway control function; SGW, signaling gateway.

for load balancing, an interrogating CSCF (I-CSCF) may be provided at the entry point to an operator's network. The I-CSCF and the S-CSCF relay on the HSS for user-related information. Networks with multiple HSSs also provide a subscription locator function (SLF) that locates the HSS handling an IMS and provides an MRFP that is able to mix, generate, and process media streams under the control of an MRFC. The MRFP can provide transcoding to allow IMS applications to interoperate with other IP-based applications employing different encoding schemes. By separating the MRFP from the MRFC, a single control function can oversee many processing functions to achieve scalability.

The IMS also provides media gateway (MGW) functions to allow IMS sessions to interwork with circuit-switched networks, including the PSTN and the ISDN, and even the CS domain of UMTS. The MGW simply transcodes the data streams to and from the format used in the external network. It is controlled by an MGCF that also handles the signaling to and from the circuit-switched network. For some types of circuit-switched networks, the MGCF is supported by a separate SGW. Finally, a BGCF determines where breakout should occur.

MULTIMEDIA SERVICES ACCESSING

Over the past years there have been major standardization activities undertaken in 3GPP and 3GPP2 for enabling multimedia services over 3G networks.[10,11] The purpose of this activity was to specify an IP-based multimedia core network that could provide a standard IP interface to wireless terminals for accessing a range of multimedia services independently from the access technology. The IMS network provides a standardized IP-based signaling for accessing multimedia services as shown in Figure 4.9.

The interface uses the SIP specified by IETF for multimedia session control. In addition, SIP is used as an interface between the IMS session control entities and the service platforms which run the multimedia applications. The initial goal of IMS was to enable mobile operators to offer to their subscribers multimedia services based on and built up Internet applications, services, and protocols. It should be noted that the IMS architecture of the 3GPP and 3GPP2 is identical, and is based on IETF specifications. Thus, IMS forms a single core network architecture that is globally available and can be accessed through a variety of access technologies, such as mobile data networks, WLANs, fixed broadband (e.g., xDSL, or x-digital subscriber line), etc. No matter what technology is used to access IMS, the user always employs the same signaling protocols and accesses the same services.

IMS IN THE NETWORK ENVIRONMENT

Many of the IMS nodes are specialized types of SIP servers. They are known as CSCF and are used to process SIP signaling packets. For example, a P-CSCF is a SIP proxy that acts as the first point of contact for an IMS terminal.

Figure 4.10 shows the IMS in the network environment. A P-CSCF is assigned to an IMS terminal during registration, and does not change for the duration of the registration. An I-CSCF is a SIP proxy located at the edge of an administrative domain.

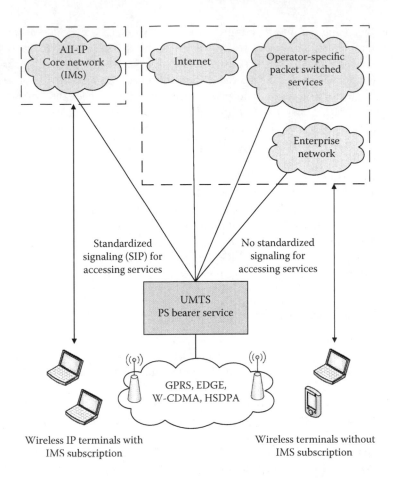

FIGURE 4.9 IP-based signaling for accessing multimedia services. EDGE, enhanced data rates for GSM evolution; W-CDMA, wideband code division multiple access; HSDPA, high-speed downlink packet access.

Its IP address is published so that remote servers can find it and use it as an entry point for all SIP packets to this domain.[22] An S-CSCF is a SIP server performing a session control role in the signaling plane. This CSCF node handles SIP registrations, which allows it to bind the user location (e.g., the IP address of the terminal) and the SIP address. The S-CSCF sits on the path of all signaling messages and can inspect every message. It decides to which application server(s) the SIP message will be forwarded to provide their services, provides renting services, and enforces the policy of the network operator. In addition to SIP functionality, IMS provides a host of common functions required for the operation of mobile networks.[23] Examples are authentication, authorization and accounting (AAA), charging, access control, and user profile databases. These functions of IMS are meant to be used by converged applications in a uniform way, so that there is no need to apply separate mechanisms to traditional circuit-switched and packet-switched voice and data services.

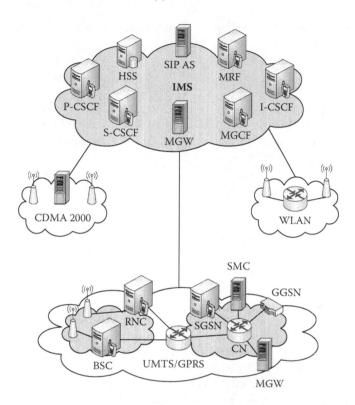

FIGURE 4.10   IMS in the network environment. RNC, radio network controller; BSC, base station controller.

WEB SERVICES IN IMS

The IMS architecture defines that incoming service invocations are carried out over SIP sessions. However, many non-SIP-based services exist in the operator domain. Such services use alternative protocols and technologies.

SIP sessions are contiguous and rather short-lived compared to the long-lived sessions and transactions used by many high-level middleware technologies like Web services (WS). WS are accessed via Web protocols and data formats such as the Hyper Text Transfer Protocol (HTTP) and the Extensible Mark-Up Language (XML). In this sense WS combine component-based development and Web technologies. WS benefit from universally accepted standards based on XML, such as Simple Object Access Protocol (SOAP) and Web Services Description Language (WSDL) for invocation and interface description. HfW SOAP is a protocol for exchange of XML-based messages over a computer network that can be used by the higher and more abstract layers to build on.

The WSDL describes the public interface exposed by the WS. It specifies message formats required to interact with WS. WS are typically stateless, which means that each invocation of the WS should contain all the information it needs to process a request, since the processing depends only on this data. This design greatly simplifies the implementation.

SERVICE DELIVERY PLATFORM

The IP multimedia subsystem is a recent idea that outlines a reference framework and architecture for building and deploying a platform that will enable a range of applications to deliver these enhanced multimedia services. Such platforms are often termed service delivery platforms.

Although the IMS architecture appears to be gaining wide acceptance as the emerging standard in service delivery, in practice a number of existing platforms have been deployed by mobile and fixed line operators using existing and alternative architectures. Many of the existing implementations were developed by focusing on the IT system integration requirements, rather than by applying a consistent implementation approach to the network aspects for multimedia delivery. Whatever the approach taken, there is a desire to merge the existing architectures with current trends in IMS delivery to maximize the effectiveness of the SDP.

There is considerable research and experimentation in the domain of multimedia, convergence, and next generation networks. The work is motivated largely by competitive forces that are driving commercial opportunities in enhanced multimedia service delivery for fixed and mobile networks. Examples include IPTV, gaming, video calls, and voice over IP (VoIP).

IMS as a basis for the service delivery platform has been suggested.[24] In Reference 25 a service delivery platform that extends IP multimedia subsystems was developed.

A test architecture that integrates SIP with a telecommunications network to deliver multimedia services has been developed. More recently there has been broader interest in IMS and service delivery characteristics.[26] Multimedia is broadly understood as the use of several forms of media, typically text, audio, graphics, and video. In addition, a service is generally defined as the nonmaterial equivalent of a good. In light of this, the term multimedia services, when used in the context of a service delivery platform, is used to denote multimedia content in the form of either content or service.

A further aberration of the multimedia service is in how this is manifested. In general, a Web-enabled application is responsible for generating the output multimedia service. When an entity (i.e., a third-party developer) creates such an application, it is the application entity that supplies the desired multimedia content or service.

The term multimedia service is also used to denote the Web application deployed to the SDP that is responsible for generating the output content or service.

EXTENDED IMS ARCHITECTURE

Functions of an IMS network can be subdivided into subscriber profile function, signaling functions, QoS policy function, media function, and charging function. Subscriber profile function is a user subscription profile database known as HSS. The signaling function of an IMS network is based on the SIP and SDP. The signaling functions consist of SIP servers mainly performing authentication, registration, location discovery, call routing, and call redirection. The signaling functions are

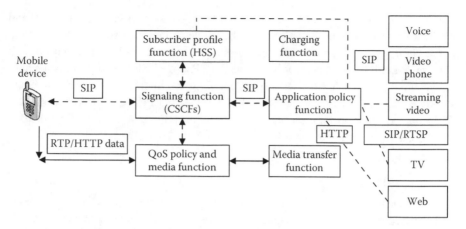

FIGURE 4.11 Extended IMS architecture. Reproduced with permission from S. Q. Khan, R. Gaglianello, and M. Luna. "Experiences with blending HTTP, RTSP and IMS," *IEEE Commun. Magazine* 45 (March 2007): 122–28.

referred to as CSCFs (proxy, serving, and interrogating call session control functions). The QoS policy function creates and enforces QoS specific policies. In some IMS infrastructures, they are referred to as policy decision function (PDF) and policy enforcement function (PEF). The charging function either performs online or offline accounting for billing purposes.

Current IMS applications are mainly voice and video telephony centric. There are a few vendor and standard initiatives to extend IMS by adding an application policy function that interfaces with Real-Time Streaming Protocol (RTSP) and HTTP-based applications, and enforces RTSP, HTTP, and WAP application-specific policies. For example, the application policy function would filter application requests and deny access to applications that are not subscribed to or are prohibited. The application policy function enables providers and subscribers to create, manage, and implement policy rules such as parental controls, privacy controls, and other application-sharing polices. In some marketing circles, this is referred to as a service delivery platform and would sit above the S-CSCF and span other network or network access technologies. Note that these application policy rules are in addition to the SIP-based filter criteria specified in 3GPP TS 23.218.[28]

The extended IMS architecture appears in Figure 4.11 and promises to provide streaming video, IPTV, and Web applications from a converged platform. Some advantages of this architecture are maintenance of a single platform, ease of service creation, reduction of redundancy, improvement of architectural scalability, and support of seamless application mobility. Some disadvantages are degraded signaling load scalability, multiple points of provisioning and support, and a degraded customer experience due to inconsistent content rendering.

Note that an element of the application policy function can be the service capability interaction manager (SCIM). This is an entity for managing interactions among SIP application servers and between SIP features and legacy signaling system components. It invokes service logic as per SIP requests. The 3GPP/3GPP2 designated the SCIM as a functional element within the IMS that sits between the S-CSCF and application servers. The SCIM also can be implemented as a standalone function. It has the opportunity to modify the messages of a session/flow as required, and can understand different protocols such as HTTP and RTSP. Thus, it can be used to enhance the underlying functionality by enabling the interaction between different standards such as HTTP and RTSP with the SIP-based IMS core. Capabilities of a SCIM include support for multiple dialog sessions, dynamic blending of application services, and personalized feature interaction management.

Policy Control in IMS

IMS provides policy control. IMS domains can specify the types of media streams that are accepted (e.g., voice streams) and the types that are not (e.g., video streams). It is possible to specify the sessions that are accepted based on characteristics of their media streams, such as the codecs used or the bandwidth requested for them.

IMS policy control and IMS charging control were implemented using separate architectures in 3GPP R5 (Release 5) and 3GPP R6. However, 3GPP R7 merged those architectures together. The policy and charging control (PCC) architecture defined in 3GGP R7 is the result of merging service-based local policy (SBLP) and flow-based charging (FBC).

Figure 4.12 shows the 3GPP R7 policy and charging control architecture. Only the AF (application function), the PCRF (policy and charging rules function), and the access gateway are relevant to this discussion. The role of the AF can be performed by the P-CSCF or by an application server.

The PCRF receives information about the offer/answer exchanges between the terminals from the AF over the Rx interface. If the characteristics of the session

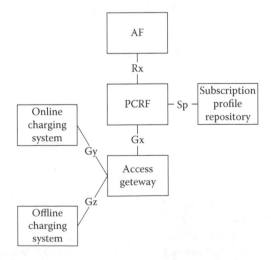

FIGURE 4.12 IMS policy and changing control architecture. Reproduced with permission from G. Camarillo et al. "Towards an imnovation oriented IP multimedia subsystem," *IEEE Commun. Magazine* 45 (March 2007): 130–36.

being established are acceptable to the PCRF (based on the domain policy), the PCRF authorizes the session on the access gateway using the Gx interface. If the characteristics of the session are not acceptable to the PCRF, it instructs the AF to terminate the session using the Rx interface. Of course, in this case the PCRF does not authorize the session on the access gateway.

One of the design principles of SIP was to enable the creation of end-to-end services without requiring the upgrading of the network elements between end points. As discussed earlier, the current approach to IMS policy control is contradictory to this principle.

Session policies provide a means for domains to communicate their policy to terminals and for terminals to provide domains with information about the sessions that they establish. There are two types of session policies: session-independent policies and session-specific policies. Session-independent policies are general policies that apply to all the sessions a terminal may attempt to establish. For example, a terminal may not be allowed to use video streams in its sessions. Session-specific policies only apply to a particular session. For example, a terminal may be required to group two of the session media streams (e.g., audio and video) into a single PDP context and transfer a third media stream over a different PDP context (e.g., instant messaging).

A domain typically requires information about the session being established by a terminal to provide it with session-specific policies for that session. The terminal informs the domain about the session and obtains the domain policies for that session. For example, a domain can use the information received from a terminal about the IP addresses it will use to open the gates of the access gateway or a pinhole in a firewall.

Some policies can be implemented as session-independent or as session-specific policies. For example, a domain may choose to provide terminals with a list of all the audio codecs the terminals are allowed to use as a session-independent policy. On the other hand, a domain could choose to be informed about the codecs a terminal intends to use in a session and inform the terminal which of the codecs are acceptable.

Some domains prefer to implement this type of policy as session-specific to avoid disclosing the entire domain policy to the terminal. These domains want to prevent competitor operators from copying their policies, which some providers believe to be a source of competitive advantage.

Session policies define a logical entity referred to as a policy server. Terminals subscribe to the policy server using a SUBSCRIBE request, and obtain information about the domain session-independent policies in NOTIFY requests.

Session-independent policies are a form of configuration information. Therefore, the event package used to transfer session-independent policies is the event package for the user agent profile delivery specified in Reference 30.

Terminals also use SUBSCRIBE requests to obtain session-specific policies. When a terminal intends to establish a session, it sends a SUBSCRIBE request to its policy server. This request uses the event package for session-specific policies defined in Reference 31. The terminal includes an XML document describing the session it is about to establish in the body of its SUBSCRIBE request.

The policy server receives the SUBSCRIBE request and, based on the XML document received and the user's profile, returns its policies in a NOTIFY request. If the terminal must change the characteristics of the session at some point, it sends a new

SUBSCRIBE within the same SUBSCRIBE-initiated dialog to the policy server. If the policy server must send additional policies to the terminal at some point, it issues a new NOTIFY request.

IMS Standardization Effort

IMS is the foundation for the next generation IP-based networks, as specified by the 3GPP/3GPP2 standards organizations and embraced by the European Telecommunication Standards Institute's Telecoms and Internet Converged Services and Protocols for Advanced Networks (ETSI TISPAN) group and International Telecommunication Union–Telecommunication Standardization Sector (ITU-T). The standards support multiple access technologies such as GSM, WCDMA, CDMA2000, WLANs, wireline broadband, and cable.

The IMS defines a control layer—on top of IP-based fixed and mobile networks—that enables seamless provisioning of multimedia services riding over the control layer. This is realized by extending IETF protocols such as the SIP and diameter for multimedia session control and authentication, authorization, and accounting (AAA). Key IMS components are the CSCF, HSS, media resource function (MRF), and ASs.

IMS aims to make Internet technologies, such as instant messaging, presence, and voice and video conferencing, available to everyone from any location over any network. It is expected to allow service providers to control the network and, in return, provide better security, QoS, and single sign-on for a combination of existing telecommunications services such as voice, SMS, MMS, and IP-based services such as IPTV, instant messaging, push to talk, and Web browsing. Switching between services will be seamless. Key benefits of IMS to service providers can be summarized as follows:

- A common access-agnostic core network supporting all applications instead of application specific networks, resulting in lower capital and operational expenditures
- Simplified creation of blended services such as combined presence, instant messaging, and telephony
- Delivery of applications across the fixed–mobile boundary
- Faster deployment of new applications based on standardized modules

As interest in fixed–mobile convergence continues to rise, IMS is emerging as the technology that enables service providers to move beyond the limitations of today's cellular mobile architectures. IMS is already being deployed in many trials and for a small number of commercial services, but it is expected to grow as service providers move from trials to full-scale deployment. Gaming, push-to-talk over cellular (PoC), and presence-based services are expected to drive IMS deployment in the mobile domain; while services such as VoIP, IPTV, and fixed/mobile convergence (FMC) are expected to drive deployment in the fixed network domain.

IMS was initially standardized by 3GPP as part of its Release 5 specifications in 2003 as a new service layer on top of IP-based 3G networks. Release 6 specifications in 2005 addressed IMS interworking with legacy circuit networks and other IP networks as well as harmonization with emerging PoC and related service enabler standards defined by the OMA. In Release 7 specifications, the IMS scope is extended to any IP networks, including fixed access networks. In addition, Release 7 addresses decreasing latency and improvements to real-time applications such as VoIP. The ongoing Release 8 addresses 3GPP long-term evolution (LTE) and system architecture evolution (SAE), including IMS-based emergency services.

The IMS standardization effort focuses primarily on the IMS core network elements and protocols, including IMS application server options: customized applications for mobile network enhanced logic (CAMEL), open services access (OSA)/ Parlay, and SIP ASs, but excludes standardization of applications. OMA investigates the applications space by standardizing service enablers on top of IMS.

IMS architecture as defined in relevant standards is complex due to the number of interfaces and definitions of functional entities. This complexity results in various challenges when deploying IMS services:

- Simplification of the architecture
- End-to-end multivendor interoperability
- End-to-end network management
- Interaction of services layer with control and transport layers
- Policy management across various market verticals to effectively provide service offerings while guaranteeing service quality
- Coexistence with and use of legacy technologies such as 3G, ATM, WLAN, Ethernet, wavelength division multiplex (WDM), and synchronous optical network (SONET)
- Use of multiple access technologies in an agnostic fashion
- Simplified and flexible billing
- Delivery of more complex and blended applications
- Length of SIP control messages, which is extremely large for wireless control channels

These challenges in architecture, protocols, and operations are being worked on in the industry. This feature topic intends to address challenges at the infrastructure and service levels.

4.4 LOCATION-BASED SERVICES

Location-based services (LBS) can be considered the most rapidly expanding field in the mobile communications market. Their first appearance, in a much more primitive form than that known today, is traced to the middle of 1990s, propelled by the advent of the Global Positioning System (GPS). Only a few years later the proliferation of the mobile/wireless Internet; the constantly increasing use of handheld, mobile devices and position tracking technologies; and the emergence of mobile computing paved the way for the introduction of this new category of services. An impressively large number of applications for the general public has come to mean solutions that leverage location information to deliver consumer applications on a mobile device. Application opportunities can be grouped into the following categories:

- Navigation and real-time traffic monitoring
- Location-based information retrieval
- Emergency assistance
- Concierge and travel services
- Location-based marketing and advertising
- Location-based billing

These categories target the widest portion of the LBS market, and will be made accessible to millions of users by large players in the telecommunication, automotive, and media industries. Apart from these services, however, an additional set of applications, focused on specialized target groups (e.g., the corporate sector, the health sector), will be developed. These include:

- Dispatch and delivery route optimization
- Monitoring person location, which includes data for health care, emergency calls, and prisoner tagging
- Third-party tracking services for enterprise resource planning (ERP) and the corporate and consumer markets (e.g., fleet management, asset or individual tracking)
- Security and theft control
- People finding

LBS cover the whole range of user needs for emergencies, as well as a large range of business needs (e.g., fleet management). For the development and provision of LBS services, a synergy of different, yet complementary, technologies and architectures is required. An overview of these technologies, along with other critical technical issues, is given in the sections below in an attempt to define the requirements and the architectural aspects of an LBS system; that is, a system focused on the delivery of location based services.

REQUIREMENTS FOR LBS DELIVERY

At first, we identify the essential components that make up the process for delivering LBS to the end users. LBS client/server architecture is shown in Figure 4.13.

There are two main entities involved in the LBS provisioning model:

- The LBS server, which is responsible for providing the location sensitive information to the LBS client. Providing such information may also require invocation or queries to other network entities.
- The LBS client, which asks the LBS server for location sensitive information and is the recipient of the response produced by the corresponding service; an LBS client may range from a notepad computer to a mobile phone or any other handheld mobile device.

We refer to the combination of these two entities as the LBS system. It is evident that the LBS provision model greatly resembles the standard client/server model.

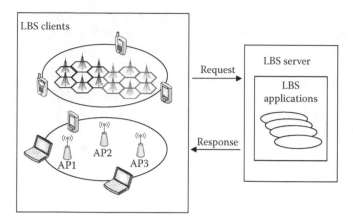

FIGURE 4.13 Client/server architecture for LBS.

There is always an entity that asks for the information and another which provides it. Although, this is true from a logical perspective, sometimes it is not very easy to physically separate the server and the client entities, as they may both reside in the same physical device.

This was the usual case until a few years ago, when the concept of an LBS system would normally bring to mind an electronic device, equipped with a GPS receiver and a display, where users could see their current position, possibly on a map. However, such proprietary LBS solutions were characterized by certain drawbacks. First, the consumer found them expensive; the cost of the electronic device was usually high. Moreover, the device was capable of providing only a certain range of LBS services and no enrichment was possible due to the lack of open interfaces.

Currently, such solutions, although not completely abandoned, are targeted to a specific class of users, and have been replaced by solutions that apply to the classical client/server model, where the entities involved are physically separate. A key factor in this development was the growth in the capabilities of mobile and handheld devices, as well as the growing maturity of the wireless infrastructure in terms of positioning technology. Moreover, the new model completely separates the service logic from the client side, thus allowing new services to be developed and accessed using the same terminal equipment.

Prime requirements that an LBS system should meet include:

- Cost-effectiveness. Service delivery should not require the end user to buy costly equipment.
- Openness. Service provisioning should support a variety of access protocols so that the service is available through different networks and different client equipment.
- Reusability. The LBS system should be capable of hosting a number of different services, with different requirements and functionality. The introduction of new services should not require changes to either the LBS server or

the client, and potential changes to the server should maintain downward compatibility (i.e., should not affect the execution of existing services).

- Security and privacy. Security in the interfaces with external entities is essential so that secure communication and privacy is achieved. This is a fundamental requirement, as the end user would not normally be willing to have personal information (e.g., location) revealed to a third party.
- Scalability. The system should be able to host a large number of services, each capable of serving numerous concurrent requests.
- Extensibility. The system should be extensible and capable of accommodating new technologies. To achieve this, there must be independence from underlying technologies. Therefore, the system should not be bound to any specific positioning or Geographical Information System (GIS) technology. This will allow it to easily adopt newly evolved technologies from both sectors, thus increasing its life expectancy.

Additional requirements, which are optional, but may greatly enhance the potential of the LBS system include the following:

- Support for many operation paradigms. This means that apart from the classic request/response functionality, the platform should support services using the push model, as well as event scheduling, which can be based on time or location of events (such as notification using SMS about the day's offers when the user enters a shop).
- Roaming across different infrastructures. Both indoor and outdoor environments should be supported.
- Support for flexible service creation processes.
- Support for service deployment and operation, which should be provided through automatic procedures.
- Portability. Independence of operating systems and hardware platforms is a characteristic of prime importance as it guarantees integration with every infrastructure.

Having established the desirable requirements, we now proceed with the LBS system.

LBS SYSTEM

An LBS system is composed of two elements: the LBS server and the LBS client. A generic LBS provisioning model is shown in Figure 4.14. The main entities involved in the LBS provisioning scheme are depicted. Apart from the core LBS system, the model also contains three additional types of systems: positioning system, spatial data (GIS) system, and supplementary systems.

The first two systems are essential to the LBS provisioning chain, as the information they provide to the LBS server is mandatory for executing the LBS application. The "supplementary systems" category includes auxiliary or optional entities (e.g., billing systems), which although not needed for the basic LBS provisioning process,

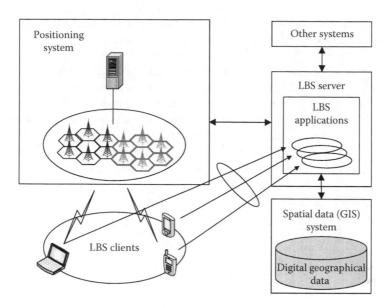

FIGURE 4.14 Generic LBS provisioning model.

can greatly enhance it and allow the implementation of advanced concepts such as service personalization, QoS differentiation, and different provisioning policies.

The presented model shows the logical separation between the different components of an LBS system. However, the LBS server may integrate the GIS system as well as the positioning system in the same physical node.

4.5 CONCLUDING REMARKS

The development of the IP-based network for IP multimedia services is exciting. Many new services will become available to make use of the higher data rates.

For IMS deployment to be successful, it must emulate the success that current IT-based solutions have experienced. To facilitate sustained growth, expansion of the number and variety of multimedia services is required.

Benefits of an enhanced IMS are as follows:

- Enables a consistent, seamless application mobility experience of SIP, RTSP, and HTTP-based applications.
- Extends the value of existing services into the IMS domain, and IMS services into the existing domain, providing accelerated return on investment for IMS investments research and development.
- Scales the IMS offering to applications and services available on the Web today.
- Avoids maintaining parallel networks mode by creating a consistent, centrally managed, policy enforcement point that is applicable to segment management, regardless of domain. Eliminates the requirement of duplicate

functions, for example, one subscriber profile function can be used for three networks—SIP, HTTP, and RTSP-based networks.

Using session policies instead of the currently specified techniques for policy control would make IMS a better platform for service innovations. Session policies remove the requirement to inspect and understand session descriptions in the network. Therefore, the use of session policies would enable service providers to offer innovative services based on new SDP extensions or other session description formats without requiring them to upgrade their networks. This way, IMS services would reach their users faster and at a lower price. The price for implementing session policies is a slightly higher session establishment delay for sessions that would have been acceptable to the network in the first place.

5 Wireless Networking Standards (WLAN, WPAN, WMAN, WWAN)

This chapter summarizes the specifications for wireless networking standards that have emerged to support a broad range of applications: wireless local area networks (WLANs), wireless personal area networks (WPANs), wireless metropolitan area networks (WMANs), and wireless wide area networks (WWANs). The standards have been developed under the auspices of the IEEE 802 LAN MAN Standards Committee of the Institute of Electrical and Electronics Engineers (IEEE). After a short presentation of IEEE 802.x standards, we deal with WLAN link layer standards. Wireless asynchronous transfer mode LAN, together with European Telecommunication Standard Institute (ETSI) BRAN HIPERLAN standard, is also included. This chapter also reviews WPAN devices and Bluetooth. Key achievements toward deployment of ultra wideband communications (UWB) are described. We continue with an overview of WMANs and WWANs. The emphasis is on the 802.16 network arrangement medium access control (MAC) protocol, as well as on the orthogonal frequency division multiplexing (OFDM) physical (PHY) layer protocol. This chapter concludes with H.264 video transmission over IEEE 802.11-based wireless networks, taking into account a robust cross-layer architecture that leverages the inherent H.264 error resilience tools.

5.1 INTRODUCTION

In the debate between standards and proprietary systems, the former seems to be winning in all moderate- to large-size markets because of the proven effectiveness of economy of scale in reducing manufacturing costs. Users are demanding interoperability so that they are free to choose equipment from different manufacturers. The global success of standards such as Global System for Mobile (GSM) or Ethernet in the LAN market are proving that standardization works.[1]

Technology underlies all developments in communication networks. However, wide-scale deployment of networks is not based directly on technology, but on standards that embody the technology, along with the economic realities. Because standards mediate between the technology and the application, they provide an observation point from which to understand the current and future technological opportunities. In the case of traditional telecommunications, such observation is obscured by a proliferation of standardization bodies, many organized on a geographical or governmental basis. However, in the case of data communications networks, standards have historically been set, and applied, on a worldwide basis, without strong geographical influence.

The information society is bringing about huge interchanges of data that must be delivered to the user in different situations for a wide range of applications.[2] The flexibility provided by wireless links can be a determinant in the rapid growth of many applications. The success of mobile communications, with mobile phones approaching or surpassing the number of fixed phones in many countries, will almost certainly be followed by the success of mobile data communications.[3]

A wireless future can be envisioned with the following features:

- Telephony will be mostly mobile, with phone numbers associated with people, instead of places. The number of mobile phones will be on the order of 90 percent of the population able to use it (everybody over age 6, with the possible exception of disabled or very old people).
- Fixed access, which will still be necessary for the home and in certain offices will be provided by microwave- and millimeter-wave lines, with point-to-multipoint standards (like Local Multipoint Distribution System, LMDS), or, in the case of high bandwidth users, point-to-point links.
- Inside the home or office, devices will communicate with one another with some of the short-range technologies (WLAN or others) that are currently being defined.

Internet access or data access in general will be provided at fixed points by the fixed access technologies, and by the third and future generations of mobile communications. Very high speed broadcast can be obtained from satellites (with digital video broadcasting, DVB, transmission) and future stratospheric platforms.[4]

Another important trend is the evaluation nature of current standards. They are not fixed once and for all, but are being updated often to add new features or to find new applications. The current speed of technological innovation is based on this fact. On many occasions, the new version of a standard can be implemented with the same hardware, by means of updating the software.

Generally speaking, standards have to be considered in the following areas:

- Networking and connectivity
- Registration and addressing
- Protocols and software, computer languages, and distributed architecture
- Format conversion and compression techniques

Standards for mobile computing are driven forward by national and international standardization organizations, by industry alliances, and by international forums. On the global level we find the international standardization organizations in the data and media worlds—for example, International Organization for Standardization (ISO), International Electrotechnical Commission (IEC), the International Telecommunication Union (ITU), and the global partnership projects, which are formed by regional and national standardization bodies. The Internet Engineering Task Force (IETF) is another organization with a global focus. Internationally, we also find a variety of industry alliances depending on specific standardization targets. Examples are the Wireless Applications Protocol (WAP) Forum, the HYPERLAN12 Global Forum, and the Bluetooth Special Interest Group.[5]

Standardization on a global basis is of fundamental importance, and even more so for mobile cellular systems because of the need for worldwide roaming. Thus, standardization is a key issue for International Mobile Telecommunication (IMT)–2000/ Universal Mobile Telecommunication system (UMTS). In this field, there exists the tradition to coordinate and define the framework standards for global wireline and wireless access for national and international telecommunication infrastructures on the Internet TV level, as well as on the regional and national levels.

Also, the standards for the mobile Internet have to be global. The fixed Internet standards, with their roots in the development of the LAN area, evolved to a global standard with the worldwide use of the Internet protocols. These standards have been set by the IETF, World Wide Web Consortium (W3C), Internet Corporation for Assigned Names and Numbers, and other Internet Protocol (IP)-related standards organizations, for example, mobile IP. These standards will be merged for UMTS in the Third Generation Partnership Project (3GPP). In addition, for data processing, data communications, television, and multimedia, ISO and IEC standards were developed and continue to be valid for all kinds of data and multimedia applications and services.

Mapping Internet content to mobile wireless devices requires new standards and innovative solutions that minimize cost and maximize efficiency. The mobile Internet must deliver information in a suitable format regardless of location and content. Wireless access has already introduced a new set of standards and protocols that add a layer of complexity to applications not necessarily compatible with the Internet (e.g., the GSM/General Packet Radio System, GPRS, standard). For UMTS these standards have to be supplemented for those services that will also be offered on the wireline network. The harmonization of the wireline terminals may also be a standardization issue. Furthermore, standardization has to specify impacts regarding addressing, which is quite different in telecommunications than it is in the IP world.

The harmonization of the UMTS standards in the IMT-2000 framework with the standards on the Internet side is necessary to make mobile multimedia happen in an international networking environment, especially for roaming use. The role of the UMTS Forum, the IP version 6 (IPv6) Forum, and other market representation partners within 3GPP is to widen the scope of the standardization and to convert its views into requirements and work items.

The individual members of these standardization bodies and forums come from the application service industry, equipment manufacturing, network operators, and Internet service providers, as well as from software companies, smart-card industries, regulatory agencies, and universities. In addition, as multimedia and mobility applications lead the way forward, other industries like the media and the automobile industry are joining these standardization activities.

This chapter addresses primarily the IEEE 802 work[14] in wireless networks, supporting low-cost products serving customers' needs for WLANs, WPANs, and WMANs.

5.2   STANDARDIZATION PROCESS IN IEEE 802

IEEE supports many technical activities, including an active program in standardization through the IEEE Standards Association (IEEE-SA). IEEE standards are
developed openly, with consensus in mind. Standards developed in any form can
produce high-quality, broadly accepted results capable of focusing companies and
industries. Project development in the IEEE-SA is normally delegated to individual
standard "sponsors," one of the most important of which is IEEE 802. The IEEE 802
process is designed for quick development of standards with broad consensus. The
demand for consensus helps to ensure that standards are technically refined and meet
market needs. The essence of the process is at two-stage balloting system, each with
multiple rounds, that seems not only to confirm consensus, but also to generate critical comments. It is sometimes said in IEEE 802 that the purpose of balloting is not
to approve the draft standard but to improve it. Experience has shown that the IEEE
802 process is extremely effective at engaging a wide variety of interested parties,
fostering comments, and implementing constructive changes. As a result, 802 drafts
are refined again. By the time a draft is ready for approval, users have confidence in
it. Yet, with careful attention and the will of the developers, it is possible to drive the
draft through the system within a reasonable time.[6,13]

The IEEE 802 wireless standards program comprises three working groups:

- IEEE 802.11 Working Group develops the IEEE 802.11 standards for
  WLANs.
- IEEE 802.15 Working Group develops the IEEE 802.15 standards for
  WPANs.
- IEEE 802.16 Working Group on Broadband Wireless Access develops the
  IEEE 802.16 standard for WMANs.

In addition, two technical advisory groups (TAGs) help coordinates activities:
the IEEE 802.18 Regulatory TAG and the IEEE 802.19 Coexistence TAG. In what
follows, we summarize the status and technology of the projects in the IEEE 802
wireless standards program.

5.3   WIRELESS LOCAL AREA NETWORKS (WLAN)

Technologies for physical media access by wireless follow several specifications,
some of which have been approved by independent standards bodies. One such suite
of specifications is IEEE 802.11. In conformity with U.S. Federal Communications
Commission (FCC) requirements, IEEE 802.11 allows both direct sequence (DS)
and frequency hopping (FH) spread spectrum. The maximum data rate offered by
the standard for either technique is 2 Mbps. However, a higher bit rate version of
IEEE 802.11 allows a data rate of up to 11 Mbps. A drawback of the 802.11 protocol
for data transfer in a home network is its overhead.[7,14]

IEEE 802 LAN/MAN Standards Committee develops LAN and MAN standards,
as well as one WPAN standard (802.15). WLANs cover single-hop or multihop communications, which can provide various network services within a limited service

area. Research and deployment of these networks has been very rapid in the past few years, leading to the development of a number of WLAN technologies like 802.11 (WiFi) and HyperLAN. Even through these technologies can provide high-speed (broadband) wireless access to IP networks, they have significant limitations, which must be overcome in order to allow seamless, scalable, and stable quality of service (QoS) for wireless mobile users.[8,9]

IEEE 802.11[15] refers to a family of specifications developed by the IEEE for WLAN technology, which operates at either the 2.4 GHz industrial, scientific, and medical (ISM) band, or the 5 GHz unlicensed national information infrastructure (UNII) band. There is a growing interest in the use of WLAN technology. Examples of applications range from standard Internet services, such as Web access, to real-time services with strict latency/throughput requirements, such as multimedia video and voice over IP (VoIP).[10] Future applications will be more demanding and may include high definition television (HDTV) and audiovisual support. With the high demands and varying requirements of these applications, there is a need to support QoS in WLAN.

OVERVIEW OF THE IEEE 802.11 STANDARDIZATION

IEEE 802.11 is an industry standard set of specifications for WLANs developed by the IEEE. In next generation wireless networks, it is likely that the IEEE 802.11 WLAN technology will play an important role and affect the style of people's daily life. The 802.11 technology provides cheap and flexible wireless access capability. It is very easy to deploy an 802.11 WLAN in campuses, airports, stock markets, offices, hospitals, and other places. Meanwhile, multimedia applications are increasing tremendously. People want voice, audio, and broadband video services through WLAN connections. Unlike traditional best-effort data applications, multimedia applications require QoS support such as guaranteed bandwidth and bounded delay/jitter. Providing such QoS support in 802.11 is challenging, as the original standard does not take QoS into account. Both the PHY layer and the MAC layer are designed for best effort data transmission.[9,15,16]

Two kinds of basic network configuration modes are provided in the 802.11 standard:

- An infrastructure mode, where transmissions of all stations (STAs) have to go through a central access point (AP) device
- An ad hoc mode, where any STA can talk to another without an AP

The 802.11 standard defines the specifications of both PHY and MAC layers to construct a WLAN using either configuration mode.

The first standard for 802.11 came out in 1977. The MAC layer was defined with three different PHY layers based on infrared, direct sequence spread-spectrum (DSSS), and frequency hoping (FH). These PHY layers supported only 1 and 2 Mbps. Then, the following extensions were developed to enhance the performance of IEEE 802.11 WLANs.

IEEE 802.11a

IEEE 802.11a operates at a data transmission rate as high as 54 Mbps and uses a radio frequency of 5.8 GHz. Instead of DSSS, 802.11a uses OFDM.[4] The general idea of OFDM is to shift a high-rate data stream into lower-rate streams which are transmitted simultaneously over a number of subcarriers. These subcarriers are orthogonal to each other in the sense that, when listening to one subscriber, they do not interfere with one another. Hence, the name orthogonal frequency division multiplexing. By using this technique, multipath delay spread and inte0-symbol interference are considerably decreased because only low bit rate streams are employed. OFDM allows data to be transmitted by subfrequencies in parallel. This modulation mode provides better resistance to interference and improved data transmission.

IEEE 802.11b

IEEE 802.11b, an enhancement to 802.11, provides standardization of the physical layer to support higher bit rates. IEEE 802.11b uses 2.45 GHz, the same frequency as IEEE 802.11 and supports two additional speeds: 5.5 and 11 Mbps. It uses the DSSS modulation scheme to provide higher data transmission rates. In direct sequence, the frequency band is divided into fewer but larger channels. To compensate for noise in a channel, a technique called chipping is implemented. Each bit is converted into a redundant bit pattern chip sequence, whereby an n-chip code spreads the signal by a factor of n. If some interference destroys part of the chip sequence, the original bit may still be recovered from the remaining chips. The bit rate of 11 Mbps is achievable in ideal conditions. In less than ideal conditions, the slower speeds of 5.5, 2, and 1 Mbps are used.[4,16]

To conclude, IEEE 802.11b specifies a DSSS system with a peak data rate of 11 Mbps. Lower rates are available for poor links with dynamic rate switching rules specified. Since this mode is a backward-compatible extension of the original DSSS system, control information is transmitted at the common 1 Mbps rate. This is one reason that actual throughput is less than ideal. The channel bandwidth is about 20 MHz, so that the typical North American and European frequency allocations (2.4 to 2.4835 GHz) provide for three non-overlapping channels. A smaller Japanese allocation is noted in the standard. These devices are used in many countries, with varying spectrum allocations and power limits.

IEEE 802.11c

IEEE 802.11c provides required information to ensure proper bridge operations. This is very important for implementation of access points, as they have to bridge between wired and WLANs.

IEEE 802.11d

When 802.11 was launched in the late 1990s, only a handful of regulatory domains (e.g., the United States, Europe, and Japan) had rules in place for the operation of 802.11 WLANs. To support widespread adoption of 802.11, the 802.11d task group has an ongoing charter to define PHY requirements that satisfy regulatory controls in additional countries.

IEEE 802.11e

The IEEE 802.11 WLAN is the dominant solution for local area wireless networking; not only is its high performance provided at low cost, but it is also easy to deploy. Most hotels, airports, office buildings, and universities are covered by an 802.11 network for Internet access.[10,17] However, applications like VoIP, videoconferencing, and online gaming with specific bandwidth, delay, and jitter requirements are not always supported adequately by 802.11 networks, making QoS a key requirement in these networks.[11,18]

For WLANs, IEEE 802.11 is designed for best-effort services. The 802.11 standard specifies two MAC mechanisms: the mandatory distributed coordination function (DCF), and the optimal point coordination function (PCF).[19] The lack of a built-in mechanism for supporting real-time services makes it very difficult to provide QoS guarantees for multimedia applications. To enhance QoS support in 802.11, the IEEE 802.11 working group worked on a new standard, IEEE 802.11e, which introduces the so-called hybrid coordination function (HCF).[20,21] HCF includes two medium access mechanisms: contention-based channel access and controlled channel access. Contention-based channel access is referred to as enhanced distributed channel access (EDCA), and controlled channel access is referred as HCF controlled channel access (HCCA).

Among various QoS issues, admission control is an important component for the provision of guaranteed QoS parameters. The purpose of admission control is to limit the amount of traffic admitted into a particular service class so that the QoS of the existing flows will not be degraded, while at the same time the medium resources can be maximally utilized.

The 802.11e standard defines a single coordination function, HCF, which combines the functions of both DCF and PCF for QoS data transmission. The major benefits offered by the 802.11e standard are as follows:[22]

- Reducing the latency through prioritizing different types of traffic packets
- Enabling access points to allocate resources based on data rate and latency requirements from each individual station
- Improving wireless bandwidth efficiency and reducing packet overheads

IEEE 802.11f

The existing IEEE 802.11 standard does not specify the communications among access points in order to support user roaming from one access point to another. To make this communication possible, IEEE 802.11 defines IEEE 802.11f that defines the rules for communication among different access points. This becomes very important to optimize the performance in Transmission Control Protocol/User Datagram Protocol (TCP/UDP) communications when mobility happens. In the absence of IEEE 802.11f, we can utilize the same vendor for access points to ensure interoperability for roaming users.

IEEE 802.11g

The goal of the IEEE 802.11 task group was to develop a higher speed extension (up to 54 Mbps) to the 802.11b PHY, while operating in the 2.4 GHz band. 802.11g will implement all mandatory elements of the IEEE 802.11b PHY standard. This also uses OFDM to increase its channel rate to 54 Mbps. The 802.11g stations use request-to-send/clear-to-send (RTS/CTS) exchange to prevent 802.11b stations from accessing the medium.[4] The RTS/CTS protocol notifies other users that the medium is expected to be busy, and therefore provides a "virtual carrier sense." The RTS/CTS mechanism is effective at addressing the hidden-terminal problem, at least with a moderate number of terminals hidden, and with some cost in the overhead. However, RTS/CTS is challenged in the wireless Internet service provider (WISP) environment by directional subscriber antennas. In this case, virtually all of the terminals are hidden from each other. Also, while responding to RTS/CTS is mandatory, generation of RTS is optional. If the terminals do not generate RTS, the mechanism does not function.

IEEE 802.11h

The IEEE 802.11h addresses the requirements of the European regulatory bodies. It provides dynamic channel selection (DCS) and transmit power control (TPC) for devices operating in the 5 GHz band. In Europe, there is a strong potential for 802.11a interfering with satellite communications, which have "primary use" designation. Most countries authorize WLANs for "secondary use" only. Through, the use of DCS and TPC, 802.11h will avoid interference with the primary user.

The IEEE 802.11 MAC Layer

The 802.11 MAC aims to provide access control functions to the wireless medium such as access coordination, addressing, frame check sequence generation, and security. There are several ongoing activities to extend the MAC layer protocols including 802.11e to enhance QoS performance; 802.11f, proposing an inter-access point (AP) protocol to allow stations to roam between multivendor access points; and 802.11i, focusing on enhanced security and authentication mechanisms.

Two medium access coordination functions are defined in the original 802.11 MAC: a mandatory DCF and a PCF.

DCF. The basic DCF uses a carrier sense multiple access with collision avoidance (CSMA/CA) mechanism to regulate access to the shared wireless medium. Before initiating a transmission, each station is required to sense the medium and perform a binary exponential backoff. If the medium has been sensed idle for a time interval called DCF interframe space (DIFS), the station enters a backoff procedure. To deal with hidden terminal problems in which some stations cannot hear each other and may transmit at the same time to a common receiver, an optional four-way handshake scheme known as request/clear to send can be associated with the basic DCF when data frame size exceeds a value called the RTS threshold.

PCF. The PCF was introduced to support multimedia transmission and it can only be used if a WLAN operates in an infrastructure mode. It is an optional MAC function because the hardware implementation of PCF was thought to be complicated at the time the standard was finalized. PCF is a polling-based contention-free access scheme, which uses an access point as a point coordinator. When a WLAN system is set up with PCF enabled, the channel access time is divided into periodic intervals called beacon intervals. A beacon interval is composed of a contention-free period (CFP) and a contention period (CP). During a CFP, the access point maintains a list of registered stations and pools them according to the list. The size of each data frame is bounded by the maximum MAC frame size (23,004 bytes). If the PHY data rate of every station is fixed, the maximum CFP duration for all stations, CFP max duration, can then be decided by the access point. The time used by an access point to generate beacon frames is called the target beacon transmission time (TBTT). The next TBTT is announced by the access point within the current beacon frame. To give PCF higher access priority than DCF, the access point waits for a starter interval called PCF interframe space (PIFS), before starting PCF.

IEEE 802.11i

The IEEE 802.11i group is actively defining enhancements to the MAC to counter the issues related to wired equivalent privacy (WEP). The existing 802.11 standard specifies the use of relatively weak, static encryption keys without any form of key distribution management. This makes it possible for hackers to access and decipher WEP-encrypted data on the WLAN. IEEE 802.11i will incorporate 802.1x and stronger encryption techniques, such as advanced encryption standard (AES).

IEEE 802.11j

The purpose of the Task Group J is to enhance the IEEE 802.11 standard and amendments in order to add channel selection for 4.9 and 5 GHz in Japan, and to conform to the Japanese rules on operational mode, operational rate, radiated power, spurious emissions, and channel sense.

IEEE 802.11k

The IEEE 802.11k standard for WLANs enables interoperability between different vendors' access points and switches, but it does not let WLAN systems access a client's radio frequency resources. Consequently, this limits administrators' ability to manage their networks efficiently. As a proposed standard for radio resource measurement, 802.11k aims to provide key client feedback to WLAN access points and switches. The proposed standard defines a series of measurement requests and reports that detail Layer 1 and Layer 2 client statistics. In most cases, access points or WLAN switches ask clients to report data, but in some cases client might request data from access points.

IEEE 802.11m

The purpose of this task group is maintenance. It will look for any editorial changes in the other 802.11 standards and will also answer any specific questions raised by implementers.

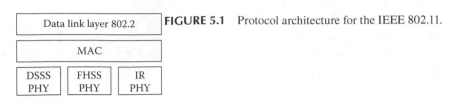

FIGURE 5.1   Protocol architecture for the IEEE 802.11.

IEEE 802.11n

The purpose of this task group is to design a MAC that will provide a base through-put of 100 Mbps at the MAC layer. This will have IEEE 802.11e as the base MAC, and the multiple input multiple output (MIMO) as its physical layer.

IEEE 802.11 GENERAL ARCHITECTURE

The IEEE 802.11 is a standard constituted by a PHY layer and a MAC layer. Over this layer, the standard foresees interfacing with the standard logic link control (LLC) layer 802.2. The protocol architecture is depicted in Figure 5.1 together with the protocol stack. PHY is chosen among three possibilities:

- Frequency hopping spread spectrum (FHSS)
- Direct sequence spread spectrum (DSSS)
- Infrared (IR)

The system is constituted by the following entities:

Station (STA). The object of the communication, in general a mobile station.

Access point (AP). A special control traffic relay station that normally operates on a fixed channel and is stationary—can be partially seen as the coordinator within a group of STAs.

Portal (PO). A particular access point that interconnects IEEE 802.11 WLANs and wired 802.x LANs. It provides the logical integration between both types of architectures.

A set of STAs, and eventually an AP, constitutes a basic service set (BSS), which is the basic block of the IEEE 802.11 WLAN. The simplest BSS is constituted by two STAs that can communicate directly. This mode of operation is often referred to as an ad hoc network because this type of IEEE 802.11 WLAN is typically created and maintained without prior administrative arrangement for specific purposes, such as transferring a file from one personal computer to another. This basic type of IEEE 802.11 WLAN is called an independent BSS (IBSS).

The second type of BSS is an infrastructure BSS. Within an infrastructure BSS, an AP which is a STA acts as the coordinator of the BSS. Instead of existing inde-pendently, two or more BSSs can be connected together with some kind of back-bone network that is called the distribution system (DS). The whole interconnected WLAN (some basic service sets and a distribution system) is identified by the IEEE 802.11 as a single wireless network called an extended service set (ESS). This sce-nario is shown in Figure 5.2.

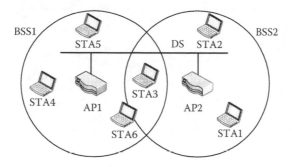

FIGURE 5.2   Extended service set (ESS).

The association between a STA and a particular BSS is dynamic. As a consequence, the setup of the system is automatic. Some basic features of the original IEEE 802.11 specification are sketched in Table 5.1.

It is important to note that IEEE 802.11 does not impose any constraint in the DS. For example, it does not specify if the DS should be data link layer or network-layer based. Instead, IEEE 802.11 specifies a set of services that are associated with different parts of the architecture. Such services are divided into those assigned to STAs, called station service (SS), and those assigned to the DS, called distribution system service (DSS). Both categories of services are used by the IEEE 802.11 MAC sublayer. The services assigned to the station are the following:

- Authentication/deauthentication
- Privacy
- MAC service data unit (MSDU) delivery to upper layer (IEEE 802.2 layer)

The services assigned to the distribution system (DS) are as follows:

- Association/deassociation
- Distribution
- Integration
- Reassociation

Both MAC and PHY layers include two management entities: MAC sublayer management entity (MLME) and PHY layer management entity (PLME). These entities

TABLE 5.1
IEEE 802.11 Original Specifications Basic Features

Spectrum	2.4 GHz
Maximum physical rate	2 Mbps
Maximum data rate, layer 3	1.5 Mbps
MAC	CSMA/CA
Fixed network support	IEEE 802 wired LANs and others

TABLE 5.2
PHY Specifications

	FH	DS	IR
Spectrum	2.4 GHz First channel at 2.402 GHz	2.4 GHz	Diffuse infrared (wavelength from 850 to 950 nm
Subcarrier	1 MHz wide	11, 13, or 14 subchannels, each of 22 MHz	
Physical rate	1 and 2 Mbps	1, 2, 5.5, and 11 Mbps	
Modulation	2GFSK, 4GFSK	DBPSK, DQPSK, CCK (apply for IEEE 802.11b)	16 pulse p1 and 2 Mbps position modulation (PPM) and 4 PPM
Other	Hop over 79 channels	11-chip Barker sequence	Nondirectional transmission

Reproduced with permission from IEEE 802.11 WG. Wireless LAN Medium Access Control (MAC) and Physical Layer (PHY) Specifications, 1999.

provide the layer management service interfaces through which layer management functions may be invoked.

The physical layer is divided into two sublayers. The first layer is the physical medium dependent (PMD) sublayer. It carries out the modulation and the encoding. The second layer belongs to the Physical Layer Convergence Protocol (PLCP), which carries out physical layer-specific functions, providing a clear channel assessment signal.

This architecture has been designed to implement under the same MAC and PHY chosen among frequency hopping (HP), DS, or IR. The PHY specifications including FH, DS, and IR are described in Table 5.2.

The MAC layer is responsible for providing the following services:

- Asynchronous data service, which provides peer IEEE 802.2 entities with the ability to exchange MAC service data units
- Security services, which in IEEE 802.11 are provided by the authentication service and the WEP mechanism
- MAC services data unit ordering, whose effect is a change in the delivery order of broadcast and multicast MSDUs, relative to direct MSDUs, originating from a single service station address

The general MAC frame format consists of the following components:

- MAC header, which comprises frame control, duration, address, and sequence control information
- A variable length frame body, which consists of information specific to the frame type

TABLE 5.3
IEEE 802.11 WLAN Link Layer Standards

	802.11	802.11a	802.11b	802.11g
Bandwidth (MHz)	300	83.5	83.5	83.5
Frequency range (GHz)	2.4–2.4835	5.15–5.25 (lower) 5.25–5.35 (middle) 5.725–5.825 (upper)	2.4–2.4835	2.4–2.4835
Number of channels	3	12 (4 pcr subband)	3	3
Modulation	BPSK, QPSK DSSS, FHSS	BPSK, QPSK MQAM, OFDM	BPSK, QPSK DSSS	BPSK, QPSK MQAM, OFDM
Coding		Convolution (rate 1/2, 2/3, 3/4)	Barker, CCK	Convolution (rate 1/2, 2/3, 3/4)
Max. data rate (Mbps)	1.2	54	11	54
Range (m)		27–30 (lower band)	75–100	30
Random access		CSMA/CA		

Reproduced with permission from IEEE P802.11e/D6.0. Wireless Medium Access Control (MAC) and Physical Layer (PHY) Specifications. Medium Access Control (MAC); Quality of Service (QoS) Enhancements, November 2003.

- A frame check sequence (FCS) which contains an IEEE 32-bit cyclic redundancy code (CRC)

There are four address fields in the MAC frame format. These fields are used to indicate the BSS identifier (BSS ID), source address, destination address, transmitting address, and receiving station address.

WIRELESS LAN LINK LAYER STANDARDS

WLANs are built around the family of IEEE 802.11 standards. The main characteristics of this standard family are summarized in Table 5.3. The baseline 802.11 standard, realized in 1997, occupies 83.5 MHz of bandwidth in the unlicensed 2.4 GHz frequency band. It specifies phase shift keying (PSK) modulation with FHSS or DSSS. Note that in FH, the transmission band is divided into different frequency channels, typically 79, and a logical channel; that is, a channel in which two or more devices communicate and that hops periodically from one frequency channel to another with a pseudorandom hopping sequence. Data rates up to 2 Mbps are supported, with CSMA/CA used for random access.

The baseline standard was extended in 1999 to create the 802.11 standard, operating in the same 2.4 GHz band using only DSSS. This standard uses variable rate modulation and coding, with binary PSK (BPSK) or quadrature PSK (QPSK) for modulation and channel coding via either the Barker sequence or complementary code keying (CCK). This leads to a maximum channel rate of 11 Mbps, with a maximum user data rate around 1.6 Mbps. The transmission range is approximately 100 m. The network architecture in 802.11b is specified as either star or peer-to-peer, although the peer-to-peer feature is not typically used. This standard has been widely used and deployed, with manufacturers integrating 802.11b WLAN cards into many laptop computers.

The IEEE 802.11a standard occupies 300 MHz of spectrum in the 5 GHz band. In fact, the 300 MHz of bandwidth is segmented into three 100 MHz subbands: a lower band from 5.15 to 5.25 GHz, a middle band from 5.25 to 5.35 GHz, and an upper band from 5.725 to 5.825 GHz. Channels are spaced 20 MHz apart, except on the outer edges of the lower and middle bands, where they are spaced 30 MHz apart. Three maximum transmit power levels are specified: 40 mW for the lower band, 200 mW for the middle band, and 800 mW for the upper band. These restrictions imply that the lower band is mostly suitable for just indoor applications, the middle band for indoor and outdoor, and the high band for outdoor. Variable-bit rate modulation and coding are used on each channel. The modulation varies over binary PSK (BPSK), QPSK, 16-quadrature amplitude modulation (QAM), and 64-QAM. On the other hand, the convolutional code rate varies over 1/2, 2/3, and 3/4. This leads to a maximum data rate per channel of 54 Mbps. For indoor systems, the 5 GHz carrier coupled with the power restriction in the lower band reduces the range of IEEE 802.11a relative to IEEE 802.11b, and also makes it more difficult for the signal to penetrate walls and other obstructions. IEEE 802.11a uses OFDM multiple access instead of FHSS or DSSS, and in that sense diverges from the original IEEE 802.11 standard (Table 5.4).

The IEEE 802.11g standard, finalized in 2003, attempts to combine the best of IEEE 802.11a and IEEE 802.11b, with data rates of up to 54 Mbps in the 2.4 GHz band for greater range. The standard is backward-compatible with IEEE 802.11b so that IEEE 802.11g access points will work with IEEE 802.11b wireless network adapters, and vice versa. However, IEEE 802.11g uses OFDM modulation and the coding scheme of IEEE 802.11a. Both access points and WLAN cards are available with all

TABLE 5.4
IEEE 802.11 Ongoing Standards Work

Standard	Scope
802.11e	Provides QoS at the MAC layer
802.11f	Roaming protocol across multivendor access points
802.11h	Adds frequency and power management features to 802.11a to make it more compatible with European operation
802.11i	Enhances security and authentication mechanisms
802.11j	Modifies 802.11a link layer to meet Japanese requirements
802.11k	Provides an interface to higher layers for radio and network measurements which can be used for radio resource management
802.11m	Maintenance of 802.11 standard (technical/editorial corrections)
802.11n	MIMO link enhancement to enable higher throughput

three standards to avoid incompatibilities. The IEEE 802.11a/b/g family standards are collectively referred to as WiFi, for wireless fidelity. Extending these standards to frequency allocations in countries other than the United States falls under the IEEE 802.11d standard. There are several standards in the IEEE 802.11 family that lack the ability to allocate specific bandwidth and delay attributes to wireless users. Admission control is essential when guaranteed QoS is desired. Also, there is a need for managing network resources in a way that ensures capacity optimization with minimal compromises in achieved QoS. The actual performance of complicated QoS-capable protocols is crucial for network planning. The random nature of MAC protocols, their distributed nature, and the fact that they are, by design, complicated, make performance evaluation challenging. Some analytic models produce the maximum capacity and delay in both saturated and unsaturated IEEE 802.11 networks. Simulations are also useful in evaluating the network protocols, but actual test beds are the only way to study the overall performance in real-life situations.

5.4 WIRELESS ATM LAN

Asynchronous transfer mode (ATM) networks embody a key technology for supporting broadband multimedia services.[23] Motivated by the growing acceptance of ATM as a standard for broadband multimedia communication, in 1996 the ATM Forum and ETSI started the extension of the current wired ATM standard to mobile wireless ATM (WATM) applications.[24] In a number of WATM programs in Europe, North America, and Japan, the needs of customers who want a unified end-to-end networking platform with high performance and robust service characteristics can be met. WATM closes the gaps between existing voice-oriented wireless systems such as GSM communications and CEST and data-oriented wireless networks such as IEEE 802.11 and high performance LAN (HIPERLAN). In the LAN scenario, WATM is normally applied for mobile users to access local information services, or to simplify wiring and dynamic configuration.[25]

The realization of WATM presents a number of technical challenges that need to be resolved. First, there is a need for the allocation and standardization of appropriate radio frequency bands for broadband communications. Second, new radio technology and access methods are required to operate at high speed. Next, location management must be capable of tracking mobile terminals as they move around the network. Fourth, handoff algorithms must be capable of dynamically reestablishing virtual circuits to new APs while ensuring in-sequence and loss-free delivery of ATM cells. Finally, WATM should provide uniform end-to-end QoS guarantees. However, providing such guarantees is difficult during periods of limited bandwidth, time-varying channel characteristics, and terminal mobility.

The operating frequency for WATM LANs is in the 5 GHz band. The 5.2 GHz band is known in Europe as the HIPERLAN band, with a bandwidth of 100 to 150 MHz (depending on national regulations). In the United States, the FCC has recently opened a 300 MHz unlicensed national information infrastructure (U-NII) band within the 5 GHz spectrum for license-except use, comprising three subbands in the frequency domains, 5.15 to 5.25 GHz, 5.25 to 5.35 GHz, and 5.725 to 5.825 GHz, with different maximum transmission power limitations.

There are two general methods to realize WATM. The first method is to design the air interface of the WLAN independent of the ATM cell format and provide protocol conversion at the wireless access point connected to an ATM-based backbone. The second method is to provide a seamless extension of the ATM services and the associated QoS control over the wireless medium. These two methods result in different locations of the ATM adaptation layer (AAL). For the first method, the AAL is located in the base station (BS). Because ATM connections are terminated in the fixed access point, requirements on the air interface can be simplified. The main disadvantage of this approach is the loss of the native ATM application programming interface (API) in the mobile terminal. Furthermore, the complexity of the BS increases with the average number of active users.

WIRELESS ATM WORKING GROUP

In spring 1996, the ATM Forum[26] created a new working group[27] to extend ATM technology to the wireless and mobile domains. During the same period, the FCC announced the allocation of 350 MHz of national information infrastructure (NII)/ SuperNET spectrum in the 5 GHz band. The charter of the working group includes the development of a set of specifications that will facilitate the use of ATM technology for a broad range of wireless network access scenarios, for public and private network access. Currently, the working group is focusing on two major work plans that include specification for mobile ATM and the radio access layer. The mobile ATM section of the specifications deals with suitable extensions to existing specifications for location management, bandwidth, routing, addressing, and traffic management. As a first step, the working group is considering modifications to ATM signaling for handoff and location management. The radio access layer section of the specification addresses the physical layer, medium access control, data link control, and radio resource control. The new radio will operate in the 5 GHz unlicensed band at 25 Mbps and is targeted toward indoor microcellular use.

The major thrust of the working group has been that of addressing architectural, mobility, and location management issues. The baseline specification includes several reference model configurations ranging from fixed WATM terminals to moving ATM switch platforms. Two architectural approaches are an integrated model, which incorporates all mobility and radio functions into the ATM switch, and an access model. The first approach places more complexity in the switch. Conversely, the latter reduces the impact of radio and mobility functions in the same switch, but requires a new AP control protocol to convey messages between the AP and the ATM switch.[26]

The aim of handoff signaling is to enable wireless terminals to move seamlessly between APs while maintaining connections with their negotiated QoS. However, in some cases QoS renegotiation and/or connection dropping may be unavoidable. The working group considered a number of proposed handoff schemes. After due consideration, the working group has adopted a simple fast virtual connection rerouting scheme as opposed to more complex schemes that argue for richer functionality. For example, complex mechanisms used for cell forwarding between the old and new APs during handoff were shown in practice to be rarely needed.[27] At the ATM

Forum meeting, the working group arrived at a consensus on the outline of a suitable handoff algorithm.[28] The proposed style of handoff is a backward handoff through the old AP. Backward handoff can be initiated through the new or old AP.

The working group has considered a number of proposals for location management. Two schemes have been actively investigated. One integrates location management with ATM signaling, and the other partitions the address space, keeping location management external to the existing signaling.

In 1997, the ETSI created a project called Broadband Radio Access Network (BRAN) to develop broadband radio local loops and other radio access systems operating in licensed and unlicensed bands at data rates between 25 and 155 Mbps.[29] The project leverages work on HIPERLAN 25 Mbps ATM LAN access in the 5.15 to 5.3 GHz band, as well as access technologies at higher frequencies (e.g., 17.1 to 17.3 GHz band) and data rates (e.g., 155 Mbps). The WATM Working Group and BRAN are combining forces to develop the radio access layer specification. BRAN will be the prime developer of new radio technologies for scalable performance, and new scheduled time division multiple access (TDMA) techniques to ensured QoS, as well as flexible error control strategies for meeting application-level requirements.

The initial strategy of the working group was to prioritize the signaling extensions and location management support in the first instance. Other important issues have been raised and marked for further consideration by the team. For example, the working group has discussed the impact of wireless technology on satellite and ad hoc networking. The major impact of these technologies on ATM is the need for a wireless private network-to-network interface (PNNI) link. Interworking issues associated with WATM and existing cellular systems (e.g., personal communication system, PCS, to ATM) have also been discussed.

ETSI BRAN HIPERLAN Standard

This section provides an overview of the ETSI BRAN HIPERLAN standard, particularly highlighting type 2.[30] We provide a brief description of its system architecture and protocol reference model. Then, attention is focused on the MAC/data link control (DLC) layer. The BRAN family is shown in Figure 5.3.

The HIPERLAN type 1 standard provides a high speed WLAN, while the HIPERLAN type 2 standard supplies short range access to networks based on IP, UMTS, and ATM. HIPERACCESS and HIPERLINK are not WLAN, although they may complement these networks.

Some manufacturers, with the purpose of accelerating the adoption of HIPERLAN1-based products worldwide, have formed the HIPERLAN Alliance. Market promotion, spectrum lobbying, and cooperation with other standardization bodies are among the objectives of the HIPERLAN Alliance.

The protocol architecture of HIPERLAN exhibits some differences between type 1 and type 2. HIPERLAN1 employs an access method called elimination yield—nonemptive priority multiple access (EY-NPMA), which constitutes a kind of CSMA/CA that splits the procedure into three phases: priority resolution, elimination, and yield. HIPERLAN2, originally based on WATM, has extended its scope for

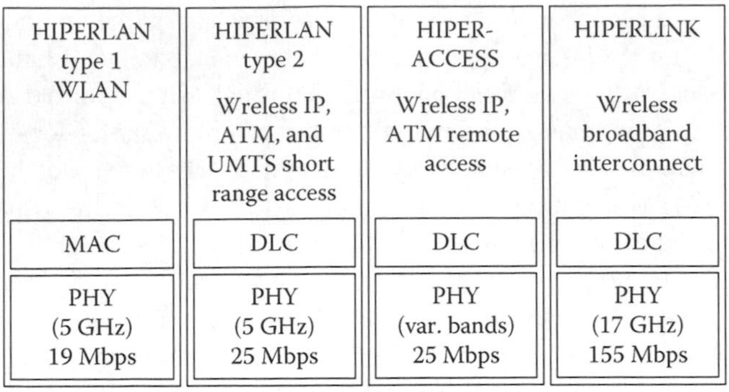

HIPERLAN type 1 WLAN	HIPERLAN type 2 Wreless IP, ATM, and UMTS short range access	HIPER-ACCESS Wreless IP, ATM remote access	HIPERLINK Wreless broadband interconnect
MAC	DLC	DLC	DLC
PHY (5 GHz) 19 Mbps	PHY (5 GHz) 25 Mbps	PHY (var. bands) 25 Mbps	PHY (17 GHz) 155 Mbps

FIGURE 5.3 The BRAN family. Reproduced with permission from ETSI TR 101 683 V1.1.1(2002-02), BRAN, HiperLAN Type 2 System Overview, 2002.

providing WLAN services and for interacting with core networks of different kinds such as IP, ATM, and UMTS.

The protocol stack of HIPERLAN2 comprises three layers, with each divided into user plane and control plane as shown in Figure 5.4. The user plane includes functions related to transmission of traffic over the established user connections, while the control plane includes functions related to the control, establishment, release, and modification of the connections.

The three basic layers of HIPERLAN2 are PHY, DLC and convergence layer (CL), which is a part of the DLC. The PHY provides a basic data transport function by means of a baseband modem. The transmission format on the PHY layer is a burst consisting of a preamble part and a data part. The modulation scheme chosen for the PHY layer is OFDM. It was chosen due to its very good performance on highly dispersive channels.

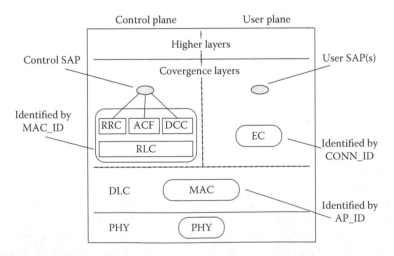

FIGURE 5.4 HIPERLAN2 protocol reference model. Reproduced with permission from ETSI TR 101 683 V1.1.1(2002-02), BRAN, HiperLAN Type 2 System Overview, 2002.

The DLC layer consists of the error control (EC), the radio link control (RLC), and the MAC functions. The DLC layer is divided into data transport and control functions. The user data transport part handles the data packets arriving from the higher layer via the user service access point (USAP). The user data transport part also contains the EC, which performs an automatic repeat request (ARQ) protocol. The DLC protocol is connection oriented and for each DLC connection, a separate EC instance is created. This allows different error controls to be performed for different connections depending on the service class. The control part contains the RLC function, which provides a transport service to the DLC connection control (DCC), the radio resource control (RRC), and the association control function (ACF).

Finally, the CL is also divided into a data transport and a control part. The data transport part provides the adaptation of the user data to the DLC layer message format. If the higher layer network protocol is other than ATM, it also contains a segmentation and reassembly function that converts higher layer packets with variable sizes into fixed size packets that are used within the DLC. The segmentation and reassembly function is an important part of the CL, because it makes possible the standardization and implementation of the DLC and PHY layers that are independent of the fixed networks to which HIPERLAN2 is attached.

HIPERLAN2 System

This system is structured in a centralized mode (CM) even though a connection between two or more mobile stations is foreseen. HIPERLAN2 centralized architecture is shown in Figure 5.5. Direct link mode (DM) can be established between two or more mobile stations so that they can directly exchange information.

Two main entities are present in the centralized system:

- The mobile terminal (MT), which is the entity that wants to be connected to others and, if necessary, to external resources.

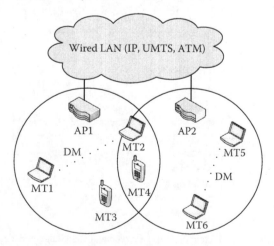

FIGURE 5.5   HIPERLAN2 centralized architecture.

TABLE 5.5
Basic HIPERLAN2 Features

Spectrum	5 GHz
Maximal physical rate	54 Mbps
Maximum data rate, layer 3	32 Mbps
MAC	Central resource control/TDMA/TDD
Fixed network support	IP/ATM/UMTS

Reproduced with permission from http://portal.etsi.org/bran/kta/hiperlan/hiperlan2.asp..

- The AP, the entity that coordinates the other MTs in its area and can control one or more sectors. Its protocol reference model differs from that of MTs for multiple MAC and RLC.

Some basic features of the HIPERLAN2 system are described in Table 5.5.

The basic transmission format on the PHY layer is a burst, constituted by pre-amble and data. The main features for the HIPERLAN PHY layer are summarized in Table 5.6.

The DLC layer represents the logical link between an AP and its associated MTs. The DLC layer implements a service policy that takes into account such factors as QoS characteristics of each connection, channel quality, number of terminal devices, and medium sharing with other access networks operating in the same area. DLC operates on a per-connection basis, and its main objective is to maintain QoS on a virtual-circuit basis. Depending on the type of required service and the channel quality, capacity, and utilization, the DLC layer can implement a variety of means such as forward error correction (FEC), ARQ, and flow packing to optimize the service provided and maintain QoS.

Two major concepts of the DLC layer are the logical channels and the transport channels. A logical channel is a generic term for any distinct data path. A set of

TABLE 5.6
HIPERLAN PHY Layer Parameters

Spectrum	5 GHz
Channel spacing	20 MHz
Subcarrier per channel	52 (48 carry data and four are pilots that facilitate phase tracking)
Guard interval	Max 800 ns; min 400 ns
Frequency selection	Single carrier with dynamic frequency selection
Forward error control	Convolutional code
Constrain length	SEVEN and generated polynomials (133, 171)
Modulation	BPSK, QPSK, 16-QAM, 64-QAM
PHY bit rate	from 6 Mbps to 54 Mbps

Reproduced with permission from http://portal.etsi.org/bran/kta/hiperlan/hiperlan2.asp.

logical channel types is defined for different kinds of data transfer services offered by the DLC layer. Each type of logical channel is defined by the type of information it conveys. The interpretation can be viewed as logical connections between logical entities and are mostly used when referring to the meaning of messages contents. DLC layer defines the following logical channels:

- **Broadcast control channel (BCCH).** It conveys downlink broadcast control channel information concerning the entire radio cell.
- **Frame control channel (FCCH).** Downlink; it describes the structure of the MAC frame. This structure is announced by resource grant messages (RGs).
- **Random access feedback channel (RFCH).** Downlink; it informs the MTs that have used the RCH in the previous MAC frame about the results of their access attempts. It is transmitted once per MAC frame per sector.
- **RLC broadcast channel (RBCH).** Downlink; it conveys broadcast control information concerning the entire radio cell. The information transmitted by RBCH is classified as: broadcast RLC messages, assignment of MAC_ID to a nonassociated MT, convergence ID information, and encryption seed. RBCH is transmitted only when necessary.
- **Dedicated control channel (DCCH).** It transports RLC messages in the uplink direction. A DCCH is implicitly established during association of an MT.
- **User broadcast channel (UBCH).** Downlink; it transmits user broadcast data from the CL. The UBCH transmits in repetition or unacknowledged mode and can be associated or unassociated with LCCHs.
- **User multicast channel (UMCH).** Downlink; it is employed to transmit user point-to-multipoint user data. The UMCH is transmitted in unacknowledged mode.
- **User data channel (UDCH).** Bidirectional; it is employed to exchange data between APs and MTs in CM, or between MTs in DM. The UDCH is associated or not associated to LCCHs.
- **Link control channel (LCCH).** Bidirectional; it is employed to exchange ARQ feedback and discard messages both in CM and in DM. The LCCH is also used to transmit resource request messages (RRs) in the uplink direction (only in CM) and discard messages for a UBCH using repetition mode. LCCHs may or may not be associated with UDCHs/UBCHs.
- **Association control channel (ASCH).** Uplink; in this case the MTs that are not associated to an AP transmit new association and handover requests.

The logical channels are mapped onto different transport channels. The transport channels provide the basic elements for constructing protocol data units (PDU) and describe the format of the various messages (e.g., length, value representation). The message contents and their interpretation are subject to the logical channels. The following transport channels are defined in the DLC layer.

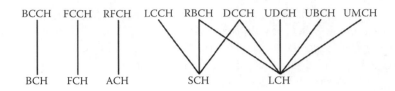

FIGURE 5.6   Mapping between logical and transport channels for the downlink.

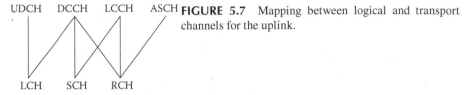

FIGURE 5.7   Mapping between logical and transport channels for the uplink.

- **Broadcast channel (BCH).** Downlink; it contains 15 bytes of radio cell information such as identification of the AP and its current transmitted power.
- **Frame channel (FCH).** Downlink; its length is a multiple of 27 octets. It contains a description of the way resources have been allocated and can also contain an indication of the empty parts of a frame.
- **Access feedback channel (ACH).** Downlink; its length is 9 octets. It contains information on access attempts made in the previous RCH.
- **Long transport channel (LCH).** Downlink and uplink; its length is 54 octets. It is used to transmit DCL user PDUs (U-PDUs) of 54 bytes with 48 bytes of payload.
- **Short transport channel (SCH).** Downlink and uplink; its length is 9 octets. It is used to exchange DLC control PDUs (C-PDU) of 9 bytes.
- **Random channel (RCH).** Uplink; its length is 9 octets. It is used for sending control information when no granted SCH is available. It carries RRs as well as ASCH and DCCH data.

Mapping between logical and transport channels for the downlink, uplinks, and direct link are shown in Figure 5.6, Figure 5.7, and Figure 5.8, respectively. Figure 5.9 shows the reference model for HIPERLAN2. A convergence layer provides connectivity with the core network.

Thus, the standard is open and it can use other networks in the future, with the only specification that of the corresponding convergence layer. The centralized

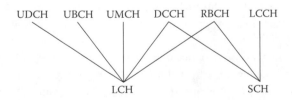

FIGURE 5.8   Mapping between logical and transport channels for the direct link.

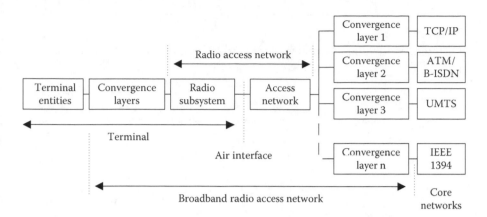

FIGURE 5.9 Reference model for HIPERLAN2. Reproduced with permission from ETSI TR 101 683 V1.1.1(2002-02), BRAN, HiperLAN Type 2 System Overview, 2002.

mode of operation allows the implementation of radio access to a fixed point. Thus, a network can be implemented to cover entire buildings through the deployment of APs in a cellular structure. It can also operate in ad hoc networks without the need for APs, but in this case one of the devices must operate as a controller which can be any of the devices in the network.

OVERVIEW OF PHYSICAL LAYERS OF HIPERLAN2 AND IEEE 802.11a

The transmission format on the physical layer consists of a preamble part and a data part. The channel spacing is 20 MHz, which allows high bit rates per channel. The physical layer for both the IEEE 802.11a and HIPERLAN2 is based on OFDM. OFDM uses 52 carriers per channel, where 48 subcarriers carry actual data and 4 subcarriers are pilots that facilitate phase tracking for coherent demodulation. The duration of the guard interval is equal to 800 ns, which is sufficient to enable good performance on channels with delay spread of up to 250 ns. An optional shorter guard interval of 400 ns may be used in small indoor environments. OFDM is used to combat frequency selective fading and to randomize the burst errors caused by a wideband fading channel. The PHY layer modes with different coding and modulation schemes are shown in Table 5.7. The MAC selects any of the available rates for transmitting its data based on the channel condition. This algorithm is called adaptation and the standard does not specify how it should be performed, thus enabling product differentiation between different vendors.

Data for transmission is supplied to the PHY layer in the form of an input PDU train or physical layer convergence procedure (PLCP) protocol data unit frame. This is then passed to a scrambler that prevents long runs of 1s and 0s in the input data. Although both 802.11a and HIPERLAN2 scramble the data with a length 127 pseudorandom sequence, the initialization of the scrambler is different. The scrambled data then passes to a convolutional encoder. The encoder consists of a 1/2 rate mother code and subsequent puncturing. The puncturing schemes facilitate the use

TABLE 5.7
Different Modulation Schemes of IEEE 802.11a and HIPERLAN2 Physical Layer

Mode Scheme	Modulation	Coding Rate	Bit Rate (Mbps)	Coded Bits/ Subcarrier	Coded Bits/OFDM Symbol	Data Bits/ OFDM Symbol
1	BPSK	½	6	1	48	24
2	BPSK	¾	9	1	48	36
3	QPSK	½	12	2	96	48
4	QPSK	¾	18	2	96	72
5	16-QAM (H/2 only)	$9/16$	27	4	192	108
5	16-QAM (IEEE only	½	24	4	192	96
6	16-QAM	¾	36	4	192	144
7	64-QAM	¾	54	6	288	216
8	64-QAM (IEEE only)	$2/3$	48	6	288	192

Reproduced with permission from http://portal.etsi.org/bran/kta/hiperlan/hiperlan2.asp.

of code rates 1/2, 3/4, 9/16 (HIPERLAN2 only), and 2/3 (802.11a only). In the case of 16-QAM, HIPERLAN2 uses rate 9/16 instead of rate 1/2 in order to ensure an integer number of OFDM symbols per PDU train. The rate 2/3 is used only for the case of 64-QAM in 802.11a. Note that there is no equivalent mode for HIPERLAN2. HIPERLAN2 also uses additional puncturing in order to keep an integer number of OFDM symbols with 54-byte PDUs. The coded data is interleaved in order to prevent error bursts from being input to the convolutional decoding process in the receiver. The interleaved data is subsequently mapped to data symbols according to a BPSK, QPSK, 16-QAM, or 64-QAM constellation. OFDM modulation is implemented by means of an inverse fast Fourier transform (FFT); 48 data symbols and four pilots are transmitted in parallel in the form of one OFDM symbol.

Two wireless network standards characteristics, IEEE 802.11 and HIPERLAN2, are presented in Table 5.8 in a summary form. They are usually intended for use as WLAN. It must be pointed out that in direct sequence spread spectrum (DSS), the channels are 22 MHz wide, but spaced 5 MHz apart, and therefore they overlap. Two overlapping channels used in one place will interfere with each other, decreasing each other's data rate. Only three non-overlapping channels are available in one place.

Note that IEEE 802.11 ad hoc mode provides direct communication of stations in the absence of an AP. Because many features of IEEE 802.11 such as QoS or power savings rely on the AP, they are not available in ad hoc mode, making this a very limited ad hoc standard, albeit with high data rates.

5.5 INFRARED STANDARDS

We can classify the new IR standards for WLANs into three main categories. The first includes data association (IrDA) developments (such as IrDA VFIR, IrDA

TABLE 5.8
Characteristics of IEEE 802.11 and HIPERLAN2

	IEEE 802.11			HIPERLAN2
	IEEE 802.11	**IEEE 802.11b**	**IEEE 802.11a**	**HIPERLAN2**
Frequency range (GHz)	2.4–2.4835 (ISM) (also IR at 850–950 nm)		5.150–5.350 5.725–5.825	5.150–5.350 5.470–5.725
Bandwidth (MHz)	83.5	300	455	
Data rate (Mbps)	1, 2	1–11	6–54	6–54
PHY (no. of channels)	FSSS (79) DSSS (11 USA, 13 EU)	DSSS (11 USA, 13 EU)	OFDM (12)	OFDM (19)
MAC	Mainly SCMA/ CA (limited QoS)			Based on Wireless ATM (high QoS support)
Comparison	Simple MAC (low processing power) Low performance (channel throughput), decreasing at higher data rates Good for low-quality applications			Complex MAC (high processing power required) High performance at any data rate Good for high end applications

Reproduced with permission from http://portal.etsi.org/bran/kta/hiperlan/hiperlan2.asp.

OBEX, or IrDA Lite).[31] The second group includes technologies for interconnecting preexisting wired networks such as the EthIR networks for Ethernet networks, or wireless access to ATM networks. The third set of developments includes new technologies for diffuse links such as the use of spread spectrum techniques.

IrDA is an open commercial association, very flexible in adapting its products to available devices, components, or other necessities.

IrDA VFIR. IrDA very fast IR is a high-speed specification that allows transmission rates of up to 16 Mbps. This represents a fourfold increase in speed from the previous maximum data rate of 4 Mbps of the IrDA fast IR version. This extension provides end users with faster throughput and is backward-compatible with equipment using the current data rate. The immediate applications for these links are the interconnection of digital cameras with personal computers, making it possible for users to download the entire contents of the camera in less than 20 s.

IrDA OBEX. IR object exchange is an industry standard of the IrDA that defines how objects can be shared among different IrDA devices. IROBEX provides the ability to put-and-get data objects simply and flexibly, thereby enabling rapid application development and interaction with a broad class of devices including personal computers, personal digital assistants (PDAs), data collectors, cellular phones, handheld scanners, and cameras. The application developer does not have to worry about the

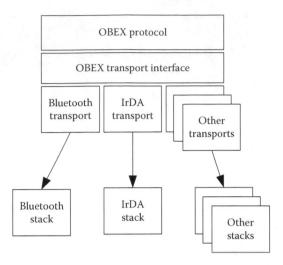

FIGURE 5.10 Structure of the multitransport OBEX protocol. Reproduced with permission from IEEE 802.15, http://grouper.ieee.org/groups/802/15.

low level IrDA functions of link discovery, set up, and maintenance, but rather can focus on higher level application development. Another IrOBEX time-saving feature enables the developer to include data object information such as object description headers, along with the data object itself. The multitransport OBEX protocol is designed to allow one or more adaptation layers to provide access to various network transports. Multitransport OBEX provides a clean, well-defined interface between the OBEX protocol component and the OBEX transport adapter module. This allows easy expansion of the types of transports supported beyond those provided for IrDA and Bluetooth. Figure 5.10 shows the structure of multitransport OBEX.

IrDA Lite. The basic strategies for IrDA Lite are described to support only the required minimal link parameters, such as 9,600 bps, data size of 643 bytes, and a 500 ms turnaround time. Also, one of the strategies is to ignore frames that do not have the correct address, as well as to use simplified algorithms for discovery. On the other hand, IrDA Lite does not support optional features such as sniffing and rate exchange, nor does it support both connectionless and connection-oriented frames. IrDA Lite implementations may range from minimal IrDA Lite through full-featured IrDA and all points in between. The weakness of the IrDA Lite implementation is that it cannot easily handle multiple independent applications running at the same time. When applications on two devices communicate, typically one of the applications (server) passively waits for the other application (client) to establish the connection. Potential uses of IrDA Lite include watches, printers, cameras, modems and terminal adapters, cell phones, instruments, and industrial equipment. On the other hand, IrDA Lite cannot easily support multiple applications running at the same time. However, the number of scenarios in which multiple applications need to run at the same time is small.

Increasing the IrDA range. According to the IrDA specification, the range is up to 1 m. In some cases, it may be desirable to increase link distance beyond the distance guaranteed by IrDA. The two ways for doing this are to increase transmitted light intensity or to increase the receiver sensitivity. To extend the link distance, both sensitivity and intensity must be increased at both ends of the IR link. To communicate with a standard IrDA device that may have minimum transmitter intensity, the receiver intensity must be increased. The standard IrDA device may also have minimum receiver sensitivity, so transmitter intensity must also be increased.

INTERCONNECTION FOR WIRELESS NETWORKS

One of the main developments for IR data communications is to provide a universal network access solution for portable devices, such as notebooks, handheld PCs, PDAs, and Internet applications such as smart phones and electronic books. The concept of providing high-speed wireless access is based on the concept of using Point-to-Point Protocol (PPP) over IR. This allows easy and completely wireless networks access with no hardware or software to install and no reconfiguration of the portable device. Connection speeds of up to 4 Mbps, when using PPP over IP, are possible with the current implementation of the IrDA standard. The ratification of the VFIR IrDA standard increases this up to 16 Mbps. These high-speed IR interfaces can be used to bridge PCs to xDSL and cable modem for high-speed Internet access. There are two main possibilities for providing IR access to Ethernet links. One is designing an IR interface for regular Ethernet boards. The other one is transforming an Ethernet connection into a wireless network AP enabling LAN, WAN, and Internet access.

TECHNIQUES FOR DIFFUSE LINKS: SPREAD SPECTRUM

The use of different wireless IR communications links for short-range indoor communications has received a lot of interest over the past year. This configuration does not need alignment. Spread spectrum techniques allow the use of low emitted power communications under the noise level and avoid undesired multipath propagation effects. Their performances have been tested in data applications in the IR field, but the final application of these techniques will probably be in the area of consumer electronics. Spread spectrum techniques offer the possibility of avoiding the intersymbol interference introduced by multipath propagation (such as in the IR indoor channel), but pose, on the other hand, higher complexities in the symbol time recovery stage of the receiver. DSSS is based on multiplying a narrowband data signal. As the spreading sequence is coded following a pseudonoise algorithm, only the receiver synchronized to that code would obtain the original narrowband signal. For another receiver using a nonsynchronized replica of the code, or a different sequence, the signal delivered to the decision stage will be almost white or colored noise. As the power of the original data signal is also spread throughout the coded-signal bandwidth, the power spectral density (PSD) of the coded signal will be much less than the PSD of the data signal, and even below the environmental PSD of noise.

As the DSSS receiver is tuned to only one code, we can establish several simultaneous communications using the same bandwidth without collision by using different codes for each link. The use of this kind of multiple access strategy (known as CDMA) is well known in other fields of application such as cellular telephony, and offers interesting possibilities in the indoor IR channel. It avoids the use of complex protocols such as CSMA/CA or carrier sense multiple access with collision detection (CSMA/CD), that are not well suited to this channel as they need to test for the presence or absence of a carrier.

5.6   WIRELESS PERSONAL AREA NETWORKS

The wireless communication environment of the future will not consist of a single technology, but will rather be a collection of different systems. On the other hand, the era of computing has shifted from traditional desktop and laptop computers to small, handheld personal devices that have substantial computing, storage, and communications capabilities. Such devices include handled computers, cellular phones, PDAs, and digital cameras. It is necessary to interconnect these devices and also connect them to desktop and laptop systems in order to fully utilize their capabilities. For example, most of these devices have personal information management (PIM) databases that need to be synchronized periodically. Such a network of devices is described as a WPAN.[32] A WPAN is defined as a network of wireless devices that are located within a short distance of each other, typically 3 to 10 m. The IEEE 802.15 standards suite aims at providing wireless connectivity solutions for such networks without having any significant impact on their form factor, weight, power requirements, cost, ease of use, or other traits.[33,34] Several systems interoperate, including wireless local loop (WLL) (Figure 5.11).

The IEEE 802.15 group adopted the existing Bluetooth standard as part of initial efforts in creating the 802.15 specifications. This standard uses 2.4 GHz transmissions to provide data rates of up to 1 Mbps for distances of up to 10 m.[34] However, this data rate is not adequate for several multimedia and bulk data transfer applications. The term multimedia is used to indicate that the information/data being transferred over the network may be composed of one or more of the following media types: text, images, audio (stored and live), and video (stored or streaming).[35] For example, transferring all the contents of a digital camera with a 128 macroblock

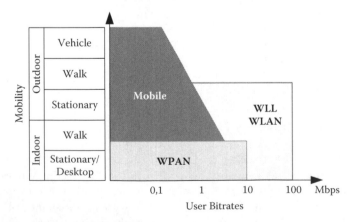

FIGURE 5.11   Several systems interoperate in the future heterogeneous communication environment.

(MB) flash card will require a significant amount of time. Other high-bandwidth demanding applications include digital video transfer from a camcorder, and music transfer from a personal music device. Therefore, the IEEE 802.15 group is examining newer technologies and protocols to support such applications.[36]

There are two new types of WPAN that are being considered. The first is for supporting low-speed, long-lifetime, and low-cost sensor networks at speeds of tens of kilobits per second. The other is for supporting the multimedia applications with higher data rates of the order of several megabits per second with better support for QoS. In an effort to take personal networking to the next level, a consortium of technology firms has been established called the WiMedia Alliance. It develops and adopts standards-based specifications for connectivity of wireless multimedia devices including applications, transport and control profiles, test suites, and a certification program to accelerate widespread consumer adoption of wire-free imaging and multimedia solutions. Even though the operations of the WPAN may resemble that of a WLAN, the interconnection of personal devices is different from that of computing devices. A WLAN connectivity solution for a notebook computer associates the user of the device with the data services available on, for instance, a corporate Ethernet-based intranet. A WPAN can be viewed as a personal communications bubble around a person, which moves as the person moves around. Also, to extend the WLAN as much as possible, a WLAN installation is often optimized for coverage. In contrast to WLAN, a WPAN trades coverage for power consumption.

WPAN Devices

Mobile communications has been based on a single integrated communication device. Also, it is anticipated that we will have several communication devices, for example, mobile phones and portable computers with WLAN. In addition, there is a plurality of user interface devices. This will ultimately immerse the user in communication.[37] Namely, the user will be able to experience a remote reality or even virtuality. The user interface devices are very diverse, ranging from high-end PDAs and cameras to low-end sensors and actuators. All these user interface and communication devices will wirelessly exchange information with one another.[38] This leads to the WPAN shown in Figure 5.12.

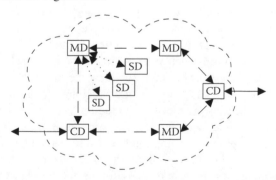

FIGURE 5.12   WPAN.

The communication devices (CD) and high-end user interface devices (called master devices or MD) communicate over an ad hoc multihop network with a data rate of at least 10 Mbps. The communication devices act as a gateway between the WPAN and other fixed or wireless networks. The low-end user interface devices behave as slave devices (SD) of an MD with a communication requirement below 1 Mbps.

Although they are very diverse with respect to complexity and communication requirements, the basic structure of all these devices in a personal area network is identical. They all have a radio that serves the communication. The trend toward heterogeneous wide-area networks puts the requirement for supporting multiple radio protocols on the radio of the communication devices. For the user interface devices, the scalability of the radio communication is a key requirement. This is due to the largely different data rates, ranging from 100 bps for temperature sensors up to 10 Mbps for video displays.

The second component of these devices is a programmable computation platform. This will typically take the form of single or multiple processors, eventually combined with programmable logic and flexible functional accelerators (e.g., MPEG codec). The idea is that multiple applications can be supported on the device, eventually downloadable over the network.

Finally, the devices have a set of sensors and actuators. These can have traditional user interface functions, like the display, microphone, loudspeakers, and keyboards. They can also have a function in the application. An example for the latter is a blood pressure sensor in a remote medical monitoring scenario.

The software architecture of these devices will consist of an operating system (OS) and libraries, a distributed Java machine, and application applets running on top of the Java machine. The two main functions of the libraries will be communication and application acceleration. The latter will be especially important for multimedia services. In addition to traditional MAC and packet formatting, the communication library will also have to support QoS aware multihop routing.

The basic structure of a WPAN device is shown in Figure 5.13. A summary of the characteristics of WPAN devices can be found in Table 5.9. The properties are provided for typical communication, master, and slave devices. Note that these are average values and that a large variety exists in each of these categories. Especially for the slave devices, many radios will be capable of even less than 1 Mbps. The complexity of the devices was estimated by studying state-of-the-art communication devices and assuming a factor 5 increase in complexity to accommodate the more complex signal processing and support for multimedia messaging.

The devices in a personal area network pose major integration challenges. Because of their large number, they must be small such that the user is hardly aware of their presence. Many of these devices have to work for a long time from a simple battery. In combination with the growing processing requirement for the radios and applications, this poses an extreme requirement on power consumption. It should be noted that not only the active power consumption but also the standby power consumption is crucial. Next to size and power consumption, cost is the third important optimization criterion. Several types of sensor devices, for instance, will be thrown away after usage and hence should be reproducible.

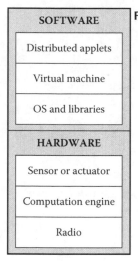

FIGURE 5.13   Basic structure of a WPAN device.

BLUETOOTH

The name Bluetooth comes from a tenth-century Viking, Harald Blatand (which translates to Bluetooth in English) who united Denmark and Norway. This is the inspiration for uniting devices through Bluetooth. Bluetooth was defined to support wireless communications. It was developed by the Bluetooth Special Interest Group (SIG) comprising mainly five major manufacturers from the computer and cellular communications fields: Nokia Mobile Phones, Ericsson Mobile Communications AB, IBM Corporation, Intel Corporation, and Toshiba Corporation. The Bluetooth specifications were published in 1999 and include a core[39] and a set of profiles[40] that are related to applications. Bluetooth wireless technology is a global specification for low-power, short-range, two-way wireless communication. Bluetooth is essentially a cable replacement technology allowing wireless data communications at ranges of about 10 m.[41] It is a cross between the Digital European Cordless Telephone (DECT) and IrDA technologies.[42-44] Bluetooth is sometimes referred to as a WPAN. A device with Bluetooth wireless technology enables data, images, sound, video, and remote control commands to be wireless, and can be exchanged with any other Bluetooth device. Interesting applications of Bluetooth are cable replacement,

TABLE 5.9
Characteristics of WPAN Devices

Device	CD	MD	SD
Functions	Multimode high-end radio	Single high-end radio	Single low-end radio
Cost	$10	$5	$0.5
Data rate	10 Mbps	10 Mbps	1 Mbps
Active power consumption	100 mW	10 mW	1 mW
Standby power consumption	10 mW	1 mW	100 µW

TABLE 5.10
Main Characteristics of Bluetooth

Parameter	Values
Band	2.45 ISM band, different channels according to the country
Carrier's separation	1 MHz
Access	FH, TDD, TDMA/FDMA
Bit rate	1 Mbps
Voice channels	64 kbps
Maximum transmit power	1 mW nominal, 100 mW with closed loop power control
Receiver sensitivity	≤70 dBm (tentative)
Other features	Authentication, encryption, power-saving functions, inter-piconet communications
Maximum physical rate	1 Mbps

wireless headphones, data transfer, and automatic synchronizer. In particular, this last application implements the concept of hidden spontaneous networking. For some applications, most notably multimedia ones such as digital imaging, Bluetooth is not suitable because of the low data rate. The range of applications is not limited nor is the profile list closed (Table 5.10).

Using Bluetooth does not involve mobile network fees. Its spectrum is in the unlicensed spectrum area (at 2.45 GHz). Data transmission speeds using Bluetooth are expected to be between 720 kbps and 1 Mbps.

Starting with Bluetooth, WPANs became a major part of what we call heterogeneous network architectures, mainly due to their ability to offer flexible and efficient ad hoc communication in short ranges, without the need of any fixed infrastructure. This led the IEEE 802 group to approve, in March 1999, the establishment of a separate subgroup, namely, IEEE 802.15, to handle WPAN standardization. Using Bluetooth as a starting point, IEEE 802.15 is working on a set of standards to cover different aspects of personal area environments. IEEE 802.15 consists of four major task groups:

- 802.15.1 standardized a Bluetooth-based WPAN as a first step toward more efficient solutions.
- 802.15.2 studies coexistence issues of WPANs (802.15) and WLANs (802.11).
- 802.15.3 aims at proposing high-rate, low-power, low-cost solutions addressing the needs of portable consumer digital imaging and multimedia applications.
- 802.15.4 investigates a low data rate solution with multimonth to multiyear battery life and very low complexity, targeted to sensors, interactive toys, smart badges, remote controls, and home automation.

The IEEE 802.15.1 standard is aimed at achieving global acceptance such that any Bluetooth device, anywhere in the world, can connect to other Bluetooth devices in its proximity. A Bluetooth WPAN supports both synchronous communication channels for telephone-grade voice communications and asynchronous communications channels for data communications. A Bluetooth WPAN is created in an ad hoc manner when devices desire to exchange data. The WPAN may cease to exist when the applications involved have completed their tasks and no longer need to continue exchanging data. The Bluetooth radio works in the 2.4 GHz unlicensed ISM band. Bluetooth belongs to the contention-free, token-based, multiaccess networks.

Bluetooth is designed to be used in a short-range radio link between two or more mobile systems. The system provides a point-to-point connection between two stations or, a point-to-multipoint connection where the medium is shared by several stations. We then have a piconet, where two or more units share the same medium.

In a piconet, one unit acts as a master, and the others as slaves. In effect, the names master and slave refer to the protocol used on the channel. Any Bluetooth unit can assume one of the two roles when required. It should be noted that all units are identical. The master is defined as the unit that initiates the connection toward one or more slave units. A piconet can have one master, and up to seven slaves can be in an active state. Active state means that a unit is communicating with a master. The station can stay in a parked state if it is synchronized to the master, but it is not active in the channel. Both active and parked stations are controlled by the master.

A slave can be synchronized with another piconet. A station that is master in one piconet can be slave in another one. In this way multiple piconets with overlapping coverage, which are not time or frequency synchronized, constitute a scatternet.

The Bluetooth specification provides low-cost connectivity by using a PHY specification with relaxed technical features compared to other systems. Also, as economies of scale are very important to reduce manufacturing costs, the first version of Bluetooth works in the ISM 2.4 GHz band (as does IEEE 802.11), which is available worldwide.

FH is employed to allow the coexistence of Bluetooth piconets with other transmissions of the same kind, or of other systems. In the scenarios for which Bluetooth is promoted, a large number of piconets can coexist in the same location without coordination. For example, inside the same room of an office, several ad hoc piconets may link the mice, keyboards, and printers with the computers. At the same time, some people may be talking with a wireless headset that is connected to their mobile phones, while others are transferring data from the PCs to the mobile phones to send it through the GSM network. One device may participate simultaneously in more that one piconet (interpiconet communications).

Figure 5.14 presents the concept of scatternet or combination of piconets that is proper to Bluetooth. The gross bit rate is 1 Mbps. The hop sequences are selected on the basis of user identity and are not orthogonal but have a low probability of persistent interference. There is a TDMA/TDD structure linked to the hopping pattern. The slot duration is 625 μs. The packet length is equal to one, three, or five slots. The carrier frequency is fixed within a packet, but changes between subsequent packets.

The hop sequence in a piconet is provided by the identity of one device, which acts as a master. It also provides the phase of the sequence. The remaining devices are slaves and coordinate their transmissions to the hop sequence and phase of the master. Any device can act as master or slave. Usually, the one that requests the service acts as master, except when the net is already established.

High-quality audio at 64 kbps can be provided through the use of a synchronous connection-oriented (SCO) link. Duplex slots (two continuous slots, one for each

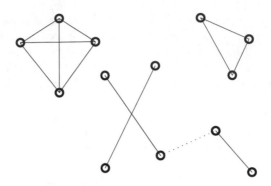

FIGURE 5.14   The combination of piconets in Bluetooth.

direction) are reserved at regular intervals. The remaining slots can be used by asynchronous connectionless (ACL) links, scheduled by the master.

With regard to the applications in WLANs, the Bluetooth core includes the object exchange (OBEX) protocol, for interoperability with IrDA. There is also a LAN access profile. This defines how Bluetooth devices can access LAN services through a LAN AP using PPP. Also, two Bluetooth devices can communicate using PPP as if they were part of a LAN. However, Bluetooth does not aim to establish a complete LAN.

5.7   ULTRAWIDEBAND COMMUNICATIONS

One of the main advances for multimedia applications in WANs is ultrawideband (UWB) communications. The potential strength of the UWB radio technique lies in its use of extremely wide transmission bandwidth, which results in desirable capacities, including accurate position location and ranging, lack of significant fading, high multiple access capability, covert communications, and possibly easier material penetration. The UWB technology itself has been in use in military applications since the 1960s, based on exploiting the wideband property of UWB signals to extract precise timing/ranging information. However, recent FCC regulations have paved the way for the development of commercial wireless communications networks based on UWB in the 3.1 to 10.6 GHz unlicensed band. Because of the restrictions on the transmit power, UWB communications are best suited for short-range communications, namely, sensor networks and WPANs. To focus standardization work in this technique, IEEE established subgroup IEEE 802.15.3a, inside 802.15.3, to develop a standard for UWB WPANs. The goals for this standard are data rates of up to 110 Mbps at 10 m, 200 Mbps at 4 m, and higher data rates at smaller distances. Based on those requirements, different proposals are being submitted to 802.15.3a. An important and open issue of UWB lies in the design of multiple access techniques and radio resource sharing schemes to support multimedia applications with different QoS requirements.

One of the enormous potentials of UWB is the ability to move between the very high data rates (HDR)/short link distance, and the very low data rates (LDR)/longer link distance applications. The trade-off is facilitated by the physical layer signal structure. The very low transmit power available invariably means multiple, low energy UWB pulses must be combined to carry 1 bit of information. In principle, trading data rate for link distance can be as simple as increasing the number of pulses used to carry 1 bit. The more pulses per bit, the lower the data rate, and the greater the achievable transmission distance. The low power and HDR of UWB systems combine to produce very high data density capabilities for UWB.

Even with the significant power restrictions, UWB holds enormous potential for wireless ad hoc and peer-to-peer networks. One of the major potential advantages in impulse radio-based systems is the ability to trade data rate for link distance by simply using more or less concatenated pulses to define a bit. Without dramatically changing the air interface, the data rate can be changed by orders of magnitude depending on the system requirements. This means, however, that HDR and LDR devices will need to coexist. The narrow time domain pulse also means that UWB offers the possibility for very high positioning accuracy. However, each device in the network must be heard by a number of other devices in order to generate a position from a delay or signal angle-of-arrival estimate. These potential benefits, coupled with the fact that an individual low-power UWB pulse is difficult to detect, offer some significant challenges for the MAC design. Figure 5.15 shows the relationship between UWB and beyond third generation (B3G) systems.

The low power restricts UWB to very short-range high data applications, or very LDR for moderate range applications, and effectively prohibits UWB from most outdoor applications. The increasing trend of users to combine both cellular and ad hoc technologies (e.g., IEEE 802.11b WLAN and 2G cellular) is a strong indicator for the inclusion of high-speed short-range wireless in a comprehensive picture of future wireless networks.

UWB systems have been targeted at very HDR applications over short distances, such as USB replacement, as well as very LDR applications over longer distances, such as sensors and radio frequency (RF) tags. Classes of LDR devices are expected to have very low complexity and very low cost. The expected proliferation of

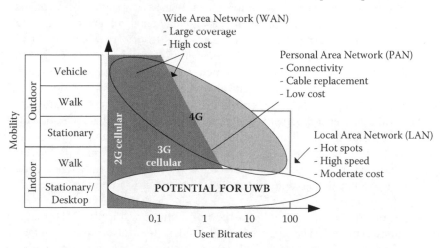

FIGURE 5.15 The area of application for UWB with respect to existing and future networks.

low-cost UWB devices means that solutions must be devised for devices to coexist, or preferably interoperate with many different types of UWB devices with different capabilities. The complexity limitations of LDR devices may mean that very simple solutions are required. HDR devices, which are expected to be higher complexity, may have much more sophisticated solutions.

ADVANTAGES OF UWB

UWB has a number of advantages which make it attractive for consumer communications applications. In particular, UWB systems

- Have potentially low complexity and cost
- Have noise-like signal characteristics
- Are resistant to severe multipath and jamming
- Have very good time domain resolution allowing for location and tracking applications

The low complexity and low cost of UWB systems arise from the essentially baseband nature of the signal transmission. Unlike conventional radio systems, the UWB transmitter produces a very short time domain pulse which is able to propagate without the need for an additional RF mixing stage. The RF mixing stage takes a baseband signal and injects a carrier frequency or translates the signal to a frequency which has desirable propagation characteristics. The very wideband nature of the UWB signal means it spans frequencies commonly used as carrier frequencies. The signal will propagate well without the need for additional up-conversion and amplification. The reverse process of down-conversion is also not required in the UWB receiver. Again, this means the omission of a local oscillator in the receiver, and the removal of associated complex delay and phase tracking loops.

Time-modulated (TM) impulse radio signal is seen as a carrierless baseband transmission. The absence of carrier frequency is a fundamental character that differentiates impulse radio and impulse radar transmissions from narrowband applications and from direct sequence (DS) spread spectrum (SS) multicarrier (MC) transmissions. TM-UWB systems can be implemented in low-cost, low-power integrated circuit processes. The TM-UWB technique also offers grating lobe mitigation in sparse antenna array systems without weakening the angular resolution of the array.

Due to the low energy density and pseudo-random (PR) characteristics of the transmitted signal, the UWB signal is noise-like, which means unintended detection is quite difficult. It appears that the low-power, noise-like UWB transmissions do not cause significant interference in existing radio systems. The interference between impulse radio and existing radio systems is one of the most important topics in current UWB research.[46]

Because of the large bandwidth of the transmitted signal, very high multipath resolution is achieved. The large bandwidth offers (and also requires) huge frequency diversity which, together with the discontinuous transmission, makes the TM-UWB signal resistant to severe multipath propagation and jamming/interference. TM-UWB systems offer good low probability of interception (LPI) and low probability of detection (LPD) properties, which make it suitable for secure and military applications.

The very narrow time pulses mean that UWB radios are potentially able to offer timing precision much better than Global Positioning System (GPS) and other radio systems. Together with good material penetration properties, TM-UWB signals offer opportunities for short-range radar applications such as rescue and anticrime operations, as well as in surveying and the mining industry. UWB does not provide precise targeting and extreme penetration at the same time, but UWB waveforms present a better choice than conventional radio systems.

UWB REGULATION

When UWB technology was proposed for civilian applications, there were no definitions for the signal. The Defense Advanced Research Project Agency (DARPA) provided the first definition for the UWB signal based on the fractional bandwidth B_f of the signal. The first definition provided that a signal can be classified as a UWB signal B_f when greater than 0.25. The fractional bandwidth can be determined using the formula:

$$B_f = 2\frac{f_H - f_L}{f_H + f} \tag{5.1}$$

where f_L is lower and f_H is higher than −3 dB point in a spectrum, respectively.[47]

In February 2002, the FCC issued the FCC UWB rulings that provided the first radiation limitations for UWB, and also permitted commercialization of the technology. The final report of the FCC First Report and Order was publicly available in April 2002. The document introduced four different categories for allowed UWB applications, as well as setting the radiation masks for them.[48,49]

The prevailing definition has decreased the limit of B_f at the minimum value of 0.2 as per Equation 5.1. Also, according to the FCC UWB rulings, the signal is recognized as UWB if the signal bandwidth is 500 MHz or more. In Equation 5.1 f_H and f_L are the higher and lower −10 dB bandwidths, respectively. The radiation limits imposed by the FCC are presented in Table 5.11 for indoor and outdoor data communication applications.

TABLE 5.11
FCC Radiation Limits for Indoor and Outdoor Communication Applications

	Frequency Range (GHz)		
	$f < 3.1$	$3.4 < f < 10.6$	$f > 10.6$
Indoor mask	−51.3 + 87 log(f/3.1)	−41.3	−51.3 + 87 log(10.6/3.1)
Outdoor mask	−61.3 + 87 log(f/3.1)	−41.3	−61.3 + 87 log(10.6/3.1)

As for the European organization, they have been influenced by the FCC decision. The recommendation for short-range devices belongs to the European Conference of Postal and Telecommunications (CEPT) working group CEPT/ERC/REC70-03.[50]

IEEE 802.15.3A

The IEEE established the 802.15.3a study group to define a new physical layer concept for short-range, HDR applications. This ALTernative PHYsical (ALT physical layer) is intended to serve the needs of groups wishing to deploy HDR applications. With a minimum data rate of 110 Mbps at 10 m, this study group intends to develop a standard to address such applications as video or multimedia links, or cable replacement. While not specifically intended to be a UWB standard group, the technical requirements very much lend themselves to the use of UWB technology. The study group has been the focus of significant attention recently as the debate over competing UWB physical layer technologies has continued. The work of the study group also includes analyzing the radio channel model proposal to be used in the UWB system evaluation.[51, 52]

The purpose of the study group is to provide a higher physical layer for the existing approved 802.15.3 standard for applications that involve imaging and multimedia. The main desired characteristics of the alternative physical layer are as follows[36]:

- Coexistence with all existing IEEE 802 physical layer standards
- Target rate in excess of 100 Mbps for consumer applications
- Robust multipath performance
- Location awareness
- Use of additional unlicensed spectrum for high rate WPANs

The IEEE 802.15.3 standard is being developed for high-speed applications including:

- Video and audio distribution
- High-speed data transfer

IEEE 802.15.4

The IEEE established the IEEE 802.15.4 study group to define a new physical layer concept for LDR applications utilizing UWB technology at the air interface. The study group addresses new applications that require only moderate data throughput, but require long battery life such as low-rate WPANs, sensors, and small networks.

UWB MEDIUM ACCESS CONTROL

UWB technology was seen as a physical layer technology with little or no protocol overhead to control the communication. MAC features are critical to achieving real benefit from UWB communication systems.

Some qualities of UWB signals are unique and may be used to produce added value beyond data communications. For example, the accurate ranging capabilities with UWB signals may be exploited by upper layers for location-aware services. Conversely, some aspects of UWB pose problems which must be solved by the MAC design. For example, when using a carrierless impulse radio system, it is cumbersome to implement the carrier-sensing capability needed in popular approaches such as carrier-sense, multiple access/collision avoidance (CSMA/CA) MAC protocols.

Another aspect that affects MAC design is the relatively large synchronization and channel acquisition time in UWB systems. CSMA/CA is used in a number of distributed MAC protocols and it is also adopted in the IEEE 802.15.3 MAC.

UWB systems have been targeted at very HDR applications over short distances, such as USB replacement, as well as very LDR applications over longer distances, such as sensors and RF tags. LDR devices are expected to be very low complexity. The potential proliferation of UWB devices means that a MAC must deal with a large number of issues related to coexistence and interoperation of different types of UWB devices with different capabilities. The complexity limitations of LDR devices may mean that very simple solutions are required. HDR devices, which are expected to be of higher complexity, may have much more sophisticated solutions.

The interaction with the MAC is critical when examining different aspects of low-complexity UWB devices. For these low-complexity/LDR devices, special solutions are required for the following purposes:

- Coexistence versus interoperation with other low-speed devices.
- Coexistence versus interoperation with high-speed devices.
- MAC enhancements for tracking and positioning. Rather than just physical/ timing jitter removal, classes of signals for timing estimation might be considered, as well as peer support for improving positioning estimation.

Strategies for ignoring or working around other devices of the same or different type based on physical layer properties will also reflect up to the MAC. Optimization of the physical layer will have implications on MAC issues such as the initial search and acquisition process, channel access protocols, interference avoidance/minimization protocols, and power adaptation protocols. The quality of the achieved channel will have implications on the link level which may necessitate active searching by a device for better conditions as happens with other radio systems.

As for strategies for working with other devices of the same type, the issue here is being able to cooperate and exchange information with devices of different data rate/ QoS class/complexity, with the emphasis on how LDRs can successfully produce limited QoS networks with HDRs. The issue for investigation focuses on the cost/ benefit trade-off of adding more complexity to very low-complexity devices.

It is possible for any single device to estimate the arrival time of a signal from another device based on its own time reference. This single data point in relative time needs to be combined with other measurements to produce a three-dimensional (3D) position estimate relative to some system reference. Exchange of timing information requires cooperation between devices. Being able to locate all devices presents a variation of the hidden node problem. The problem is complicated for positioning, as multiple receivers need to detect the signal of each node to allow 3D location to be determined. Tracking requires that each device be able to be sensed (measured at a suitable rate to allow a resonable update rate for a small number of devices, but difficult for an arbitrarily large number of devices). Information exchange between devices of timing and position estimates of neighbors (ad hoc modes) requires coordination. Calculation of actual position needs to be performed elsewhere, centralized or distributed, and results fed to the information sink. Finally, it is important to have the received signal as unencumbered by multiple access interference as possible in order to allow the best estimation of time of arrival, since every 3 ns error translates to at least 1 m extra position error.

All of these issues (information exchange, device sampling rate, node visibility, signal conditioning) require MAC support, and are significant obstacles to existing WLAN and other radio systems offering reliable positioning/tracking when added on to the MAC post-design.

The UWB MAC considerations described are only partially supported by the IEEE 802.15.3 MAC.

A WPAN is a wireless ad hoc data communications system that allows a number of independent data devices to communicate with one another. A WPAN is distinguished from other types of data networks in that communications are normally confined to a person or object that typically covers about 10 m in all directions and envelopes the person or a thing whether stationary or in motion.

The group of devices in the IEEE 802.15.3 MAC is referred to as a piconet, illustrated in Figure 5.16. A piconet includes a piconet controller (PNC) and associated devices (DEVs).

The piconet is centrally managed by the coordinator of the network, referred to as the PNC. The PNC always provides the basic timing for the WPAN. Additionally, the PNC manages the QoS requirements of the WPAN.

A DEV willing to join a piconet first scans for an existing PNC. The presence of a PNC is detected by the reception of a beacon sent from the PNC on a periodic basis. If no PNC is found, and if the device is capable of the task, the new device becomes a PNC and starts a piconet itself. The PNC periodically sends network information via a beacon. Other devices that can receive the beacon may associate with the PNC. Associated devices can exchange data directly between one another, that is, without

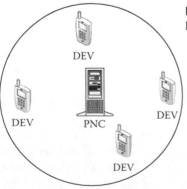

FIGURE 5.16 Piconet with the group of devices: PNC and DEVs.

using the central node as a relay, but resources are managed centrally by the PNC. The role of PNC can be handed over to an elected device according to a predetermined protocol.

The main characteristics of the IEEE 802.15.3 MAC include the following:

- **High rate WPAN.** Means HDR (currently up to 110 Mbps, to be increased by 100 to 800 Mbps) short range (minimum 10 m, up to 70 m possible).
- **Dynamic topology.** Means that mobile devices often join and leave piconet; short time to connect (<1 s).
- **Ad hoc network with multimedia QoS provisions.** Comprises TDMA for streams with time based allocations and peer-to-peer connectivity.
- **Multiple power management modes.** Designed to support low power portable devices.

Centralized and connection-oriented ad hoc networking topology is responsible for the coordinator (PNC) maintenance network synchronization timing. It also performs admission control and assigns time for connection among IEEE 802.15.3 devices.

- **Support for multimedia QoS.** Comprises TDMA superframe architecture with channel time allocation period (CTAP).
- **Authentication, encryption, and integration**
- **Multiple power saving modes**
- **Simplicity.** PNC handles only channel time requests.
- **Robustness.** This characteristic allows for the existence of dynamic channel selection PNC handover.

A compliant physical layer may support more than one data rate. In each physical layer there is one mandatory base rate. In addition to the base rate, the physical layer may support rates that are both faster and slower than the base rate. A DEV will send a frame with a particular data rate to a destination DEV only when the destination DEV is known to support that rate.

There are two main candidates for the IEEE UWB standard, neither of which has yet led to concrete proposals at the MAC level. The first is a multiband approach which is based on the flexible utilization of a number of UWB channels, each of approximately 500 MHz. With this technique, the generated UWB signal is reshaped according to country-specific regulatory constraints and the presence of known narrowband interferers (such as WLAN devices). The second is based on impulse radio- or pulse-based UWB techniques. In both cases, a flexible, dynamic MAC protocol must be developed to manage the system.

UWB has a significant role to play as an extension of or complement to cellular technology in future mobile systems. However, to reach this goal, much work is still to be done to develop the higher layers which will take advantage of what is being offered by the physical layer.

5.8   WIRELESS METROPOLITAN AREA NETWORKS

The area of broadband Internet has gained a lot of interest in recent years due the exciting business opportunities enabled by high-speed Internet connectivity to homes and businesses. Content owners (e.g., movie studios and record labels) and content providers/distributors (e.g., music and MPEG4 download services) see broadband Internet to the home as crucial to providing the next source of revenue, as it solves the difficult last mile access problem.

The IEEE Working Group 802.16 on Broadband Wireless Access (BWA) has developed the WMAN air interface standard to support the development and the deployment of WMAN.[53] The Working Group is primarily addressing applications of wireless technology to link commercial and residential buildings to high-rate core networks and thereby provide access to those networks.

It is with this background that WMANs based on the 802.16 technology have recently gained a lot of interest among vendors and Internet service providers (ISPs) as the possible next development in wireless IP, and as offering a possible solution for the last mile access problem. With the theoretical speed of up to 75 Mbps and with a range of several miles, 802.16 broadband wireless offers an alternative to cable modem and digital subscriber line (DSL technologies), possibly displacing them in the future.

The Working Group's first air interface project was completed with the April 2002 publication of IEEE Standard 802.16 Air Interface for Fixed Broadband Wireless Access Systems. As specified in this standard, a WMAN provides network access to buildings through exterior antenna communications with central radio base stations (BSs) in a point-to-multipoint topology. The WMAN offers an alternative to cable access networks, such as fiber-optic links, coaxial systems using cable modems, and DSL links. Because wireless systems have the capacity to address broad geographic areas without the costly infrastructure development required in deploying cable links to individual sites, the technology may prove less expensive to deploy and may lead to more ubiquitous broadband access.

In this scenario, with WMAN technology bringing the network to a building, users inside the building will connect to the WMAN with conventional in-building networks such as, for data, Ethernet (IEEE 802.3) or WLANs (IEEE 802.11). However, the fundamental design of the standard may eventually allow for the efficient extension of the WMAN networking protocols directly to the individual user. For instance, a central BS may someday be exchanging MAC protocol data with an individual laptop computer in a home. The links from the BS to the home receiver ,and from the home receiver to the laptop, would likely use quite different physical layers, but design of the WMAN MAC could accommodate such a connection with full QoS. The standard has already begun to evolve to support nomadic and increasingly mobile users.

IEEE Standard 802.16 was designed to evolve as a set of air interfaces based on a common MAC protocol, but with physical layer specifications dependent on the spectrum of use and the associated regulations. The published base standard addresses frequencies from 10 to 66 GHz, where extensive spectrum is currently available worldwide, but at which the short wavelengths introduce significant challenges, including propagation that is essentially limited to line of sight. The 10 to 66 GHz air interface is designated WMAN-SC because it uses single-carrier modulation. The base station basically transmits a TDM signal, with individual subscriber stations allocated time slots serially. Access in the uplink directions is by TDMA. Both time-division duplexing (TDD) and frequency-division duplexing (FDD) are handled in a common burst fashion. Half-duplex FDD subscriber stations, which may be less expensive since they do not simultaneously transmit and receive, are easily supported in this framework. Both the TDD and FDD alternatives support adaptive burst profiles in which modulation and coding options are dynamically assigned on a burst-by-burst basis.

IEEE Project 802.16a is developing a draft addressing frequencies in the 2 to 11 GHz band, including both licensed and license-exempt bands. Compared to the higher frequencies, such bands offer the opportunity to reach many more customers less expensively, though at generally lower data rates. This suggests that such services will be oriented toward individual homes or small- to medium-size enterprises. Design of the 2 to 11 GHz physical layer is driven by the need for non-line-of-sight (NLOS) operation. This is essentially to support residential applications, since rooftops may be too low for a clear line of sight to a BS antenna, possibly due to obstruction by trees. Therefore, significant multipath propagation must be expected. Furthermore, outdoor-mounted antennas are expensive due to both hardware and installation costs.

The current document specifies that systems implement one of three air interface specifications, each of which provides for interoperability:

- WMAN-SC (this uses a single-carrier modulation format)
- WMAN-OFDM (this uses orthogonal frequency division multiplexing with a 265-point transform; access is by TDMA)
- WMAN-OFDM (this uses orthogonal frequency division multiplexing with a 2,048-point transform; in this system, multiple access is provided by addressing a subset of multiple carriers to individual receivers)

Because of the propagation requirements, the use of advanced antenna systems is supported.

To accommodate the more demanding physical environment and different service requirements found at frequencies between 2 and 11 GHz, the 802.16a project upgrades the MAC to provide ARQ. Also, an optional mesh topology is defined to expand the basic point-to-multipoint architecture.

IEEE 802.16 Working Group is turning its attention to enhancing the standard, primarily by adding support for mobile user devices. To assist the successful transition from standardization to wide-scale deployment, the working group is also placing a great deal of attention on compliance and interoperability. The group has begun rapidly developing test documentation. The Worldwide Interoperability Microwave Access (*WiMAX*) Forum, an industry consortium, has arisen to help define and carry out interoperability assurance tests.[54]

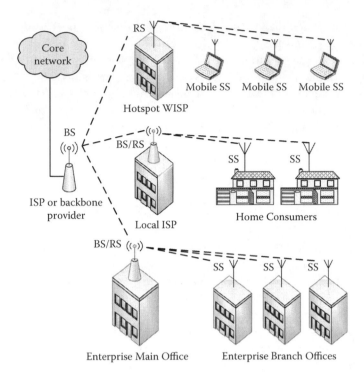

FIGURE 5.17 An example of the IEEE 802.16 network topology.

IEEE 802.16 Network Management

The basic arrangement of an 802.16 network or cell consists of one (or more) BSs and multiple subscriber stations (SSs). Depending on the frequency of transmission, the SS may or may not need to be in the line-of-sight of the BS antenna. In addition to base stations and subscriber stations, there might also be other entities within the network, such as repeater stations (RSs) and routers, which provide connectivity of the network to one or more core or backbone networks.

The 802.16 network topology is shown in Figure 5.17. The BS has a number of tasks within the cell, including management of medium access by the SS, resource allocation, key management, and other security functions.

An implementation of an 802.16 network will typically deploy a fixed antenna for the SS, with the BS using either a sectored or omni-directional antenna. The BS would be installed in a location that can provide the best coverage, which would usually be the rooftops of buildings and other geographically elevated locations. Although a fixed SS would use a fixed antenna, with the future development of the mobile subscriber station (MSS), it is possible that an SS could be using an omni-directional antenna. In practice, the cell size would be about 5 miles or less in radius. However, given suitable environmental conditions and the use of OFDM, the cell radius can reach 20 or even 30 miles. To increase the range of a given implementation, a mesh topology can also be used instead of the point-to-point topology.

Frequency Range of IEEE 802.16

In an answer to the broad frequency range of 802.16 (namely, the 2 to 66 GHz band), the standard specifies the support for multiple physical layer specifications. Since the electromagnetic propagation in this broad range is not uniform all over, the 802.16 standard splits the range into three different frequency bands, each to be used with a different physical layer implementation as necessary. The three frequency bands are as follows:

- 10 to 66 GHz (licensed bands): Transmission in this band requires line of sight between BS and SS. This is because within this frequency range the wavelength is very short, and thus fairly susceptible to attenuation (e.g., due the physical geography of the environment or interference). However, the advantage of operating in this frequency band is that higher data rates can be achieved.
- 2 to 11 GHz (licensed bands): Transmission in this band does not require line of sight. However, if line of sight is not available, the signal power may vary significantly between the BS and SS. As such, retransmissions may be necessary to compensate.
- 2 to 11 GHz (unlicensed bands): The physical characteristics of the 2 to 11 GHz unlicensed bands are similar to the licensed bands. However, because they are unlicensed, there are no guarantees that interference may not occur due to other systems or persons using the same bands.

IEEE 802.16 Medium Access Control Protocol

The scope of the IEEE 802.16 standard comprises the MAC and the PHY layer as illustrated in Figure 5.18. The MAC includes a service-specific convergence sublayer that interfaces with higher layers. The MAC common part sublayer carries the key functions and resides below the security sublayer.

The service specific convergence sublayer (CS) provides any transformation or mapping of external data received through the CS service access point (SAP). This includes classifying external network service data units (SDU) and associating them with the proper service flow identified by the connection identifier (CID). A service flow is a unidirectional flow of packets that is provided with a particular QoS.

The MAC common part sublayer (CPS) provides system access, bandwidth allocation, connection establishment, and connection maintenance. It receives data from various CSs classified to particular CIDs. QoS is applied to transmission and scheduling of data over the PHY layer.

IEEE 802.16 is optimized for point-to-multipoint (PMP) configurations where several SSs are associated with a central BS.

The system supports a frame-based transmission, in which the frame can adopt variable lengths. The frame structure of the OFDM PHY layer operating in TDD mode is shown in Figure 5.19.

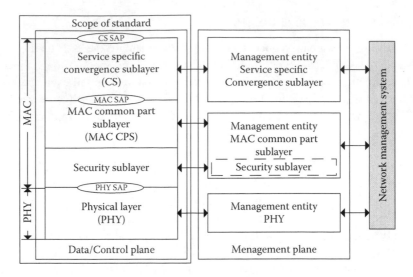

FIGURE 5.18 The scope of the 802.16 protocol layering. Reproduced with permission from http://www.ieee802.org/15/pub.

Each frame consists of a downlink (DL) subframe and an uplink (UL) subframe, with the DL subframe always consisting of only one DL PHY transmission burst starting with a long preamble (2 OFDM symbols) used for synchronization.

The frame control header (FCH) contains the DL frame prefix (DLFP) and occupies one OFDM symbol. The DLFP specifies the location as well as the modulation and coding scheme (PHY mode) of up to four DL bursts following the FCH. The mandatory modulation used for the FCH is BPSK with code rate 1/2.

The FCH is followed by one or multiple DL bursts, which are ordered by their PHY mode. While the burst with the most robust PHY mode, for example, BPSK ½, is transmitted first, the last burst is modulated and coded using the highest PHY mode, that is, 64-QAM 3/4. Each DL burst is made up of MAC PDUs scheduled for DL transmission. Optionally, a DL burst might start with a short preamble (1 OFDM symbol) to allow for an enhanced synchronization and channel estimation of SSs. MAC PDUs transmitted within the same DL burst might be associated with different connections and/or SSs, but they are all encoded and modulated using the same PHY mode. In the DL as well as in the UL direction, the burst length is an integer number of the OFDM symbol length so that burst and OFDM symbol boundaries match each other.

The UL subframe consists of contention intervals scheduled for initial ranging and bandwidth request purposes and one or multiple UL PHY transmission bursts, each transmitted from a different SS. The initial ranging slots allow an SS to enter the system by requesting the basic management CIDs, by adjusting its power level and frequency offsets, and by correcting its timing offset. The bandwidth request slots are used by SSs to transmit the bandwidth request header.

Each UL PHY transmission burst contains only one UL burst and starts with a short preamble (1 OFDM symbol). For better synchronization and channel estimation

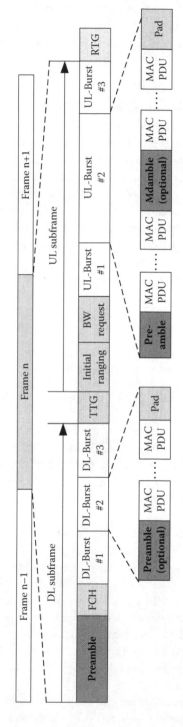

FIGURE 5.19   IEEE 802.16 MAC frame in TDD mode. Reproduced with permission from http://www.ieee802.org/15/pub.

optional midambles (1 OFDM symbol) might be periodically included in the UL burst. All MAC PDUs of a UL burst are transmitted by a single SS using the same PHY mode. DL and UL subframes are separated by the receive/transmit transition gap (RTG) and the transmit/receive transition gap (TTG), respectively.

IEEE 802.16 OFDM PHY LAYER

The IEEE 802.16 physical layer uses OFDM with a 265 point transform, designed for both LOS and NLOS operation in frequency bands below 11 GHz, both licensed and license exempt. TDM and FDM variants are defined. Channel bandwidths vary from 1.25 to 28 MHz. Additional air interfaces based on OFDM access with a 2,048-point transform and based on single-carrier (SC) modulation are specified.

The OFDM-based transmission specified in the IEEE 802.16 standard has been standardized in close cooperation with the ETSI, whose standard is named HiperMAN. Both the HiperMAN standard and the OFDM-based transmission mode of IEEE 802.16 form a basis for WiMAX certified technology. The WiMAX is an industry-led, nonprofit corporation formed to promote and certify compatibility and interoperability of broadband wireless products such as IEEE 802.16 and HiperMAN.[57]

EXAMPLE 5.1

The main advantage of fixed broadband wireless access (FBWA) technologies over wired systems like DSL and cable modems results mainly from the high costs of the labor-intensive deployment of cables. A 200-square-kilometer service area costs a DSL provider more than $11 million. The same area can be served wirelessly for about $450,000.

Because a single harmonized frequency band is not present, it is recommended that the frequency bands 3.4 to 3.6 GHz, 10.15 to 10.3 GHz, and 10.5 to 10.65 GHz should be identified as preferred bands for FBWA.[58] In addition to the 5 and 3.5 GHz bands, WiMAX targets the licensed 2.5 GHz bands. The WiMAX Forum announced its intention to advance the allocation of licensed and license-exempt spectrum in lower frequency bands.[57]

Link distances, that is, cell sizes, vary strongly based on the frequency bands used, the environment, propagation conditions, and antenna gain. The system targets distances between 2 and 4 km for NLOS, and up to 15 km for LOS.

The delay spread is due to multipath scattering. To avoid intersymbol interference, a cyclic prefix (CP) is introduced in front of every data part of an OFDM symbol.[59] Table 5.12 lists common maximum delay spread values in different types of environments. These delay spread values remain unchanged for any operating frequency above 30 MHz, since the wavelengths become much smaller than man-mode architectural structures. Recent measurements do confirm the values for frequency bands between 800 MHz and 6 GHz.[59]

The IEEE 802.16 frame structure, and therewith the system capacity, depends on the number of SSs and on the number of different PHY modes in use.

TABLE 5.12
Maximum Delay Spread Values in Different Environments

Type of Environment	Maximum Delay Spread (µs)
In-building (house, office)	<0.1
Large building (factory, malls)	<0.2
Open area	<0.2
Suburban area LOS	0.2–1.0
Suburban area NLOS	0.4–2.0
Urban area	1.0–3.0

5.9 WIRELESS WIDE AREA NETWORKS

IEEE 802.20 Mobile Broadband Wireless Access (MBWA) Working Group, the establishment of which was approved by IEEE standard board in December 2002, aims to prepare a formal specification for a packet-based air interface designed for IP-based services named as MobileFi.

IEEE 802.20 will be specified according to a layered architecture, which is consistent with other IEEE 802 specifications. The scope of the working group consists of the PHY, MAC, and LLC layers. The air interface will operate in bands below 3.5 GHz and with a peak data rate of over 1 Mbps.

The goals of 802.20 and 802.16e, the so-called *mobile WiMAX*, are similar. A draft 802.20 specification was balloted and approved on January 2006.

The baseline specifications that have been proposed for this specification aim considerably higher than those available on our current mobile architecture. The draft standard's proposed benefits are

* IP roaming and handoff (at more than 1 Mbps)
* New MAC and PHY with IP and adaptive antennas
* Optimization for full mobility up to vehicular speeds of 250 km/h
* Operation in licensed bands (below 3.5GHz)
* Utilization of packet architecture
* Low latency

In June 2006, the IEEE-SA standard board directed that all activities of the 802.20 working group be temporarily suspended. In September 2006, the IEEE-SA approved a plan to enable the IEEE 802.20 working group to move forward with its work to develop a mobile broadband wireless access standard and provide the best opportunity for its completion and approval.

Remember that three main wireless data technologies exist, namely, the IEEE 802.11, IEEE 802.16, and IEEE 802.20 technologies. The IEEE 802.11 WLAN technology, so-called WiFi, has been widely deployed in a range within 100 m. The IEEE 802.16 WMAN technology, so-called *WiMAX*, only supports fixed broadband wireless access systems in which subscriber stations are in fixed locations. *WiMAX* provides data rates of about 10 Mbps over longish distances, competing with wired DSL broadband services. The emerging IEEE 802.16e standard enhances the original IEEE 802.16 standard with mobility, so that mobile subscriber stations can move during services. *WiMAX* for longer-range broadband services has been recently deployed.

However, WiFi and *WiMAX* are limited by the range of their coverage areas from a few hundred feet to 30 miles. The IEEE 802.20 MBWA, so-called MobileFi, may change the direction of wireless networking. MobileFi sits on existing cellular towers, promising the same coverage area as a mobile phone system with the speed of a WiFi connection. MobileFi is for truly mobile, high-speed data, and truly mobile rates of 20 Mbps are possible. Both IEEE 802.16e and IEEE 802.20 aim to combine the benefits of mobility, standardization, and multivendor support. MobileFi will increase the coverage or mobility compared with WLAN and WiMAX. IEEE 802.11, IEEE 802.16, and IEEE 802.20 all define MAC protocols with several different PHY layer specifications dependent on the spectrum of use and the associated regulations.

The IEEE 802.20 MBWA is an efficient packet-based air interface that is optimized for the transport of IP-based services to enable worldwide deployment of affordable, ubiquitous, always-on, and interoperable multivendor MBWA networks. This will be an incredible boost for users of laptops and PDAs, as well as other wireless network devices. It also opens the field for integrating applications, such as targeted advertising and video on demand, throughout MBWA networks.

The IEEE 802.20 standard draft began to be developed after an evaluation by the IEEE working group concluded that five necessary criteria would be able to be met by the IEEE 802.20 standard. The five criteria are broad market potential, compatibility, distinct identity, technical feasibility, and economic feasibility.[60]

To achieve the goal of compatibility, the standard must conform to IEEE 802 ID (MAC Bridges) and 802.IF (Virtual LAN Bridges). Distinct identity is achieved due to the fact that IEEE 802.20 is much different from any other standard in that no other current IEEE 802 standard supports full vehicular mobility at high sustained data rates. The technical feasibility has been proved in small-scale proprietary systems that use widely available components such as modems, radios, and antennas. Finally, economic feasibility was proved to the working group because cost factors for mobile services and components are well documented. Furthermore, investors are willing to invest large sums of money in such projects because they believe that the return on investment will be good. Since IEEE 802.20 is a packet-based access network, its designers believe that it is optimal for data services, which are characterized by high peak demand but bursty requirements overall.

WLAN standards still exist numerous times. In the United States and Europe, a number of other standards for high-speed wireless: a high-speed wireless U.S. IEEE 802.11 standards (including 802.11a and 802.11b) and HomeRF standard. IEEE 802.lib standard maximum data transmission rate can reach 11 Mbps require a 2.4 GHz frequency bands. The standard is very popular in North America. In another high-speed wireless standard, IEEE 802.11a, the data transmission rate for 54Mbps requires a 2.4 GHz bandwidth, faster than the current IEEE 802.11b technology by nearly five times. Here the term refers to the WWAN equipment across wide areas using a single mobile wireless business products and services. We can see from Figure 5.20 that 3G and IEEE 802.16 belong to MAN. IEEE 802.20 also has a distinct advantage—the great coverage.

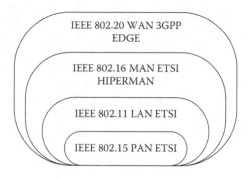

FIGURE 5.20 Coverage of the IEEE 802.20 standard.

The main technical characteristics of IEEE 802.20 are as follows:

- Full support for real-time and non-real-time operations
- Maintenance of continuous connectivity
- Frequency reunification may be resumed
- Air interface for QoS supports integration with the core network QoS levels end line
- Support of IPv4 and IPv6 with the agreement
- Rapid allocation of the necessary resources
- Environmental change on the basis of access
- Automatic selection of optimal data transmission rate
- Authentication mechanisms between terminals and networks provided
- Existing cellular system with a common operation, with the network air interface between the open interfaces

IEEE 802.20, IEEE 802.11, and IEEE 802.16 are strongly complementary; combined they can provide a very good solution for entire network coverage for true wireless access to meet the needs of the market. IEEE 802.11, IEEE 802.16 WLAN, and MAN technology are mainly targeted at livestock TourType wireless access, providing walk rate movements.

MISSION, SCOPE, AND PURPOSE

The mission of the IEEE 802.20 working group is to develop a specification for an efficient packet-based air interface optimized for IP-based services, with the goal of enabling worldwide deployment of affordable, ubiquitous, always-on, and interoperable multivendor mobile broadband wireless access networks.

The scope of IEEE 802.20 is to define the PHY and MAC layers for interoperable mobile broadband wireless access systems, operating in licensed bands below 3.5 GHz, optimized for IP-data transport, with peak data rates larger than 1 Mbps per user. IEEE 802.20 supports various vehicular mobility classes up to 250 km/hour in a MAN environment and targets spectral efficiencies, larger sustained user data rates, and larger numbers of active users.[61]

The IEEE 802.20 technology will fill the gap between cellular networks (low bandwidth and high mobility) and other IEEE 802 wireless networks (high bandwidth and low mobility) currently in use, such as IEEE 802.11 WLANs and IEEE 802.16 WMANs. It will provide seamless integration between the three domains of work, home, and mobile,[62] allowing users to have a single connection that provides for their networking needs wherever they go, by laying the groundwork for a worldwide network that is cost-effective, spectrum efficient, always on, and interoperable. More specifically, it will offer transparent support of real-time and non-real-time applications, always-on connectivity, universal frequency reuse, support of inter-technology roaming and handoff (e.g., from MBWA to WLANs), seamless intercell and intersector handoff, fast resource allocation for both uplink and downlink, QoS that is policy based and supports both IPv4 and IPv6, as well as the ability to be codeployed with existing cellular systems.

TECHNICAL ASPECTS RELATED TO IEEE 802.20 SYSTEM

Technical aspects related to an IEEE 802.20 system include application support, user expectations, and QoS, as well as security. In connection with this are peak user data rate, peak aggregate data rate, spectral efficiency, bandwidth maximum operating frequency, and so on. Table 5.13 shows the data rates available to the end user for the available bandwidths of an MBWA base station.[63] Both 1.25 and 5 MHz channel bandwidth are supported. Peak data rates and the aggregate data rates delivered to the user are shown in Table 5.13.

The IEEE 802.20-based air interface will support applications such as video, web browsing with full graphical capability, e-mail, file uploading and downloading without size limitations, streaming video and audio, virtual private network (VPN) connections, VoIP, instant messaging, and online multiplayer gaming, all while maintaining an always-on connection similar to cable modem or xDSL. With regard to voice services, the MBWA will support VoIP services by leveraging QoS features that deal with latency, jitter, and packet loss.

The MAC layer should be able to handle more than 100 simultaneous active sessions per sector. An active session is defined as a time duration during which a user can receive and/or transmit data with a short delay of less than 25 ms with a

---

TABLE 5.13
Available Bandwidth at MBWA Base Station

	1.25 MHz	5 MHz
Peak user data rate (downlink)	>1 Mbps	>4 Mbps
Peak user data rate (uplink)	>300 kbps	>1.2 Mbps
Peak aggregate data rate per cell (downlink)	>4 Mbps	>16 Mbps
Peak aggregate data rate per cell (uplink)	>800 kbps	>3.2 Mbps

Reproduced with permission from IEEE 802.20 WG PD-06. Introduction to IEEE 802.20: Technical and Procedural Orientation, March 2003.

TABLE 5.14
Degradation Expected for a Typical MBWA Implementation

	3 km/h	120 km/h
Peak user data rate (downlink)	>1 Mbps	>4 Mbps
Peak user data rate (uplink)	>300 kbps	>1.2 Mbps
Peak aggregate data rate per cell (downlink)	>4 Mbps	>16 Mbps
Peak aggregate data rate per cell (uplink)	>800 kbps	>3.2 Mbps

Reproduced with permission from W. Bolton, Y. Xiao, M. Guizani. "IEEE 802.20: Mobile broadband wireless access," IEEE Wireless Communications 14, no. 2 (February 2007): 84–95.

probability of at least 0.9. Certain applications such as VoIP, however, will have to be given higher priority with respect to delay in order to satisfy QoS requirements.

Velocities, both on foot (or even stationary) and at vehicular speeds, will undoubtedly play a large role in the throughput that the user can expect to see. As a user's speed increases, average throughput available to the user degrades. Table 5.14 illustrates the amount of degradation that the IEEE 802.20 working group expects to see for a typical MBWA implementation.

With regard to QoS, the IEEE 802.20 MAC and PHY layers are the primary components responsible for providing efficient QoS to users. The system should be intelligent enough to recognize that a user may be using several different applications with differing QoS requirements at the same time. For example, the user may be browsing the web and participating in a video conference that has separate audio and video streams associated with it. Clearly, the two services differ enough that they need separate QoS negotiations. The system should be able to recognize and categorize various kinds of IP traffic based on specific packet flows associated with each, such as delay, bit rate, error rate, and jitter. The bit rate, or data rate, should scale from the lowest allowable data rate to the maximum rate supported by the MAC/PHY. Delivery delay, also known as latency, should be in a range from 10 ms to 10 s. It should be noted, however, that 10 ms is the targeted objective for round-trip delay time, and that even 50 ms is considered by the IEEE 802.20 working group to be way too high. The error rate, after corrections have been made in the MAC and PHY, should be in the range from 10^{-8} to 10^{-1}. Delay variation, also known as jitter, should fall in the range from 0 to 10 s.[64]

Wireless networking is inherently less secure than wired networking. This is a proven fact that has been validated many times over by security experts and is simply a result of the nature of the medium. Because wireless systems pass information through the air, they are especially vulnerable to the threat of interception, by which a malicious third party is able to gain access to information that is being passed over the network. The IEEE 802.11-based systems in particular have been criticized by security experts due to their lackluster security enforcement. Recognizing this, and knowing that a successful commercial deployment of MBWA systems will depend in no small part on the level of trust people have in the security of the system, mechanisms that provide for an optimal level of security are being thoroughly studied and built into the IEEE 802.20 specification.

Security in an MBWA system is centered on three major factors: protecting against theft of service on behalf of the service provider, protecting the privacy of the user, and deterring denial-of-service attacks. The system should provide for the following four combinations of privacy and integrity: encryption and message integrity, encryption and no message integrity, message integrity and no encryption, and no message integrity and no encryption.

However, encrypting data is only one piece of the larger puzzle of securely transmitting data in a network environment. Although encryption is an important mechanism for transmitting data securely, equally important is for both parties to be confident that they are actually exchanging data with whom they think they are. In the particular case of an MBWA system, the base station needs to be protected against unauthorized access by a rogue mobile station, and the mobile station needs to be sure it is talking to an actual base station and not a third party. This process is known as authentication and is the basis on which trust is formed in a networked environment.

AIR INTERFACE CHARACTERISTICS

The air interface should support multiple MAC protocol states with fast and dynamic transitions among them. The purpose of the dynamic switching capability is to improve system capacity and provide for a seamless networking experience on the part of the user, which is good for end-to-end TCP/IP performance. System capacity is increased because the dynamic switching capability is similar to an operating system that efficiently manages the process.

With regard to resource allocation, the air interface should support fast resource assignment and release procedures on both the uplink and downlink channels. This is especially well suited to the bursty requirements of IP applications, and in combination with the dynamic transitioning between MAC protocol states, should allow for the maximum system utilization.

Both intersector and intercell handover procedures for a stationary user and all the way up to full vehicular speeds should be supported. The concepts of cell and sector are similar to those in cellular networks. The handover procedures should be designed and implemented in a way that minimizes both packet loss and latency and provides for an interrupt IP packet transmission.

An MBWA system is able to distinguish different types of packets and categorize them into certain classes, each of which has distinct QoS requirements in order to function properly. The air interface is designed to operate in the portion of the licensed spectrum below 3.5 GHz. The MBWA system frequency plan should provide for both paired and unpaired channel plans with multiple bandwidths, that is, 1.25 MHz or 5 MHz. The air interface should automatically select the optimized data rate that the user can achieve based on various network loading factors. An IEEE 802.20 system is designed to use OFDM to multiplex data signals together. One of the major advantages of using an OFDM approach is that there is no intracell interference. Also, OFDM allows much higher system capacities than CDMA.

TABLE 5.15
MAC-Related Air Interface Characteristics

Parameter	Proposed Value
Number of active users per sector/cell	>100
Transition from active On to active Hold state	<100 ms
Transition time from active Hold state to active on state	<50 ms
Transition time from active Hold state to inactive sleep state	<100 ms
Access time from inactive Sleep state to active on state	<200 ms
Paging signal periodicity	<100 ms
Paging signal duration	<1 ms
Minimum scheduling interval	<2 ms
UL request time	<10 ms
Intersector/cell handoff time	<200 ms

Reproduced with permission from W. Bolton, Y. Xiao, M. Guizani. "IEEE 802.20: Mobile broadband wireless access," IEEE Wireless Communications 14, no. 2 (February 2007): 84–95.

While traditional TDMA and FDMA systems achieve the desired characteristic of users being orthogonal to one another for in-cell interference, their performance suffers with regard to out-of-cell interference. CDMA, on the other hand, exhibits the inverse behavior. Although it does not perform well with regard to in-cell interference, it does provide for average-case out-of-cell interference characteristics. Thus, although OFDM is more complex and has more overhead associated with it, it does provide for good in-cell and out-of-cell interference behavioral patterns that are deemed important in a working MBWA system.

Table 5.15 gives a summation of the characteristics that the MAC layer is responsible for in the air interface.[66] The air interface should be given the capability to support multiple MAC protocol states with fast transitions between them. Three states should be supported: On, Hold, and Sleep. In the On state, the user is actively using system resources to transmit and receive data. To make better use of the system resources and provide for higher system efficiency, the Hold state should be initiated when the user is temporarily not using the system. When a mobile user is completely inactive, the Sleep state should be initiated.

User states are extremely beneficial, since a significant amount of air-link resources related to power control, timing control, and traffic requests, are required to enable users (i.e., users that are in the On mode) to actively send and receive traffic. The fewer air-link resources that are consumed by individual users, the more users that can be added to a particular cell, ultimately decreasing total system costs for both the user and the operator, as well as increasing total system efficiency.

The number of users that are inactive and in the Sleep state can be almost unlimited. To be utilized effectively and to minimize noticeable effects on behalf of the user, transitions between the states should be both fast and dynamic. A mechanism for waking users up from the Sleep state and bringing them into an active On state must be implemented. The mechanism that the designers of the MBWA standard are choosing to use is known as paging. The primary advantage that paging has to offer lies in its ability to allow a mobile station to conserve energy by way of the Sleep state, and still allow the mobile station to receive incoming packets as needed. This is especially important for real-time applications such as voice and instant messaging in which the station needs to be responsive at all times to incoming packets. To reduce the delay associated with waking a user up, the MBWA air interface should support the ability to send paging signals as often as once every 100 ms. The paging signals themselves should be very brief, however; they should last no longer than 1 ms, so that a mobile system in Sleep state can listen for a page very briefly and then be able to go back to sleep again. This is especially important for a battery-powered mobile device that seeks to retain as much energy as possible, as it allows the device to expend the least amount of energy possible during the Sleep state.

The air interface should support both intersector and intercell handoff at a rate that is comparable to the state transition time of 200 ms. This requirement is crucial for allowing minimal packet loss and latency, two properties that allow for seamless IP packet transmission.

IEEE 802.20 SYSTEM ARCHITECTURE

The IEEE 802.20 standard is designed using a modular approach. This means that there should be a clear separation of functionality in the system between the user, data, and control. The PHY and MAC layers should each have a set of clearly defined responsibilities that are encapsulated within the respective layers. The use of a layered architecture is consistent with the design methodologies employed in the current IEEE 802-based systems. The MAC layer should be optimized to support a specific PHY implementation. If more than one PHY implementation is to be used, the MAC layer should be designed so that it has a PHY-specific layer as well as a more general part.

The cellular environment, in which MBWA system has to operate, requires support for macrocell, microcell, and picocell sizes. The system provides for non-line-of-sight outdoor to indoor usage, as well as outdoor coverage. Service attributes that a system should provide for include, but are not limited to, the following: low end-to-end latency, a high-frequency reuse network, high capacity per sector/per carrier, and a full mobile broadband user experience. The implementation at the mobile station should be optional and determined by the market, based on economic factors. The base station should also support antenna diversity, but just as in the case of multiantenna technology, implementation at the mobile station should not be mandatory.

We can summarize the PHY and MAC features as follows:

- The physical (PHY) layer specification consists of two different duplexing modes (TDD and FDD); two different forward link hopping modes (SymbolRateHopping and block-hopping); two different synchronization modes (semisynchronous and asynchronous); and two different multicarrier modes (MultiCarrierOn and MultiCarrierOff).
- Modulation uses OFDM with QPSK, 8-PSK, 16-QAM, and 64-QAM modulation formats.

- The MAC layer includes session, convergence, security, and lower MAC functions. The lower MAC sublayer controls operations of data channels: Forward traffic channel and reverse traffic channel all. It includes control channel MAC protocol, access channel MAC protocol, shared signaling MAC protocol, forward traffic channel MAC protocol, reverse control channel MAC protocol, and reverse traffic channel MAC protocol. Forward- and reverse-link transmissions are divided into units of superframes, which are further divided into units of PHY frames. FDD and TDD superframe timing is used.
- An FDD forward-link superframe consists of a superframe preamble followed by several (e.g., 24) forward frames; and an FDD reverse-link superframe consists of several (e.g., 24) reversed frames.
- A TDD forward-link superframe consists of a superframe preamble and several forward frames, and a TDD reverse-link superframe consists of several reversed frames.
- The default access channel MAC protocol provides an access terminal to transmit by initial access or handoff via sending an access probe.

The emerging 802.16e and 802.20 standards will both specify new mobile air interfaces for wireless broadband. Although the two standards seem very similar, there are some important differences between them. For example, 802.16e will add mobility in the 2 to 6 GHz licensed bands, whereas 802.20 aims for operation in the licensed band below 3.5 GHz. More importantly, the 802.16e specification will be based on an existing standard (802.16a), while 802.20 is starting from scratch.

The IEEE approved the 802.16e standards effort with the intent of increasing the use of broadband wireless access by taking advantage of the inherent mobility of wireless media. The 802.20 interface seems to provide real-time data transmission rates in WWAN to speeds that rival DSL and cable connections (1 Mbps or more) based on cell ranges of up to 15 km or more. It plans to deliver those rates to mobile users even when they are traveling at speeds up to 250 km/hour. This makes 802.20 an option for deployment in high-speed trains. The 802.16e project authorization request specifies only that it will support user terminals moving at vehicular speeds. 802.16 is looking at the mobile user walking around with a PDA or laptop, whereas 802.20 will address high-speed mobility issues. Some argue that 802.20 is a direct competitor to 3G wireless cellular technologies.

STANDARD IEEE 802.20 RELATIONSHIP TO 802.16E AND 3G

The IEEE 802.20 standard represents an attempt to bridge many of the technical differences between IEEE 802.16e and 3G networks, combining them in a powerful and accessible manner for the end user (Figure 5.21). This standard is an attempt to combine the global mobility and roaming support of 3G with the high-bandwidth and low-latency characteristics of IEEE 802.16e. Key differences among IEEE 802.20, IEEE 802.16e, and 3G cellular networks are shown in Table 5.16. The standard 802.16e handles low mobility such as walking with a PDA or laptop in the 2 to 6 GHz licensed bands, whereas IEEE 802.20 handles high-speed mobility in bands

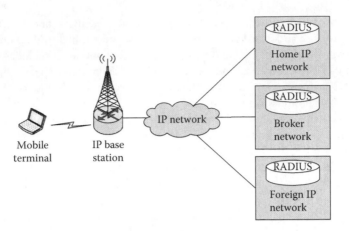

FIGURE 5.21   IP reference model of IEEE 802.20 architecture.

below 3.5 GHz. IEEE 802.20 should be optimized for data mobility and broadband wireless networks.

3G is fundamentally a voice-based technology and is not optimized for packet-based delivery services. Many technical differences for carrying data exist including spectral efficiency, latency, and the single-threaded call model structure of the

TABLE 5.16

Key Differences Among IEEE 802.20, IEEE 802.16e and 3G

Network	802.16e	802.20	3G
Mobility	High-data rate with adjunct mobility service	Fully mobile, high-throughput data user	Voice user requiring data services
Data pattern	Symmetric data services	Symmetric data services	Highly asymmetric data services
Services	Low-latency data and real-time voice	Low-latency data	Lack of support for low-latency services
Mobility and roaming support	Local/regional	Global	Global
MAC/PHY	Extensions to 802.16a MAC and PHY	New PHY and MAC optimized for packet data	W-CDMA, CDMA2000
Technology	Optimized for and backward compatible with fixed stations	Optimized for full mobility	Evaluation of GSM or IS-41
Bands	2–6 GHz	<3.5 GHz	<2.7 GHz
Channel bandwidth	>5 MHz	<5 MHz	<5 MHz

Reproduced with permission from W. Bolton, Y. Xiao, M. Guizani. "IEEE 802.20: Mobile broadband wireless access," *IEEE Wireless Communications* 14, no. 2 (February 2007): 84–95.

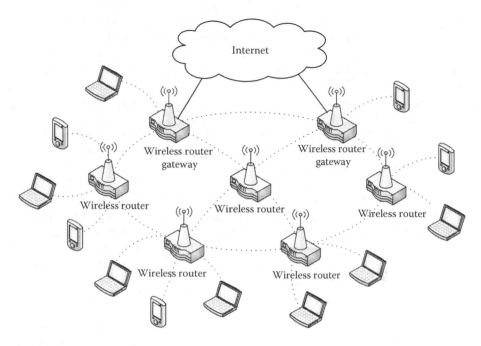

FIGURE 5.22   An example of wireless mesh network. Reproduced with permission from W. Bolton, Y. Xiao, M. Guizani. "IEEE 802.20: Mobile broadband wireless access," *IEEE Wireless Communications* 14, no. 2 (February 2007): 84–95.

interface. Other issues include circuit-switched uplink, voice-centric channel parameters, contention-based access, intracell interference, and insufficient support for QoS.

IEEE 802.20 WIRELESS MESH NETWORKS

A wireless mesh network includes routers and mobile wireless stations. Wireless mesh network architecture is shown in Figure 5.22. Wireless mesh networks combine wired networks and wireless networks with wireless routers as backbones and mobile stations as users.

Wireless routers communicate with each other, form a backbone of the wireless network, connect wired networks, and conduct multihop communications to forward mobile wireless stations' traffic to/from wired networks. Mobile users' traffic travels over wireless routers and reaches wired networks such as the Internet. Each mobile station also acts as a router, forwarding packets for other mobile stations. Wireless routers may have multiple different wireless interfaces to connect heterogeneous wireless networks.

People love the convenience of not being constrained by a wired line. Therefore, we have TV remotes, cordless phones, cellular phones, and so on. In a wireless mesh network, a mobile user can reach the Internet via only a few hops of connections, which is much more reliable and practical than using ad hoc networks. For example, a wireless mesh network can be built on roofs of buildings to provide inexpensive broadband Internet access, while outdoor cellular WiFi cells also form a wireless mesh network. In a wireless mesh network, users can connect to the Internet as long as they are in range of another device that somehow connects to the Internet. A wireless mesh network for vehicles enables people to access traffic information and location-based services.

The IEEE 802 working groups are working on several wireless mesh standards such as IEEE 802.11s, IEEE 802.15.5, IEEE 802.16a, and IEEE 802.20. The IEEE 802.11s extends the IEEE 802.11 standards to enable access points to form backbones of wireless mesh networks. The IEEE P802.15.5 Mesh Network Task Group was formed in November 2003 to define technologies to support mesh networks of IEEE 802.15 WPANs. The IEEE 802.15.5 mesh networks increase WPANs' coverage with shorter-distance radio transmissions and higher throughput, especially for UWB systems. The IEEE 802.16a supports optional mesh networks with a TDMA-based MAC layer, so that all subscriber stations can have direct links with each other. Either the base stations can control resource allocations of these direct links to subscriber stations, or multiple base stations can coordinate subscriber station transmissions within one or two hops. The IEEE 802.20 adopts a cellular architecture with macrocells, microcells, and picocells. IEEE 802.20 addresses high-speed mobility up to 250 km/hour.

Mobile stations in IEEE 802.20 can form a wireless mesh network. Furthermore, an IEEE 802.20 mesh network can connect IEEE 802.16a, IEEE 802.11s, IEEE 802.15.5, 3GPP, and 3GPP2 technologies to form heterogeneous wireless mesh networks. The IEEE 802.20 systems must be designed to provide ubiquitous mobile broadband wireless access in a cellular architecture. The system architecture must be one of the following architectures: point-to-multipoint topology, mesh network topology or a hybrid of both mesh and point-to-multipoint. The requirements for a mesh network are completely different from those for point-to-multipoint networks.

There are many challenges for building a large-scale and high-performance multihop wireless mesh network, such as compatibility, and coexistence, scalability, security, QoS. The first and the most important challenge is compatibility. Designing wireless networks should always take into account compatibility with old/current technologies in terms of radio transmissions, MAC, security, routing, and so on. Coexistence should consider all aspects of technologies, which pose a great challenge, including rapid technologies, MAC layer, routing, security, QoS, smooth handoff, and so forth. This task also paves a road to fourth generation (4G) cellular networks.

The third challenge is scalability. The performance of a wireless mesh network should not be degraded significantly when the number of mobile stations and the network size increase significantly. The current MAC protocols are not scalable, as evidenced by the fact that throughput is significantly decreased as the number of mobile stations increases. The current routing protocols are not scalable, as evidenced by the fact that performance is significantly degraded as the number of hops increases. However, totally new MAC and routing protocol designs should also be limited by the compatibility issue.

The fourth challenge is security. Although security mechanisms and services are designed for different wireless networks, there is a great challenge when these networks coexist. Security is especially difficult when compatibility, efficiency, and QoS issues are also considered at the same time.

The fifth challenge is QoS. QoS becomes a very tough task in a heterogeneous wireless mesh network, and it involves many layers, such as MAC, routing, and application layers, as well as compatibility with different architectures and technologies under different wireless networks.

Furthermore, there are many other issues, such as high-speed radio interfaces, system resource management, range extension, new MAC and routing protocols (constrained by compatibility), cross-layer design and optimization, mobility, power management, energy efficiency protocols, topology management, and applications.

IEEE 802.20-BASED ARCHITECTURE: AN EXAMPLE

Figure 5.23 illustrates an example of the IEEE 802.20-based network architecture. This standard is applied for the wireless network infrastructure in the broadband railway digital network (BRDN) based on the mobile broadband Internet access technology.[68,69]

A WLAN is deployed in a train composition with a typical configuration. The Ethernet (IEEE 802.3) cable is wired through all carriages via a multivehicle bus (MVB). In each carriage, WiFi access points are deployed with complete coverage of the area and are hooked up to the Ethernet. End users access this WLAN via their mobile terminals, such as laptops, PDAs, mobile IP phones, and so on. An IEEE 802.20 client, interfaced to the Ethernet over the router, connects the WLAN to the

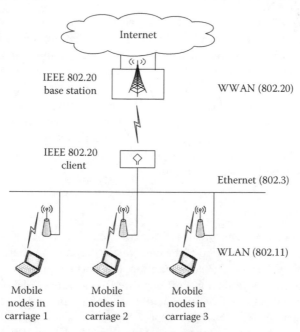

FIGURE 5.23   An example of IEEE 802.20-based network architecture.

base stations of the IEEE 802.20 on the ground, which jointly offer the transparent services on the physical layer and data link layer between the Internet and the mobile nodes. Because of the availability of network address translation (NAT) technology, the whole network may only need a single IP address. A server can automatically assign each mobile node of the network a dynamic IP address. An interconnection scheme of WWAN between the IEEE 802.20 base stations has several choices: IEEE 802.20, IEEE 802.16, the existing cellular communication systems infrastructure, and the cable, because these base stations are stationary.

5.10   H.264 VIDEO TRANSMISSION OVER IEEE 802.11 WIRELESS LAN NETWORKS

One of the driving forces of the next WLAN generation is the promise of high-speed multimedia services. Providing multimedia services to mobiles and land lines through wireless access can be a reality with the development of the following:[70–72]

- Two high-speed PHY layers: IEEE 802.11g (54 Mbps) and IEEE 802.11n (100 Mbps)
- The IEEE 802.11e QoS-based MAC layer

However, wireless channel characteristics such as shadowing, multipath, fading, and interferences still limit the available bandwidth for the deployed applications. Consequently, video compression techniques are a crucial part of multimedia applications over WLAN.

The H.264/AVC video coding standard,[73] proposed by both the Joint Video Team (JVT) of the ITU-T and the MPEG, achieved a significant improvement in compression efficiency over the existing standards. For instance, digital satellite TV quality was reported to be achievable at 1.5 Mbps, compared to the current operation point of MPEG-4 Part 2 video codec at around 3 Mbps. Additionally, the H.264 standard introduces a set of error resilience techniques such as slice structure, data partitioning, and flexible macroblock ordering (FMO). However, these techniques are insufficient because the resource management and protection strategies available in the lower layers (PHY and MAC) are not optimized explicitly considering the specific characteristics of multimedia applications.[74]

H.264 OVERVIEW

H.264 consists of two conceptually different layers. First, the video coding layer (VCL) contains the specification of the core video compression engines that achieve basic functions such as motion compensation, transform coding of coefficients, and entropy coding. This layer is transport unaware, and its highest data structure is the video slice—a collection of coded macroblocks (MBs) in scan order. Second, the network abstraction layer (NAL) is responsible for the encapsulation of the coded slices into transport entities of the network. In this H.264 overview, we particularly focus on the NAL layer features and transport possibilities.

The NAL defines an interface between the video codec itself and the transport world. It operates on NAL units (NALUs) that improve transport abilities over almost all existing networks. An NALU consists of a 1-byte header and a bit string that represents the bits constituting the MBs of a slice. The header byte itself consists of an error flag, a disposable NALU flag, and the NALU type. Finally, the NAL provides a means to transport high-level syntax (i.e., syntax assigned to more than one slice, e.g., to a picture or group of pictures or an entire sequence).

One very fundamental design concept of the H.264 codec resides in its ability to generate self-contained packets, making mechanisms such as header duplication and the MPEG-4 header extension code (HEC) unnecessary. The way this is achieved is to decouple information relevant to more than one slice from the media stream. This higher-layer information should be sent reliably and asynchronously, and also before transmitting video slices. Here, provisions for sending this information in-band are also available for applications that do not have an out-of-band transport channel appropriate for the purpose. The combination of higher-level parameters is called the parameter set concept (PSC). The PSC contains information such as picture size, display window, optional coding modes employed, MB allocation map, and so on. To be able to change picture parameters without necessarily retransmitting PSC updates, the video codec can continuously maintain a list of parameter set combinations to switch on. In this case each slice header would contain a codeword that indicates the PSC to be used.

The H.264 standard includes a number of error resilience techniques. Among these techniques, DP (data partition) is an effective applications-level framing technique that divides the compressed data into separate units of different importance. Generally, all symbols of MBs are coded together in a single bit string that forms a slice. However, DP creates more than one bit string (partition) per slice, and allocates all symbols of a slice into an individual partition with a close semantic relationship. In H.264, three different partition types are used:

- **Partition A**, containing header information such as MB types, quantization parameters, and motion vectors. This information is the most important because, without it, symbols of the other partitions cannot be used.
- **Partition B** (intra partition), carrying intra coded block pattern (CBP) and intra coefficients. The type B partition requires the availability of the type A partition in order to be useful at the decoding level. In contrast to the inter information partition, intra information can stop further drift and hence is more important than the inter partition.
- **Partition C** (inter partition), contains only inter CBPs and inter coefficients. Inter partitions are the least important because their information does not resynchronize the encoder and decoder. Their use requires the availability of the type A partition, but not of the type B partition.

Usually, if the inter or intra partitions (B or C) are missing, the available header information can still be used to improve the efficiency of error concealment. More specifically, due to the availability of the MB types and motion vectors, a comparatively high reproduction quality can be achieved as only texture information is missing.

DISTRIBUTED COORDINATION FUNCTION

The distributed coordination function (DCF) is the basic mechanism for IEEE 802.11. It employs CSMA/CA as the access method. Before initiating a transmission, each station is required to sense the medium. If the medium is busy, the station defers its transmission and initiates a backoff timer. The backoff timer is randomly selected between 0 and contention window (CW). Once the station detects that the medium has been free for a duration of DCF interframe spaces (DIFS), it begins to decrement the backoff counter as long as the channel is idle. If the backoff timer expires and the medium is still free, the station begins to transmit. In case of a collision, indicated by the lack of an acknowledgment, the size of the CW is doubled until it reaches the CW_{max} value. Furthermore, after each successful transmission, the CW is initialized with CW_{max},

$$CW = (CW_{min} \, x2^i)-1 \qquad\qquad (5.2)$$

where i is the number of transmission attempts.

Under DCF, all stations compete for channel access with the same priority. There is no differentiation mechanism to provide better service for real-time and multimedia applications.

The need for a better access mechanism to support service differentiation has led Task Group e of IEEE 802.11 to propose an extension of the actual IEEE 802.11 standard. The 802.11e draft introduces the hybrid coordination function (HCF) that concurrently uses a contention-based mechanism and a pooling-based mechanism, EDCA, and HCF controlled channel access (HCCA), respectively. Like DCF, EDCA is very likely to be the dominant channel access mechanism in WLAN because it features a distributed and easily deployed mechanism. QoS support in EDCA is realized with the introduction of access categories (ACs). Each AC has its own transmission queue and its own set of channel access parameters. Service differentiation between ACs is achieved by setting a different CW_{min}, CW_{max}, arbitrary interframe space (AIFS), and transmission opportunity duration limit (optional). If one AC has a smaller arbitrary interframe space (AIFS) or CW_{min} or CW_{max}, the AC traffic has a better chance of accessing the wireless medium earlier. Generally, AC3 and AC2 are reserved for real-time applications (e.g., voice or video transmission), and the others (AC1, AC0) for best-effort and background traffic.

VIDEO COMMUNICATION OVER WLAN

Error control is one of the most popular application level approaches dealing with packet loss and delay in multimedia communication over bandwidth-limited fading wireless channels.

Two classes of communication protocols are used in practice to reliably communicate data over packet networks: synchronous and asynchronous. Asynchronous communication protocols such as ARQ operate by dividing the data into packets and appending a special error check sequence to each packet for error detection purposes. The receiver decides whether a transmission error occurred by calculating the check sequence. For each intact data packet received in the forward channel, the receiver sends back an acknowledgment. While this model works very well for data communication, it is not suitable for multimedia streams with hard latency constraints. The maximum delay of the ARQ mechanism is unbounded, and in the case of live streaming it is necessary to interpolate late arrival or missing data rather than insert a delay in the stream playback.

In synchronous protocols (i.e., FEC-based protocols), data are transmitted with a bounded delay but generally not in a channel adaptive manner. FEC codes are designed to protect data against channel erasures by introducing parity packets. No feedback channel is required. If the number of erased packets is less than the decoding threshold for the FEC code, the original data can be recovered perfectly. The FEC mechanism represents a lack of efficiency since FEC does not adapt to variable error channel conditions: either a waste of bandwidth may occur when the radio channel is in good condition, or insufficient error protection may exist when it becomes bad.

It should be pointed out that approaches based on these two mechanisms (FEC, ARQ) are implemented and supervised at the application layer, and consequently do not have access to lower layers' transmission parameters.

Cross-layer architecture is an interesting alternative to the above-mentioned mechanisms for robust H.264 video transmission over WLAN. Cross-layer architectures for video transport over wireless networks can be classified into the following categories:[75,76]

- **Top-down approach.** The higher layer optimizes the parameters and the strategies at the next lower layer.
- **Bottom-up approach.** In this architecture, the lower layer isolates the higher layers from losses and bandwidth variations.
- **Application-centric approach.** The application layer optimizes the lower-layer parameters one at a time in either a bottom-up (starting from the PHY layer) or top-down manner, based on its requirements.
- **MAC-centric approach.** In this cross-layer technique, the application layer passes its traffic information and requirements to the MAC, which decides which application layer packets/flows should be transmitted and at what QoS level.
- **Integrated approach.** The strategies to design a cross-layer architecture are determined jointly by all the open system interconnection (OSI) layers.

Transport of H.264 video is expected to be an important component of many wireless multimedia services such as video conferencing, real-time network gaming, and TV broadcasting. However, due to wireless channel characteristics and lack of QoS support, the basic 802.11-based channel access procedure is nearly sufficient to deliver non-real-time traffic. The delivery should be augmented by appropriate mechanisms to better consider different QoS requirements and ultimately adjust the medium access parameters to the video data content characteristics.

PROPOSALS FOR FUTURE WORK ON H.264 STANDARD

The subject matter and activities of the Video Coding Experts Group (VCEG) in July 2006 consisted of work on video coding.[76] The primary purpose was to consider proposals for future work on H.264, namely

- Progression of work on revision of H.264 for scalable video coding
- Considerations and progression of work on revision of H.264 for 4:4:4 video coding
- Consideration and progression of work on revision of H.264 for 3D/multi-view video coding
- Maintenance of H.26x
- Consideration of proposals for supplemental enhancement information for use with H.264
- Consideration of last-call remarks as necessary relating to H.262, H.264, and H.271
- Consideration of future work proposals for revision of H.264, H.264.1, and H.264.2
- Consideration of proposals and organizational work toward eventual development of an H.265
- Collection of nonnormative content to aid in the study and implementation of H.264
- Study and coordination relating to use of video coding in systems
- Coordination and communication with other organizations
- Planning for future work of question Q.5 and Joint Video Team (JVT)

It should be understood that current work is not intended to imply any need for near-term planning to create additional enhancements of H.264 that are not already under way, or to start drafting of an H.265 or H.266, and so on, but is rather to determine whether and when work on such things should begin in earnest. When work is begun on an H.265, computational efficiency should be one serious and concentrated goal of the effort, in addition to coding efficiency. In principle, encoder as well as decoder computational efficiency is worthy of consideration.

In the ITU-R Q.1/16 VCEG meeting in January 2007, the subject matter of the activities consisted of work on video coding as follows:[77,78]

- Consideration of last-call remarks as necessary on color space amendment to H.262 and new profiles amendment to H.264
- Consideration of proposals for new enhancements of Recomm. H.264, H.264.1, H.264.2, H.271, and T.851
- Consideration of proposals and organizational work toward eventual development of an H.265
- Maintenance of H.26x, H.271, and T.8x video and image coding standards
- Collection of nonnormative content to aid in the study and implementation of H.264
- Study and coordination relating to use of video and image coding in systems

- Review, planning, and coordination for work of Q.6 and JVT
- Coordination and communication with other organizations on video and image coding related topics

Most or all contributions to this meeting seem more in the direction of an H.264+ as opposed to an H.265. When beginning work on an H.265, computational efficiency should be one serious and concentrated goal of the effort (obviously, along with coding efficiency and other considerations). In principle, we consider encoder as well as decoder computational efficiency to be worthy of consideration.

5.11 CONCLUDING REMARKS

With implementation of WLAN data communication, services in residential and office environments have undergone significant changes. WLAN products are available from a range of vendors. Depending on the transmission scheme, products may offer bandwidth ranging from about 1 Mbps up to 54 Mbps.

WPANs will play a significant role in the multimedia landscape. They will become integrated with a hybrid multimedia network architecture, initially providing services based on the Bluetooth technology, and later based on the high-speed WPAN technology which satisfies the requirements of the digital consumer electronic market. One of the main advances for multimedia applications in WPANs is UWB communication. UWB has significantly higher data rates than Bluetooth. It also occupies significantly more bandwidth and has the strongest power restrictions which prevent it from interfering with primary band users. This is only suitable for short-range indoor applications. UWB only defines a link layer technology, so it requires a compatible MAC protocol as well as higher layer protocols to become part of a wireless network standard. UWB is likely to become the link layer technology for wireless networks supporting imaging and multimedia applications. The potential strength in the UWB radio technique lies in the use of extremely wide transmission bandwidths, which results in desirable capabilities, including accurate position location and ranging, as well as high multiple access capability.

WMANs based on the IEEE 802.16 technology garnered a lot of interest among vendors and Internet service providers as the development for wireless IP offering a possible solution for the last mile access problem. Its OFDM mode targets frequency bands below 11 GHz.

A new standard, IEEE 802.20, should be developed and optimized for data mobility and broadband wireless networks. The following necessary criteria are important for this standard: broad market potential, compatibility, distinct identity, and technical and economic feasibility. The technical details of the standard include QoS parameters, data rates available to end users, application support, and security. IEEE 802.20 can be successfully applied in a broadband network by adopting IEEE 802.20 and 802.11 technologies to implement the function of physical and data link layers. The physical layer provides reliable data frame transmission between neighboring hosts using bit transmission function packets by the physical layer.

Finally, the H.264 video transmission standard over IEEE 802.11 based wireless networks takes into account a robust cross-layer architecture that leverages an inherent H.264 error resilience tool, that is, data partitioning. Transport of H.264 video is expected to be an important component of many wireless multimedia services such as video conferencing and TV broadcasting.

6 Advances in Wireless Video

This chapter seeks to provide information concerning advances in wireless video. We start with introducing error robustness support using the H.264/AVC coding standard that makes it suitable for wireless video applications. When errors cannot be avoided, we introduce some essential system design principles. Also, error concealment and limitation of error propagation are considered. Next, we move to error resilience video transcoding for wireless communications. Transcoding is used to reduce the rate and change the format of the originally encoded video service to match network conditions and terminal capabilities. We provide an overview of the error resilience tools, including benefits according to category (localization, data partitioning, redundant coding, and concealment-driven techniques). Efficiently utilizing energy is a critical issue in designing wireless video communication systems. This chapter highlights recent advances in joint source coding and optimal energy allocation, including joint source-channel coding and power adaptation. Finally, multipath transport is analyzed, together with general architecture for multipath transport of video streams. Multipath transport provides a new degree of freedom in designing robust multimedia transport systems. It is a promising technique for efficient video communications over ad hoc networks.

6.1 INTRODUCTION

In the last two decades two distinct communication technologies have experienced unequaled rapid growth and commercial success: mobile communications and multimedia communications. The great success in both areas has inspired many electrical engineers and computer scientists. Today, 50 years after Shannon's[1] pioneering contributions to information theory, the communications community has grown in both the quality and quantity of communications services. Researchers continue scanning the horizon for new ideas in order to approach information theoretical performance prediction. Burst-by-burst adaptive turbo-coded, turbo-equalized single-carrier, multicarrier, and code division multiple access (CDMA) schemes constitute just a few examples of the potential techniques that facilitate the approach toward Shannon's predictions over wireless channels.

Against a backdrop of emerging third generation (3G) wireless personal communications standards and broadband access network standard proposals, this chapter is dedicated to a range of topical wireless multimedia and video communications aspects. The transmission of multimedia information over wireline-based links can now be considered a mature area, where a range of interactive and distributive services is offered by various providers across the globe.

The rapid growth in the numbers of mobile phones and Internet users has resulted in spectacular strides being made in wireless communication systems. The increase in data rates provided by 3G wireless systems will facilitate an even wider variety of applications, including real-time multimedia. Users will be able to receive the same services on their wireless systems that they currently receive over wireline networks, including such bandwidth demanding applications as interactive multimedia, video games, and video conferencing. Although the demand for networked and wireless media services involving video is ever-increasing, the high error rates in these channels and stringent delay constraints still remain significant obstacles to enabling these services.

Following the rapid growth of wireless networks and the great success of Internet video, wireless video services are expected to be widely deployed in the near future. Toward this end, both wireless communication and video coding technologies have advanced substantially. On the wireless side, new broadband wireless technologies such as WiFi, UWB, and *WiMAX*, in conjunction with new paradigms such as peer-to-peer communication and 3G/Super3G technologies have evolved to provide a platform viable for many video applications (e.g., video streaming, video sharing, and video home networking). On the video coding side, International Organization for Standardization (ISO), Motion Picture Experts Group (MPEG), and International Telecommunication Union–Telecommunication Standardization Sector (ITU-T) have made excellent progress in advanced video coding benefiting the whole gamut of applications from wireless streaming to high-definition television (HDTV).

The rapid growth of wireless communication and networking protocols will ultimately bring video to our lives anytime, anywhere, and on any device. Until this goal is achieved, wireless video delivery faces numerous challenges, among them highly dynamic network topology, high error rates, limited and unpredictably varying bit rates, and scarcity of battery power. Unsurprisingly, then, the exploration of advanced video coding and delivery techniques in conjunction with wireless network technologies is an active area of investigation.

Media coding over wireless networks is governed by two dominant rules. One is the well-known Moore's law, which states that computing power doubles every 18 months. Moore's law has been at work for codec evolution, and there have been huge advances in technology in the 10 years since the adoption of MPEG-2. The second governing principle is the huge bandwidth gap (one or two orders of magnitude) between wireless and wired networks. This bandwidth gap demands that coding technologies achieve efficient compact representation of media data over wireless networks. Obviously, the most essential requirement for wireless video is coding efficiency.

Unfortunately, Moore's law continues to increase transistor budgets. We all know that clock speed is a key means of increasing CPU performance, but every time the clock speed goes up, so does the power. If media coding power dissipation increases beyond a modest 100 mW (which corresponds to wireless local area network, or WLAN, IEEE 802.11 receiver power consumption), it will be hard to implement the media application in portable devices.

In addition to coding efficiency and power dissipation, error resilience is an important issue, since mobile networks generally cannot guarantee error-free communication during fading periods. A functional block diagram for wireless video communication with respect to error control consists of four layers:

- The layer ½ transport, which correspond to physical and media access control (MAC) layers
- An end-to-end transport layer such as Transmission Control Protocol/Internet Protocol (TCP/IP) or Real-Time Protocol/User Datagram Protocol/IP (RTP/UDP/IP)
- An error-resilience tool and network adaptation layer such as H.264 NAL
- A source coding layer

To provide error robustness, we can take a number of approaches, combining the four layers. The reconstructed video quality depends on redundancy in the data. This is required to protect and recover the transmitted video at the four layers. Intuitively, error control in wireless video transmission is concerned with how and what redundancy is allocated at each layer.

6.2   ERROR ROBUSTNESS SUPPORT USING H.264/AVC

H.264/AVC features can be used exclusively or jointly for error robustness purposes, depending on the application. It is necessary to understand that most codec-level error resilience tools decrease compression efficiency. Therefore, the main goal when transmitting video goes along with the spirit of Shannon's famous separation principle: combine compression efficiency with link layer features that completely avoid losses such that the two aspects, compression and transport, can be completely separated. Nevertheless, if errors cannot be avoided, the following system design principles are essential:

- **Loss correction below the codec layer.** Minimize the amount of losses in the wireless channel without completely sacrificing the video bit rate.
- **Error detection.** If errors are unavoidable, detect and localize erroneous video data.
- **Prioritization methods.** If losses are unavoidable, at least minimize loss rates for very important data (e.g., control).
- **Error recovery and concealment.** In case of losses, minimize the visual impact of losses on the actual distorted image.
- **Encoder-decoder mismatch avoidance.** Limit or completely avoid encoder and decoder mismatches resulting in annoying error propagation.

ERROR CONTROL

Error control such as forward error correction (FEC) and retransmission protocols are the primary tools to provide quality of service (QoS) in mobile systems, especially on the radio access part. QoS methods are essential in good system designs, as minimizing or eliminating transmission errors has many advantages for applications. However, usually the trade-off of reliability versus delay has to be considered.

Nevertheless, to compensate for the shortcomings of non-QoS-controlled networks (e.g., the Internet or some mobile systems), as well as address total blackout periods caused by, say, network buffer overflow or a handover of transmission cells, advanced transport protocols provide features that allow error control to be introduced at the application layer.

For point-to-point services, selective application layer retransmission schemes can be used to retransmit RTP packets. For many applications it can be assumed that at least a low-bit-rate feedback channel from the receiver to the transmitter exists that allows general back-channel messages to be sent. For example, RTP is accompanied by Real-Time Transport Control Protocol (RTCP) providing control and management messages. Media receivers can send receiver reports, including instantaneous and cumulative loss rates, as well as delay and jitter information. RTCP has recently been extended with the extended report packet type, which allows the loss or reception of each RTP packet to be indicated by the receiver to the sender.

ERROR CONCEALMENT

The severity of residual errors can be reduced if error concealment techniques are employed to hide visible distortion as much as possible. If data partitioning and strong error protection for the motion vector are used, one might rely on the transmitted motion vectors for motion-compensated concealment.[2] If motion vectors are lost, they can be reconstructed by appropriate techniques, for example, by spatial interpolation of the motion vector field,[3] which can be enhanced by additionally considering the smoothness of the concealed macroblock along the block boundaries.[4,5] The interested reader is referred to Reference 6 for a comprehensive overview of concealment techniques. All the feedback-based error control approaches discussed in the signal benefit similarly from better concealment. It suffices to select one technique.

Despite error control techniques, error resilience in the video is still necessary whenever the video decoder observes residual losses. These problems mainly occur in conversational applications due to the delay constraints, as well as in multicast/broadcast situations due to the missing feedback link. Slice structured coding typically allows the encoder to choose between two slice coding options, one with a constant number of macroblocks within one slice, but arbitrary size of bytes; and one with the slice size bounded to some maximum S_{max} in bytes, resulting in an arbitrary number of macroblocks per slice. The latter is especially useful to introduce some QoS, as in general the slice size determines loss probability in wireless systems due to the processing.

H.264/AVC decoders should detect losses of slices by keeping a record of which slices of a picture have been received and decoded. Entirely lost reference pictures should be detected based on gaps in the sequence number for reference pictures, or prediction from missing pictures in the reference picture buffer (when a bit stream may include subsequences).

As soon as the erroneous macroblocks are detected, error concealment for all of them should be invoked. For example, in the H.264/AVC test model software two types of error concealment algorithms have been introduced, one exploiting spatial information only, suitable mainly for intraframes, and one exploiting temporal information. It is important to select the appropriate error concealment technique, spatial or temporal, adaptively to obtain reasonably good visual quality.

Using *a priori* knowledge about image/video signals, it is possible to include error concealment capabilities in decoders so that the severity of artifacts resulting from transmission errors is minimized. Error concealment is an extremely important component of any error robust video codec. Examples of spatial and temporal interpolations utilized in error concealment methods include maximally smooth recovery,[9] projection onto convex sets,[10] and various motion vector and coding mode recovery methods such as motion compensated temporal prediction.[11] Like the error resilience tools,[12] the effectiveness of an error concealment strategy is highly dependent on the performance of the resynchronization scheme. If the resynchronization methods can accurately localize the error, then the error concealment problem becomes much more tractable. Simple concealment strategies based on copying blocks from previous frames, instead of displaying corrupted blocks from a current frame, can be very effective.

Error detection and localization are usually achieved by checking if the information decoded is legal given the syntax of the bit stream. When reversible variable length codes (RVLCs) are used, the decoder has the additional capability of error detection by cross-checking of the forward and backward results. In this approach,[13,14] the variable length code words are designed such that they can be read both in the forward as well as the reverse direction. Intelligently designed, RVLCs and corresponding decoding methods can significantly improve the error robustness of the bit stream with little or no loss of coding efficiency.

LIMITATION OF ERROR PROPAGATION

Despite error control and concealment techniques, packet losses still result in imperfect reconstruction of pictures. Thus, the effects of spatiotemporal error propagation resulting from the motion prediction can be severe. Therefore, the decoder has to be provided with other means that allow error propagation to be reduced or completely stopped.

The most common way to accomplish this task is the reduction of temporal prediction in the encoding process by encoding image regions in intramode. The straightforward method of inserting instantaneous decoding refresh (IDR) frames is quite common for broadcast and streaming applications, as these frames are also necessary to randomly access the video lures, which can be undesirable, especially in conversational applications. Therefore, more subtle methods are frequently used to synchronize encoder and decoder reference frames.

In early work it was proposed to introduce ultracoded macroblocks using a constant update pattern, either randomly, or adaptively based on a cost function. The selection of an appropriate update ratio depends on different parameters such as the sequence characteristics, transmission bit rate, and, most important, channel characteristics. Most suitably, the selection of coding modes can be incorporated in the operational encoder control, taking into account the influence of the lossy channel. The encoder control is modified such that the expected decoder distortion is used instead of the encoding distortion. In addition to limiting the error propagation with macroblock intra updates, encoders can also guarantee that macroblock intra updates result in gradual decoding refresh (GDR), that is, entirely correct output pictures after a certain period of time.

The availability of the feedback channel in conversational and unicast streaming applications has led to different standardization and research activities on interactive error control (IEC) in recent years. If online encoding is performed, the slice loss information of the decoder can be directly incorporated in the encoding process to reduce, eliminate, or even completely avoid error propagation. The basics of these methods all come under the term error tracking.

The error tracking approach uses the INTRA mode for some macroblocks to stop interframe error propagation, but limits its use to severely affected image regions only. During error-free transmission, the more effective INTER mode is used, and the system therefore adapts effectively to varying channel conditions. This is accomplished by processing the negative acknowledgments (NACKs) from a feedback channel in the coding control and reconstructing the resulting error distribution in the current frame as described below. The coding control of a forward-adaptive encoder can then effectively stop interframe error propagation by selecting the INTRA mode whenever a macroblock is severely distorted. On the other hand, if error concealment was successful and the error of a certain macroblock is small, the encoder may decide that INTRA coding is not necessary. For severe errors, however, a large number of macroblocks are selected, and the encoder may have to use a coarser quantizer to maintain constant frame and bit rates. In this case, the overall picture quality decreases with a higher frequency of NACKs. Unlike retransmission techniques such as automatic repeat request (ARQ), error tracking does not increase the delay between encoder and decoder. It is therefore particularly suitable for applications that require a short latency.

The syntax of H.264/AVL permits incorporating methods for reduced or limited error propagation in a straightforward manner. Similar to operational encoder control for error-prone channels, the delayed decoder state can also be integrated in modified encoder control.

6.3   ERROR RESILIENCE VIDEO TRANSCODING

Because many video sources are originally coded at a high rate without considering the different channel conditions that may be encountered later, a means to reproduce this content for delivery over a dynamic wireless channel is needed. Transcoding is typically used to reduce the rate and change the format of the originally encoded video source to match network conditions and terminal capabilities. Given the existence of channel errors that can easily corrupt video quality, there is also the need to make the bit stream more resilient to transmission errors.

HISTORY

Many researchers have studied ways to provide error resilience transmission of the bit streams created by both discrete cosign transform (DCT)-based and wavelet-based video compression algorithms. Boyd et al.[15] integrate error detection into arithmetic coding, while Elmasry[16] considers embedding block codes. This task is done by dividing [0,1] vector space, operated on by the arithmetic coder, into more intervals than there are symbols input from the source coder. For a given symbol, bit errors now cause the output to fall into an incorrect interval. Because of the structure chosen for these intervals, a small number of bit errors can be corrected. Wen and Villasenor[13] describe a class of reversible variable length codes (RLVC). The basic idea here is that after a loss of synchronization due to a bit error, the decoder will resynchronize at an appropriate boundary. Instead of losing all pixels between these two points, the variable length code can be decoded in the reverse direction to recover them.

Layered coding is an effective method for wireless video transport. Ayanoglu et al.[18] describe a two-layer compression system for MPEG-2 video transmission in wireless asynchronous transfer mode (ATM) local area networks (LANs). Reed-Solomon channel coding is used, and they find that compared to a one layer system, performance degrades more gracefully. Mathew and Arnold[19] add a drift correction signal to a layered H.261 encoder to improve the quality of the lower layer. Fixed length packetization is also useful, as shown by Cherriman and Hanzo[20] for H.263 transmission. Swann and Kingsbury[21] transcode MPEG-2 compressed video, so that an enhanced version of error resilient entropy coding (EREC) can be used.[22]

Benzler[23] proposes a subband system where the input video is decomposed into four subbands. The base layer is coded similar to the MPEG-2 scalable profile (motion compensated DCT coding). The motion vectors are computed in the base layer and used on the enhancement layer. Gharavi and Ng[24] develop an H.263 compatible subband coder, with three-dimensional (3D) variable length codes and unequal error protection of the subbands. Van Dyck and Ganti[25] use concatenated convolutional and redundant slice (RS) channel coding, along with fixed-length packets (two ATM cells), in a two-layer subband system. The base layer (motion vectors and the low-frequency wavelet coefficients) uses the concatenated code, whereas the enhancement layer uses only a high-rate RS code. Bit errors in the enhancement layer cause only slight blurring of the imagery.

A well-known theorem of Shannon's states that an optimal communications system can be constructed by considering, separately, the source and channel coder designs. Shannon's theorem has been used as a design principle, motivating the concatenation of separately optimized source and channel coders. However, the theorem assumes that the source coder is an optimal one that removes all source redundancy. Moreover, it assumes that, for rates below channel capacity, the channel coder corrects all errors. Such optimal systems can be achieved in general only by allowing limitless encoding/decoding complexity and delay. Practical systems must limit complexity and delay, and thus sacrifice performance. In many cases, it is not reasonable to assume that the conditions for the separation theorem hold even approximately. For example, for several noisy, fading channels, block channel codes (without lengthy interleaving) may fail to reduce errors, and may in fact increase the bit error rate (BER).[26] Moreover, the output of practical source encoders may contain a significant amount of redundancy, especially for the sources with memory such as speech[27] and images/video.[28] In these cases, one can improve performance by considering the source and channel design jointly. Joint source-channel (JSC) coding can take several forms. Some JSC approaches do not assume any knowledge of the channel. However, if such knowledge is known *a priori*, this information can be incorporated within the design to improve performance.

ERROR RESILIENCE TRANSCODER

In a typical video distribution scenario, video content is captured and then immediately compressed and stored on a local network. At this stage, compression efficiency of the video signal is most important as the content is usually encoded with relatively high quality independent of any actual channel characteristics. We note that the heterogeneity of client networks makes it difficult for the encoder to adaptively encode video contents to a wide degree of different channel conditions; this is especially true for wireless clients. Subsequently, for transmission over wireless or highly congested networks, the video bit stream first passes through a network node, such as a mobile switch/base station or proxy server, which performs error resilience transcoding. In addition to satisfying rate constraints of the network and display or computational requirements of a terminal, the bit stream is transcoded so that an appropriate level of error resilience is injected into the bit stream. The optimal solution in the transcoder is one that yields the highest reconstructed video quality at the receiver.[29] Figure 6.1 shows video transmission architecture with error resilience transcoding.[30]

It should be noted that the process of error resilience video transcoding is not achieved by the addition of bits into the input bit stream to make the output bit stream more robust to errors. Such an approach is closer to conventional channel coding approaches in which some overhead channel bits are added to the source payload for protection and possible recovery. Rather, for the video source, a variety of strategies exist that affect the bit stream structure at different levels of the stream (e.g., slice vs. block level). The different techniques are to localize data segments to reduce error propagation, partition the stream so that unequal error protection can be applied, or add redundancy to the stream to enable a more robust means of decoding.

Figure 6.2 illustrates the high-level operation of a typical error resilience transcoder. From the source side, characteristics of the video bit stream are usually extracted to understand the structure of the encoded bit stream and begin building the end-to-end rate-distortion model of the source, while from the network side, characteristics of the channel are obtained. Both the content and channel characteristics,

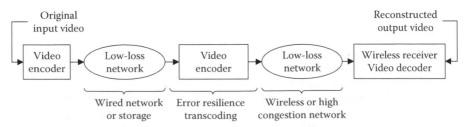

FIGURE 6.1   Video transmission architecture with error resilience transcoding. Reproduced with permission from A. Vetro, Y. Xin, and H. Sun. "Error resilience video transcoding for wireless communications," *IEEE Wireless Commun.* 12 (August 2005): 14–21.

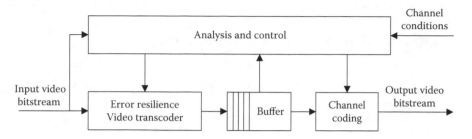

FIGURE 6.2 Error resilience transcoding of video based on analysis of video bit stream, channel measurement, and buffer analysis. Reproduced with permission from A. Vetro, Y. Xin, and H. Sun. "Error resilience video transcoding for wireless communications," *IEEE Wireless Commun.* 12 (August 2005): 14–21.

as well as the current state of the buffer, are used to control the operation of the error resilience transcoder.

It should also be noted that the transcoding of stored video is not necessarily the same as that for live video. For instance, preanalysis may be performed on stored video to gather useful information that may be used during the transcoding process.

While coding efficiency is the most important aspect in the design of any video coding scheme, the transmission of compressed video through noisy channels has always been a key consideration. This is evidenced by the many error resilience tools available in today's video coding standards.

Table 6.1 provides a summary of the tools covered and the key benefits associated with each class of tools. The different strategies, which are not mutually exclusive, that may be employed during coding are as follows:

- **Localization.** Remove the spatial/temporal dependency between segments of the video to reduce error propagation.
- **Data partitioning.** Group coded data according to relative importance to allow for unequal error protection or transport prioritization.
- **Redundant coding.** Code segments of the video signal or syntactic elements of the bit stream with added redundancy to enable robust decoding.
- **Concealment-driven.** enable improved error concealment after decoding using additional information embedded into the coded stream, or uniquely ordering segments of the video signal.

All these strategies for error resilience indirectly lead to an increase in bit rate and loss of coding efficiency, where the overhead with some is more than with others. In the following we describe each tool in terms of the benefits it provides for error-resilient transmission, as well as its impact on coding efficiency.

Localization techniques break the prediction coding loop so that if an error does occur, it is not likely to affect other parts of the video. A high degree of localization leads to lower compression efficiency. There are two methods for localization of errors in a coded video: spatial and temporal localization. Illustration of spatial and

TABLE 6.1

Benefits of Error Resilience Tools According to Category

Category	Benefit	Tools
Localization	Reduces error propagation	Resynchronization marker
Adaptive intra refresh		
Reference picture selection		
Multiple reference pictures		
Data partitioning	Enables unequal error protection and transport prioritization	Frequency coefficients
Motion, header, texture		
Redundant coding	Enables robust decoding	Reversible variable-length coding
Multiple-description coding		
Redundant slice		
Concealment-driven	Enables improved error concealment	Concealment motion vectors
Flexible macroblock order		

Reproduced with permission from A. Vetro, Y. Xin, and H. Sun. "Error resilience video transcoding for wireless communications," IEEE Wireless Commun. 12 (August 2005): 14–21.

temporal localization to minimize error propagation within a frame and over time, respectively, is shown in Figure 6.3.

Spatial localization considers the fact that most video coding schemes make heavy use of variable-length coding to reach high coding performance. In this case, even if one bit is lost or damaged, the entire bit stream may become undecodable due to the loss of synchronization between the decoder and the bit stream. To regain synchronization after a transmission error has been detected, resynchronization markers are added periodically into the bit stream at the boundary of particular macroblocks in a frame. This marker would then be followed by essential header information that is necessary to restart the decoding process. When an error occurs, the data between the synchronization point prior to the error and the first point where synchronization is reestablished is typically discarded. For portions of the image that have been discarded, concealment techniques could be used to recover the pixel data.

For resynchronization markers to be effective in reducing error propagation, all predictions must be contained within the bounds of the marker bits. The restriction on predictions results in lower compression efficiency. In addition, the inserted resynchronization markers and header information are redundant, and lower the coding efficiency. The spatial localization technique is supported in MPEG-2 and H.264/AVC using slices, and in MPEG-4 using video packets.

Although resynchronization marker insertion is suitable to provide spatial localization of errors, the insertion of intracoded macroblocks is used to provide **temporal localization** of errors by decreasing the temporal dependency in the coded video sequence. While this is not a specific tool for error resilience, the technique is widely adopted and recognized as useful for this purpose. A higher percentage of

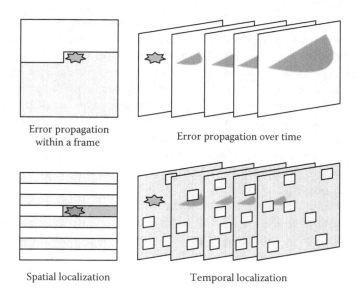

Error propagation within a frame

Error propagation over time

Spatial localization

Temporal localization

FIGURE 6.3 Spatial and temporal localization to minimize error propagation within a frame and over time, respectively. Reproduced with permission from A. Vetro, Y. Xin, and H. Sun. "Error resilience video transcoding for wireless communications," *IEEE Wireless Commun.* 12 (August 2005): 14–21.

intrablocks used for coding the video will reduce the coding efficiency in order to reduce the impact of error propagation on successively coded frames. In the most extreme case, all blocks in every frame are coded as intrablocks. In this case there will be no temporal propagation of errors, but a significant increase in bit rate could be expected. The selection of intracoded blocks may be cyclic, in which these blocks are selected according to a predetermined pattern; the intracoded blocks may also be randomly chosen or adaptively chosen according to content characteristics.

Another form of temporal localization is reference picture selection, which was introduced in the H.263 and MPEG-4 standards for improved error resilience. Assuming a feedback-based system, the encoder receives information about corrupt areas of the picture from the decoder (e.g., at the slice level), and then alters its operation by choosing a noncorrupted reference for prediction, or by applying intracoding to the current data. In a similar spirit, support for multiple reference pictures in H.264/AVC could achieve temporal localization as well. We should note that multiframe prediction is not exclusively an error resilience tool as it can also be used to improve coding efficiency in error-free environments.

Data partitioning techniques have been developed to group coded bits according to their importance to the decoding such that different groups may be more effectively protected using unequal protection techniques. For example, during the bit stream transmission over a single-channel system, more important partitions can be better protected with stronger channel codes than less important partitions. Alternatively, with a multichannel system, more important partitions could be transmitted over the more reliable channel.

In MPEG-2, data partitioning divides the coded bit stream into two parts: a high-priority partition and a low-priority one. The high-priority partition contains picture type, quantization scale, motion vector ranges, and so on, without which the rest of the bit stream is not decodable. It may also include some macroblock header fields and DCT coefficients. The low-priority partition contains everything else.

In MPEG-4, data partitioning is achieved by separating the motion and macroblock header information from the texture information. This approach requires that a second resynchronization marker be inserted between motion and texture information, which may further help localize the error. If the texture information is lost, the motion information may be used to conceal these errors. That is, the texture information is discarded due to errors, while the motion information is used to motion compensate the previously decoded picture.

The **redundant coding** category of techniques tries to enhance error resilience by adding redundancy to the coded video. The redundancy may be added explicitly, such as with the RSs, or implicitly in the coding scheme, as with RVLC and multiple description coding (MDC).

RVLC was developed for the purpose of data recovery at the receiver. By using this tool, variable-length codes are designed so that they can be read in both the forward and reverse directions. This allows the bit stream to be decoded backward from the next synchronization marker until the point of error.

Because this tool is designed to recover from bit errors, it is not a suitable tool for transmission over packet-erasure channels.

MDC encodes a source with multiple bit streams such that a basic-quality reconstruction is achieved if any one of the bit streams is correctly received, while enhanced-quality reconstructions are achieved if more than one of them is correctly received. In MD coding, the redundancy may be controlled by the amount of correlation between descriptions. Generally, MD coded video streams are suitable for delivery over multiple independent channels in which the probability of failure over one or more channels is high.

Redundant slice is a new tool adopted in the H.264/AVC standard that allows for different representations of the same source data to be coded using different encoding parameters.[31] For instance, the primary slice may be coded with fine quantization, and the RS with coarse quantization. If the primary slice is received, the RS is discarded, but if the primary slice is lost, the RS is used to provide a lower level of reinstructed quality. In contrast to MDC, the two slices together do not provide improved reconstruction.

Concealment-driven techniques refer to error resilience coding tools that help with error concealment at the decoder. Although such techniques do add redundancy to the coded bit stream, they are different from the above redundant coding tools in that they aim to improve the concealment of errors after decoding.

Concealment motion vectors are motion vectors that may be carried by intra-macroblocks for the purpose of concealing errors. According to the MPEG-2 standard, concealment motion vectors for a macroblock should be appropriate for use in the macroblock that lies vertically below the macroblock in which the concealment motion vector is carried. In other words, when an error occurs in a given macroblock, the concealment motion vector from the macroblock above it could be used to form prediction from the previous frame.

In the H.264/AVC video coding standard, flexible macroblock ordering (FMO) has been adopted to enable improved error concealment. The idea of FMO is to specify a pattern that allocates the macroblocks in a picture to one or several slice groups, not in normal scanning order, but in a flexible way. Thus, the spatially consecutive macroblocks may be assigned to different slice groups. Each slice group is transmitted separately. If a slice group is lost, the image pixels in spatially neighboring macroblocks that belong to other correctly received slice groups can be used for efficient error concealment. The allowed patterns of FMO range from rectangular patterns to regular scattered patterns, such as checkerboards, or completely random scatter patterns. Further information on the use of this tool in different application and delivery environments may be found in References 8 and 32.

ERROR RESILIENCE TRANSCODING TECHNIQUES

One of the earliest error resilience transcoding schemes, which was based on MPEG-2 video, is referred to as error resilience entropy coding (EREC). In this method, the incoming bit stream is reordered without adding redundancy such that longer VLC blocks fill up the spaces left by shorter blocks in a number of VLC blocks that form a fixed-length EREC frame. Such fixed-length EREC frames of VLC codes are then used as synchronization units, where only one EREC frame, rather than all the codes between two synchronization markers, is dropped should any VLC code in the EREC frame be corrupted due to transmission errors.

Some years later, a rate-distortion framework with analytical models that characterized the error propagation of a corrupted video bit stream subjected to bit errors was proposed.[33] The models were used to guide the use of spatial and temporal localization tools (synchronization markers and intra-refresh) to compute the optimal bit allocation among spatial error resilience, temporal error resilience, and the source rate. One drawback of this method is that it assumes the actual rate-distortion characteristics of the video source are known, which makes the optimization difficult to realize practically. Also, the impact of error concealment is not considered.

The work in Reference 34 proposes an error resilience transcoder for General Packet Radio Service (GPRS) mobile access networks with the transcoding process performed at a video proxy that can be located at the edge of two or more networks. Two error resilience tools, adaptive intra-refresh (AIR) and feedback control signaling (FCS), are used adaptively to reduce error effects while preserving the transmission rate adaptation feature of the video transcoders.

In Reference 35, optimal error resilience insertion is divided into two subproblems: optimal mode selection for macroblocks and optimal resynchronization marker insertion. In Reference 36, a method to recursively compute the expected decoder distortion with pixel-level precision to account for spatial and temporal error propagation in a packet loss environment is proposed for optimal macroblock coding mode selection. Compared to previous distortion calculation methods, this method has been shown to be quite accurate on a macroblock level. In both of these methods, interframe dependency is not considered, and optimization is only conducted on a macroblock basis.

In an effort to exploit MDC for error resilience transcoding, an MD-FEC-based scheme, which uses the $(N, i, N_{-i} + 1)$ Reed-Solomon erasure-correction block code to protect the ith layer of an N-layer scalable video, was proposed in Reference 37. The MD packetization method is specially designed to allow the ith layer to be decodable when i or more descriptions arrive at the decoder.

Considering networking-level mechanisms, the scheme in Reference 38 proposes to implement an ARQ proxy at the base station of a wireless communication system for handling ARQ requests and tracking errors in order to reduce retransmission delays, as well as to enhance the error resilience. The ARQ proxy resends important lost packets (e.g., packets with header information and motion vectors) detected through the retransmission requests from wireless client terminals, while dropping less important packets (e.g., packets carrying DCT coefficients) to satisfy bandwidth constraints. A transcoder is used to compensate for the mismatch error between the front-end video encoder and the client decoders caused by the dropped packets.

6.4 ENERGY-EFFICIENT WIRELESS VIDEO

The allocation of energy affects the quality or level of distortion of the received video sequence, as well as the required delay before the sequence may be displayed. In addition, it affects the level of interference in a multiple-user environment.

The energy needed to send a packet of L bits with transmission power P is given by $E = PL/R$, where R is the transmission rate in source bits per second. These three quantities can be adapted in a variety of ways in an actual system. For example, power adaptation can be implemented by power control at the physical layer. The change of the transmission rate R can be implemented by selecting different modulation modes or channel rates, or allowing a waiting time for each packet before transmission. In addition, it can be implemented by selecting different rate channel codes at the link or physical layers (as in an IEEE 802.11 system).

In wireless systems, one channel parameter that can be specified is the transmission power used to send each packet. For a fixed transmission rate, increasing the transmission power will increase the received signal-to-noise ratio (SNR) and result in a smaller probability of packet loss. This can be modeled by letting the packet loss probability of the kth packet be given by $Pk = f(Pk, \theta k)$, where Pk, is the transmission power used for the packet, and θk represents the available channel state information (e.g., the fading level and average noise power). In many systems the transmitter is able to estimate the channel state (e.g., using a pilot signal or feedback from the receiver). The specific channel state information needed to relate the effect of resource adaptation on the probability of packet loss depends on which channel model is used. The function f could be determined empirically or modeled analytically.

In an energy-efficient wireless video transmission system, transmission power needs to be balanced against delay to achieve the best video quality. For example, for a fixed transmission power, increasing the transmission rate will increase the BER but decrease the transmission delay needed for a given amount of data (or allow more data to be sent within a given time period). Furthermore, the amount of transmission energy required to achieve a certain level of distortion typically decreases with increased delay. For example, in a wireless system, the transmission energy required to maintain a fixed probability of error can be reduced by increasing the transmission time and decreasing the transmission power.[39] This observation is used in Reference 40 to provide energy-efficient packet transmission over wireless links. Therefore, to efficiently utilize resources such as energy and bandwidth, those two adaptation components should be jointly designed.

ENERGY-EFFICIENT VIDEO CODING

To show how the source coding and channel parameters can be jointly selected to achieve energy-efficient video coding, we deal with some techniques like JSC coding and power allocation, JSC coding and power adaptation, and joint source coding and data rate adaptation.

JSC coding and power allocation techniques deal with the varying error sensitivity of video packets by adapting the transmission power per packet based on the source content and the channel state information (CSI). In other words, these techniques use transmission power as part of an unequal error protection (UEP) mechanism. In this case, the channel coding parameter is the power level for each video packet. Video transmission over CDMA networks using a scalable source encoder, along with error control and power allocation, is considered in Reference 41. A scheme for allocating source data and transmission power under bandwidth constraints is considered in Reference 42. In Reference 43 optimal mode and quantization selection are considered jointly with transmission power allocation.

JSC coding and power allocation shows how the source coding and channel parameters can be jointly selected to achieve energy-efficient video coding and transmission.[44] The channel parameters consist of both channel coding and power allocation. Error-resilient source coding is achieved by mode selection and the use of rate-compatible convolution (RCPC) channel codes, with the power assumed to be adjustable in a discrete set of the physical layer.

Source-channel coding and power adaptation can also be used in a hybrid wireless/wireline network, which consists of both wireless and wired links as shown in Figure 6.4.

In this case different channel codes can be used to combat different types of channel errors: packet dropping in the wireline network and bit errors in the wireless link. Reed-Solomon codes are used to perform interpacket protection at the link layer, and

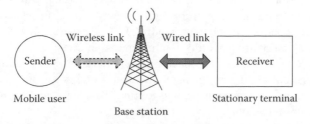

FIGURE 6.4   An example of video transmission over a hybrid wireless/wireline network.

RCPC codes are used to perform intrapacket protection at the physical layer. The selection of channel codes is jointly considered with source coding parameter selection and power adaptation to achieve energy-efficient communication. Cross-layer design is a powerful approach for dealing with different types of channel errors in such a hybrid wireline/wireless network.

JSC coding and data rate adaptation have been studied as a means to provide energy-efficient video communications. To maintain a certain probability of loss, the energy consumption increases as the transmission rate increases. Therefore, to reduce energy consumption, it is advantageous to transmit at the lowest rate possible. In addition to affecting energy consumption, the transmission rate also determines the number of bits that can be transmitted within a given period of time. Thus, as the transmission rate decreases, the distortion from source coding increases. JSC coding and transmission rate adaptation techniques adapt the source coding parameters and the transmission rate to balance energy consumption against end-to-end video quality. In Reference 45, the authors consider optimal source coding and transmission rate adaptation. Stochastic dynamic programming is used to find an optimal source coding and transmission policy based on a *Markov chain* channel model. A key idea is that the performance can be improved by allowing the transmitter to suspend or slow transmissions during periods of poor channel conditions, as long as the delay constraints are not violated.

6.5   MULTIPATH VIDEO TRANSPORT

In video communications a receiver usually displays the received video continuously. Such continuous display requires timely delivery of video data, which further translates to stringent QoS requirements (e.g., delay, bandwidth, and loss) on the underlying network. For the successful reconstruction of received video, the path used for the video session should be stable for most of the video session period. Furthermore, packet losses due to transmission errors and overdue delivery caused by congestion should be kept low such that they can be handled by error control and error concealment techniques. However, this situation does not hold true in ad hoc networks. These are wireless mobile networks without an infrastructure within which mobile nodes cooperate with each other to find routes and relay packets. There is a demonstrable need for providing video service for users of ad hoc networks, such as first responders and military units. In ad hoc networks, wireless links are frequently broken and new ones reestablished due to mobility. Furthermore, a wireless link has a high transmission error rate because of shadowing, fading, path loss, and interference from other transmitting users. Consequently, for efficient video transport, traditional error control techniques, including FEC and ARQ, should be adapted to take into consideration frequent link failures and high transmission errors. In addition, one should take a holistic approach in video transport system design, by jointly considering and optimizing mechanisms in various layers, including video coding, error control, transport mechanisms, and routing. This approach is often referred to as cross-layer optimization.

Among various mechanisms, multipath transport, by which multiple paths are used to transfer data for an end-to-end session, is highly suitable for ad hoc networks, where

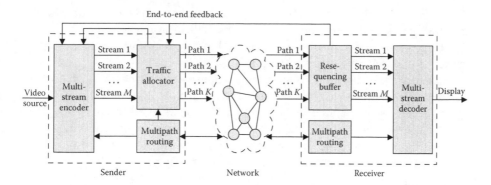

FIGURE 6.5 The general architecture for the multipath transport of video streams in the case of point-to-point communications. Reproduced with permission from S. Mao et al. "Multipath video transport over ad hoc networks," *IEEE Wireless Commun.* 12 (August 2005): 42–49.

a mesh topology implies the existence of multiple paths for any pair of source and destination nodes. Multipath transport has been applied in various settings for data.[46]

There has been considerable research on using multipath transport for real-time multimedia applications.[47–50] For example, multipath transport has been combined with MDC,[51,52] ARQ,[53] and FEC[54] for video transport. It has been shown that, when combined with appropriate source and/or channel coding and error control schemes, multipath transport can significantly improve the media quality over traditional schemes based on shortest path routing. This also inspired previous and ongoing standardization efforts for multipath transport protocols in the Internet Engineering Task Force (IETF).[55]

GENERAL ARCHITECTURE

The general architecture for multipath transport of video streams is shown in Figure 6.5. At the sender, raw video is first compressed by a video encoder into M streams. When $M > 1$, we call the coder a multistream coder. Then the streams are partitioned and assigned to K paths by a traffic allocator. These paths are maintained by a multipath routing protocol. When the streams arrive at the receiver, they are first put into a resequencing buffer to restore the original order. Finally, the video data is extracted from the resequencing buffer to be decoded and displayed. The video decoder is expected to perform appropriate error concealment if any part of a substream is lost.

In general, the quality of the paths may change over time. We assume that the system receives feedback about network QoS parameters. Although not necessary, such feedback can be used to adapt the coder and transport mechanisms to network conditions (e.g., the encoder could perform rate control based on feedback information in order to avoid congestion in the network). In practice, it is desirable for the sender to use a predesigned multistream coder that always produces a fixed number of streams (say, two to four). On the other hand, the number of available paths, as well as their bandwidths, may vary over time due to network topology changes and congestion.

Therefore, it is likely that $M \neq K$, and the traffic allocator will be responsible for distributing the video packets from the M streams to the K available paths.[56]

The point-to-point architecture can be used for two-way conversational services as well as one-way streaming services.

An important advantage of using multipath transport is the inherent path diversity (i.e., the independence of loss processes of the paths). As a result, the receiver can always receive some data during any period, except when all the paths are down simultaneously, which occurs much more rarely than single path failures. One may jointly design the source encoder, multipath routing algorithm, and traffic allocator to explore path diversity to optimize overall system performance. In addition, multipath transport provides a larger aggregate rate for a video session, thus reducing the distortion caused by lossy video coders. Finally, multipath transport distributes traffic load in the network more evenly, resulting in low congestion and delay in the network.

These advantages come at the cost of higher coding redundancy, higher computational complexity, and higher control traffic overhead in the network. In general, using more streams and paths will increase the robustness to packet losses and path failures, and reduce network congestion due to better load balancing. However, more streams may increase the video bit rate for the same video quality, as well as incur higher computation overhead and delay during traffic partitioning and resequencing. Maintaining multiple paths in a dynamic ad hoc network environment involves higher control traffic overhead and more complex routing/path selection algorithms.

For multipath transport to be helpful for sending compressed video, one must carefully design the video coder to generate streams so that the loss in one stream does not adversely affect the decoding of other streams. However, this relative independence between the streams should not be obtained at great expense in coding efficiency. Therefore, a multistream encoder should strive to achieve a good trade-off between coding efficiency and error resilience. In addition one must consider what is feasible in terms of transport layer error control when designing the source coder.

6.6   CONCLUDING REMARKS

Error resilience transcoding is a key technology that enables robust streaming of stored video content over noisy channels. It is particularly useful when content has been produced independent of the transmission network and/or under dynamically changing network conditions. This chapter has reviewed a number of error resilience coding tools, most of which can be found in the latest video coding standards. Additionally, we have outlined a number of error resilience transcoding techniques.

A key factor determining the effectiveness of a mobile device for wireless video transmission is its energy management strategy. This chapter offers an overview of energy-efficient system design for video transmission over an uplink wireless channel. The goal is to achieve the best video delivery quality with the minimum energy consumption.

Jointly adapting components from different communication layers requires more effective communications between those layers. In the traditional layered protocol stack, each layer is independently optimized or adapted to the changing network conditions. The adaptation, however, is very limited due to the limited interactions between layers. Therefore, more efficient adaptation requires cross-layer design not only from the video applications side, but also from the network protocols side.

Although there has been a fair amount of work in the area of error resilience transcoding, there seem to be several promising directions for further exploration that aim to maximize reconstructed video quality. For one, modeling error propagation for the most recent and emerging video coding formats poses several interesting challenges. While H.264/AVC offers superior compression efficiency compared to previous video coding standards, its prediction model is much more complex. To achieve optimal or near-optimal bit allocation in practical implementations, accurate low-complexity models that characterize the performance in error-prone transmission environments for these formats are required. It is also worth noting that many of the new coding tools adopted for H.264/AVC have yet to be explored in the context of error resilience transcoding. Finally, combining channel coding techniques and unique aspects of different networking environments with error resilience transcoding of the source remains an open research problem.

As for multipath video transport, we describe a framework for transport over wireless ad hoc networks, and examine its essential components, including multistream video coding, multipath routing, and transport mechanisms. Multipath transport combined with appropriate video coding techniques can lead to substantial gain over a system using a single path.

7 Cross-Layer Wireless Multimedia

The layered architecture is a reasonable way to operate wireless networks. In an attempt to improve the performance of wireless networks, there has been increased interest in protocols that rely on interactions among different layers. Studies on the cross-layer design for efficient multimedia delivery with quality of service (QoS) assurance over wireless networks utilize the differentiated service architecture to convey multimedia data. This chapter is organized as follows. First, we deal with cross-layer design and describe a modeling system. Next, we describe cross-layer architecture for video delivery over wireless channels. We continue with a cross-layer optimization strategy. Information is exchanged between different layers, while end-to-end performance is optimized by adapting to this information at each protocol layer. After that, a short overview of cross-layer design approaches for resource allocation third generation (3G) code-division multiple access (CDMA) networks is provided. Then, we move to the problem of cross-layer resource allocation for integrated voice/data traffic in wireless cellular networks. The analysis shows the effectiveness of the cross-layer approach. This, chapter concludes with the resource allocation and scheduling for orthogonal frequency division multiplexing (OFDM)-based broadband wireless networks.

7.1 INTRODUCTION

Research on cross-layer design and engineering is interdisciplinary in nature, involving several areas such as signal processing, adaptive coding and modulation, channel modeling, traffic modeling, queuing theory, network protocol design, and optimization techniques.

If one looks at the literature in the area of cross-layer design, several observations can be made. First, there are several interpretations of cross-layer design. This is probably because the cross-layer design effort has been made independently by researchers from different backgrounds who work on different layers of the stack. Second, although there are many cross-layer design proposals in the literature, including those that build on top of other cross-layer design proposals, some more fundamental issues (coexistence of different cross-layer design proposals, when cross-layer design proposals should be invoked, what roles the layers should play, etc.) are not addressed directly. Third, the synergy between the performance viewpoint and implementation concerns is weak; most proposals focus and elaborate on the performance gains from cross-layer design, although there are some ideas on how cross-layer interactions may be implemented. Finally, the wireless medium allows richer modalities of communication than wired networks. For example, nodes can make use of the inherent broadcast nature of the wireless medium and cooperate with each other. Employing modalities like node cooperation in protocol design also calls for cross-layer design.

Cross-layer design breaks away from traditional network design in which each layer of the protocol stack operates *independently*. In an effort to improve the performance of wireless networks there has been increased interest in protocols that rely on interactions between different layers. We discuss key parameters used in cross-layer information exchange, along with the associated cross-layer adaptation and optimization strategies. An in-depth understanding and comparative evolution of these strategies is necessary to effectively access and enable the possible trade-offs in multimedia quality power, consumption, implementation complexity, and spectrum utilization that are provided by various layers. This opens the question of cross-layer optimization and its effectiveness.[3]

Due to their flexible and low cost infrastructure, wireless local area networks (WLANs) are balanced to enable a variety of delay-sensitive multiple applications. However, existing wireless networks provide only limited time-varying QoS for delay-sensitive, bandwidth intense, and loss-tolerant multimedia applications. Fortunately, video transport can cope with a certain amount of packet losses depending on the sequence characteristics and error concealment strategies. Consequently video transmission does not require complete insulation from packet losses, but rather that the application layer cooperate with the lower layers to select the optimal wireless transmission strategies to maximize multimedia performance. This chapter focuses on statistical QoS cross-layer optimization and interaction between layers.

Wireless networks typically have time-varying and nonstationary lines due to the following factors: fading effects coming from path loss, roaming between heterogeneous mobile networks, and the variations in mobile speed, average received power, and surrounding environments. Consequently, the quality of the wireless link varies. This can be measured by the variation of the signal-to-noise ratio (SNR) or the bit error rate (BER). These variations result in time-varying available transmission bandwidth at the link layer, which also leads to time varying delay of arrival of video packets at the application layer, especially when retransmission is employed at the link layer. Since the buffer size at the link layer is typically limited, the time-varying channel service rate can induce buffer overflow, and therefore video packet loss, due to the bit rate mismatch between the transmitting video packet and the channel service rate. At the application layer, due to variation in arrival time of video packets, some packets may become useless during playback if their arrival times exceed a certain threshold.

Existing wireless networks provide only limited, time-varying QoS for delay-sensitive, bandwidth-intense, and loss-tolerant multimedia applications. Fortunately, multimedia applications can cope with a certain amount of packet loss depending on the sequence characteristics and error concealment strategies available at the receiver. Consequently, unlike file transfers, real-time multimedia applications do not require complete insulation from packet losses, but rather that the application layer cooperate with the lower layers to select the optimal wireless transmission strategies that maximize multimedia performance.[4]

For video streaming, high bandwidth requirements are coupled with tight delay constraints, as packets need to be delivered in a timely fashion to guarantee continuous media play out. When packets are lost or arrive late, the picture quality suffers, as decoding errors tend to propagate to subsequent partitions of the video. Due to the high bit rate requirements of video, a media stream may significantly congest the network. Hence, it is imperative to account for the potential impact of each video user on the network statistics and guarantee that the network is not operating beyond the capacity. While protocol layering is an important abstraction that reduces network design complexity, it is not well suited to wireless networks since the nature of the wireless medium makes it difficult to decouple the layers.

A cross-layer approach to networking enhances the performance of a system by *jointly designing* multiple protocol layers. This allows upper layers to better adapt their strategies to varying link and network conditions. These concepts are useful for supporting delay-constrained applications such as video. In such a structure, each layer is characterized by some *key parameters* which are passed to the adjacent layers to help them determine the operation modes that will best send the current channel, network, and application conditions. In such a design, each layer is not oblivious of the other layers, but interacts with them to find its optimal operational point. The difficulty in this cross-layer approach is characterized by parameters representing the channel capacity, such as signal-to-interference plus-noise ratio (SINR), link layer state information such as bit error rate (BER), or supported data rates. Similarly, the network and medium access control (MAC) layers must exchange the requested traffic rates and supportable link capacity.[5-7]

Wireless networks are poised to enable a variety of existing and emerging multimedia streaming applications. As the use of WLANs spreads beyond simple data transfer to bandwidth-intense, delay-sensitive, and loss-tolerant multimedia applications, addressing quality of service issues will become extremely important. Currently, a multitude of protection and adaptation strategies exists in the different layers of the open systems interconnection (OSI) stack. Hence, an in-depth understanding and comparative evaluation of these strategies is necessary to effectively assess and enable the possible trade-offs in multimedia quality, power consumption, implementation complexity, and spectrum utilization that are provided by the various OSI layers. This further opens the question of cross-layer optimization and its effectiveness in providing an improved solution with respect to the above trade-offs.

Due to their flexible and low-cost infrastructure, WLANs are poised to enable a variety of delay-sensitive multimedia applications (videoconferencing, emergency services, surveillance, telemedicine, remote teaching and training, augmented reality, and entertainment).

Fortunately, multimedia applications can cope with a certain amount of packet loss depending on the sequence characteristics and error concealment strategies available at the receiver. Consequently, unlike file transfers, real-time multimedia applications do not require complete insulation from packet losses, but rather that the application layer cooperate with the lower layers to select the optimal wireless transmission strategy that maximizes multimedia performance.

7.2   CROSS-LAYER DESIGN

A block diagram of a cross-layer design with information exchange between the different layers is shown in Figure 7.1. At the link layer, adaptive technologies are used to maximize the link rates under varying channel conditions. Each point of this region indicates a possible assignment of the different link capacities. Based on link state information, the MAC selects one part of the capacity region by assigning time slots, codes, or frequency bands to each of the links. The MAC layer operates jointly with the network layer to determine the set of network flows that will minimize congestion. To find a jointly optimal solution for capacity assignment and network flows, successive suboptimal solutions are exchanged iteratively between these two middle layers. At the transport layer, congestion-distortion optimized scheduling is performed to control the transmission and retransmission of video packets. The application layer determines the most efficient coding rate.

MODELING SYSTEM

Consider a wireless network which must provide high rate and high QoS communications and control information as detailed in Figure 7.2. The branch units (users) communicate to the headquarters through a gateway. Multiple users are connected to the gateway over wireless channels, using time-division multiplexing/time-division multiple-access (TDM/TDMA). We focus on the downlink, although this can be extended to the uplink as well.

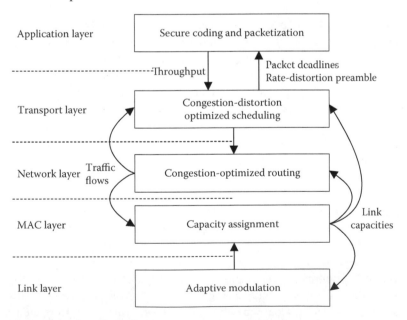

FIGURE 7.1   Cross-layer design for real-time video streaming. Reproduced with permission from E. Salton et al. "Cross-layer design for ad hoc networks for real-time video streaming," *IEEE Wireless Commun.* 12 (August 2005): 59–65.

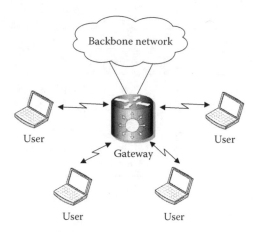

FIGURE 7.2   Cross-layer modeling system diagram.

The wireless link from the gateway to each user is shown in Figure 7.3. A finite-length buffer is implemented at the gateway for each user and operates in a first-in/first-out (FIFO) mode. The adaptive modulation and coding (AMC) controller follows the queue at the gateway (transmitter) and the AMC selector is implemented at each user (receiver). At the data link layer, the processing unit is a packet comprising multiple information bits.[8] We can assume that the queue has a finite length (capacity of K packets per user). The customers of the queue are packets, served by the AMC module at the physical layer.

At the same physical layer, the processing unit is a frame consisting of multiple transmitted symbols. We assume that multiple transmission modes are available to each user, with each mode representing a pair of a specific modulation format and FEC code, as in the HIPERLAN/2 and the IEEE 802.11a standards. Based on channel estimates obtained at the receiver, the AMC selector determines the modulation-coding pair (mode), which is sent back to the transmitter through a feedback channel for the AMC controller to update the transmission mode. Coherent demodulation

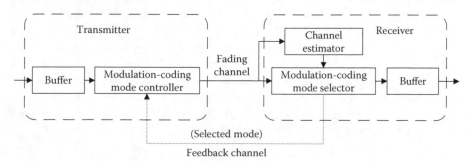

FIGURE 7.3   Wireless link from the gateway to each user. Reproduced with permission from Q. Liu, Sh. Zhak, and G. B. Giannakis. "Cross-layer scheduling with perceived QoS guarantees in adaptive wireless networks," *IEEE J. Selected Areas Commun.* 23 (Mary 2005): 1056–66.

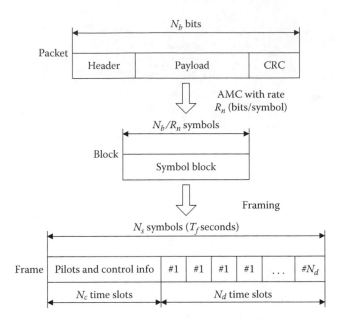

FIGURE 7.4 Processing units at each layer. Reproduced with permission from Q. Liu, Sh. Zhak, and G. B. Giannakis. "Cross-layer scheduling with perceived QoS guarantees in adaptive wireless networks," *IEEE J. Selected Areas Commun.* 23 (Mary 2005): 1056–66.

and maximum-likelihood (ML) decoding are employed at the receiver. The decoded bit streams are mapped to packets, which are pushed upward to the data link layer.

Now, we describe the packet and frame structures, with processing units at each layer as shown in Figure 7.4. At the data link layer, each packet contains a fixed number of bits N_b, which include packet header, payload, and cyclic redundancy check (CRC) bits. After modulation and coding with mode n and rate R_n at the gateway, each packet is mapped to a symbol-block containing N_b/R_n symbols. After the physical layer, the data are transmitted frame by frame through the wireless link, where each frame contains a fixed number of symbols (N_s). Given a fixed symbol rate, the frame duration (T_f) is constant, and represents the time-unit. With TDM, each frame is divided into time $N_c + N_d$ slots. As a consequence, each time-slot contains a fixed number of N_b/R_1 symbols. Each time slot can transmit exactly $R_n/R_1 = 1$ packets with transmission mode n. One time slot can accommodate R_n/R_1 packets with mode $n = 1$, $R_2/R_1 = 2$ packets with mode $n = 2$, and so on. The N_c time slots contain control information and pilots. The N_d time slots convey data, which is scheduled for different users dynamically with TDMA. Each user is allocated a certain number of time slots during each frame.

ARCHITECTURE FOR VIDEO DELIVERY OVER WIRELESS CHANNEL

The cross-layer architecture for video delivery over wireless networks is shown in Figure 7.5. This is an end-to-end delivery system for a video source which includes a source video encoding module, cross-layer mapping and adaptation module, link

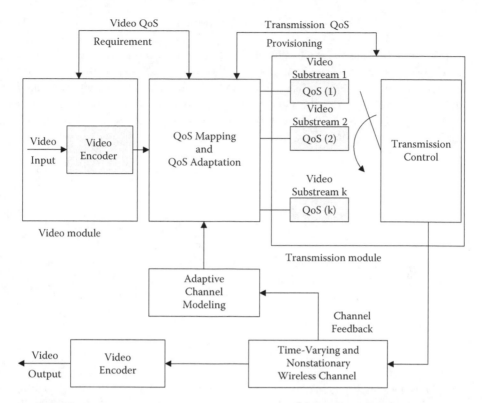

FIGURE 7.5 Cross-layer architecture for video delivery over wireless channel. Reproduced with permission from W. Kumviaisek et al. "A cross-layer quality of service mapping architecture for video delivery in wireless networks," *IEEE J. Selected Areas Commun.* 21 (December 2003): 1685–98.

layer packet transmission module, wireless channel (time varying and nonstationary), adaptive wireless channel modeling module, and video decoder/output at the receiver. Because the main challenge here is the time-varying and nonstationary behavior of the wireless link, we describe its modeling first.

The wireless channel is modeled at the link layer instead of at the physical layer because link layer modeling is more amenable to analysis (delay bound or packet loss rate). The wireless link is expected to be fading, time-varying, and nonstationary. This provides time-varying available transmission bandwidth for video service. Although the wireless channel is expected to be time-varying and nonstationary, it can be assumed that within each small time interval, say, g, the channel rate is stationary and time-varying. Furthermore, within each small time interval g, it can be assumed that the service rate for the time-varying wireless channel can be modeled by a first-order L-state Markov model.[9]

Denote $X_c(n)$ as the state of the channel at time n and $X_c(n) \in \{1,2,\ldots,L\}$. Each state $X_c(n) = i$ corresponds to a channel link condition, which can be characterized by an achievable channel transmission rate of r_i. The achievable channel transmission rate at state i is

$$r_i = R \log_2(1 + \gamma_i) \ [\text{bits/s}] \tag{7.1}$$

Here, R is the transmission bandwidth in hertz, while γ_i represents the SNR value of the wireless channel condition at state i. For the L-state discrete-time Markov chain, denote p_{ij} as the state transition probability from state i (at time $n-1$) to state j (at time n) with a transition time interval of 1 time unit and $1 < g$. That is, $p_{ij} = P\{X_c(n) = j/X_c(n-1) = i\}$. Thus, the L-state Markov chain can be completely characterized by the $L \times L$ state transition matrix, i.e.,

$$P_{transition} = \begin{matrix} p_{11} & p_{12} & \cdots & p_{1L} \\ \cdot & \cdot & \cdot & \cdot \\ p_{L1} & p_{L2} & \cdots & p_{LL} \end{matrix} \tag{7.2}$$

Using the state transition matrix, we can calculate the state probability for the Markov model within the time interval g,[19] which we denote as $p_1, p_2, ..., p_L$. The expected link layer transmission rate r_{channel}, during the time interval g is

$$r_{channel} = \sum_{i=1}^{L} r_i \, p_i \tag{7.3}$$

where r_i is the achievable link layer transmission rate, previously defined in Equation 7.1. At the end of each time interval g, the state transition matrix in Equation 7.2 will be updated by the adaptive channel modeling module to reflect the nonstationary nature of the wireless environment.

In the link-layer transmission control module, we employ a class-based buffering and scheduling mechanism to achieve differentiated services. We maintain K quality of service priority classes with each class of traffic maintained in separate buffers. A priority scheduling policy is employed to serve packets among the classes. That is, packets in a higher priority queue will always be sent first. On the other hand, packets in the lower priority queue will be sent only if there are no packets in the priority queues. Also, packets within the same class queue are served in a FIFO manner. A packet that experiences excess queuing delay (i.e., will miss its scheduled playback time) will be discarded without being sent over the wireless channel. In that way, based on class-based buffering and strict priority scheduling mechanisms, each QoS priority class will have some sort of statistical QoS guarantees in terms of probability of packet loss and packet delay. Statistical QoS guarantees of multiple priority classes can be translated into rate constraints. The calculated rate constraints will specify the maximum data rate that can be transmitted reliably with a statistical QoS guarantee over the time-varying wireless channel. This will classify video substreams into classes and allocate transmission bandwidth for each class. Figure 7.6 shows a queuing system for a time-varying service rate and channel service rate. The accumulated amount of data generated by the source from time 0 to time t is a random variable of the form

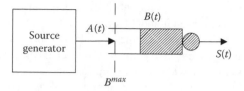

FIGURE 7.6 Queuing system for time-varying source rate and channel service rate.

$$A(t) = \int_0^t \alpha(u)\, du \qquad (7.4)$$

where $\alpha(u)$ is the source data generation rate.

The amount of data $A(t)$ will be stored in the buffer of size $B_{\max}$ awaiting transmission. The accumulated channel service from 0 to t is of the form

$$S(t) = \int_0^t \alpha^{c(u)}(u)\, du \qquad (7.5)$$

where $\alpha^{c(u)}$ is the channel service rate at time u. Note that the time-varying channel service rate has been modeled by an L-state discrete-time Markov chain, where $\alpha^{c(u)} \in (r_1, r_2, \ldots, r_L)$. The stochastic behavior of the accumulated channel service $S(t)$ can be described by the concept of effective capacity which can be presented in the form

$$(\Theta) = -\frac{\Lambda^{(c)}(\Theta)}{\Theta} \qquad (7.6)$$

Here, $\Lambda^{(c)}(\Theta)$ is the asymptotic log-moment generating function $S(t)$, defined as

$$\Lambda^{(c)}(\Theta) = \lim_{t \to \infty} \frac{\log E\ e^{-\Theta S(\Theta)}}{t}$$

where Θ is called the QoS exponent corresponding to the effective capacity (Θ).[20] The parameter Θ is related to the statistical QoS guarantee (e.g., packet loss probability) of the time-varying channel. The statistical QoS guarantee in terms of packet loss probability can be derived as a function of Θ. Namely,

$$P\{B(t) > B_{\max} | \Theta\} \approx \xi e^{-\Theta B_{\max}} \qquad (7.7)$$

where $B(t)$ is the buffer occupancy at time t, $B_{\max}$ is the maximum buffer size, ξ is the probability that the buffer is not empty, and $\xi e^{-\Theta B_{\max}}$ is the approximate packet loss probability guarantee. Intuitively, this says that the effective capacity in Equation 7.6

imposes a limit on the maximum amount of data that can be transmitted over the time-varying channel with the statistical QoS guarantee in Equation 7.7.

The adaptive wireless channel modeling module and link-layer transmission module are application independent. They are installed at the wireless end system as a common platform to support a wide range of applications, not limited to video delivery. The advantages of such a design are universal applicability, modularity, and economy of scale.

The mapping and adaptation module is application specific. It is designed to optimally match the video application layer and the underlying link-layer. Since the QoS measure at the video application layer (distortion and uninterrupted video service perceived by the end user) is not directly related to the QoS measure in the link layer (packet loss/delay probability), a mapping and adaptation mechanism must be in place to maximize application layer QoS with the time-varying available link layer transmission bandwidth. At the video application layer, each video packet is characterized based on its loss and delay properties, which contribute to the end-to-end video quality and service. These video packets are then classified and optimally mapped to the classes of a link transmission module under the rate constraint. The video application layer and link layer are allowed to interact with each other and adapt along with the wireless channel condition. The objective of these interactions and adaptations is to find a satisfactory QoS trade-off so that each end user's video service can be supported with available transmission resources.

7.3   CROSS-LAYER OPTIMIZATION

In recent years the research focus has been to adapt existing algorithms and protocols for multimedia compression and transmission to the rapidly varying and often scarce resources of wireless networks.[21] However, these solutions often do not provide adequate support for multimedia applications in crowded wireless networks, when interference is high, or stations are mobile. This is because the resource management, adaptation, and protection strategies available in the lower layers of the stack—the physical (PHY), MAC, and network/transport layers—are optimized without explicitly considering the specific characteristics of multimedia applications. Conversely, multimedia compression and streaming algorithms do not consider the mechanisms provided by the lower layers for error protection, scheduling, resource management, and so on. This *layered optimization* leads to a simple independent implementation, but results in suboptimal multimedia performance (objective and/or perceptual quality). Alternatively, under adverse conditions, wireless stations need to optimally adapt their multimedia compression and transmission strategies jointly across the protocols stack to guarantee a predetermined quality at the receiver.[11]

A layered architecture is a good candidate for a baseline design. Several optimization opportunities do present themselves through increased interaction across layers. Cross-layer design proposals explore a much richer *interaction between parameters across layers*.

In evaluating these proposals, the trade-off between performance and architecture needs to be fundamentally considered. As noted above, the *performance metrics* of the two are different. The former is shorter term, the latter longer term. Thus, a particular cross-layer suggestion may yield an improvement in throughput or delay performance. To be weighed against this are longer-term considerations.

The layered architecture and controlled interaction enable designers of protocols at a particular layer to work without worrying about the rest of the stack. Once the layers are broken through cross-layer interactions, this luxury is no longer available to the designer. The interaction can affect not only the layers concerned, but also other parts of the system. In some cases, the implementation itself may introduce dependencies that are not really essential to providing the functionality. It is important to consider the effect of the particular interaction on a remote, seemingly unrelated part of the stack. There could be disastrous unintended consequences on overall performance.

CROSS-LAYER WIRELESS TRANSMISSION

Numerous solutions have been posed for efficient multimedia streaming over wireless networks. Potential solutions for robust wireless multimedia transmission over error-prone networks include application-layer packetization, (rate-distortion optimized) scheduling, joint source-channel coding, error resilience, and error concealment mechanisms.[12-14]

Transport issues for wireless multimedia transmission have been examined in References 15 through 17. At the PHY and MAC layers, significant gains have been reported by adopting cross-layer optimization, such as link adaptation, channel aware scheduling, and optimal power control. However, these contributions are aimed at improving throughput or reducing power consumption without taking into consideration multimedia content and traffic characteristics. Explicit consideration of multimedia characteristics and requirements can further enhance the important advances achieved in cross-layer design at the lower layers. Possible solutions and architectures for cross-layer optimized multimedia transmission have been proposed in References 5 and 11.

OPTIMIZATION STRATEGIES

In wireless networks, multimedia streaming uses Real-Time Transport Protocol (RTP) and User Datagram Protocol (UDP). In that way, the transport layer is less important for error protection and bandwidth adaptation. This can easily be extended to include other layers. Consider PHY, MAC, and application (APP) layers. Let N_p, N_M, and N_A denote the number of adaptation and protection strategies available at the PHY, MAC, and APP layers, respectively. For example, the strategies PHY_i, , $i \in \{1, 2, ..., N_p\}$, may represent the various modulation and channel coding schemes existing for a particular WLAN standard. The strategies MAC_i, $i \in \{1,2,..., N_M\}$ correspond to different packetization, automatic repeat request (ARR), scheduling, admission control, and forward error correction (FEC) mechanisms. The strategies APP_i, $i \in \{1,2,..., N_A\}$, may include adaptation of video compression parameters packetization, traffic shaping, traffic prioritization, scheduling, ARQ and FEC mechanisms. We define the joint cross-layer strategies as

$$S = (PHY_1 \ldots, PHY_{Np}, MAC_1, \ldots, MAC_{NM}, \ldots) \qquad (7.8)$$

From Equation 7.8, it can be seen that there are $N = N_P \times N_M \times N_A$ possible joint design strategies. The cross-layer optimization problem seeks to find the optimal composite strategy represented by

$$S^{opt}(x) = \arg_S \max Q\ S(x) \qquad (7.9)$$

This strategy results in the best (perceived/objective) multimedia quality Q subject to the following wireless station constraints:

$$Delay\ S(x) \le D_{max} \qquad (7.10)$$

$$Power\ S(x) \le Power_{max} \qquad (7.11)$$

as well as overall system constraints, such as fairness strategies and bandwidth allocation. Given the instantaneous channel condition $x = $ (SNR, contention), maximum tolerable delay D_{max}, and maximum power Power$_{max}$, we need to solve Equation 7.9 subject to the wireless station and system constraints.

The conceptual block scheme of the proposed cross-layer optimization is given in Figure 7.7. Deriving analytical expressions for Q, Delay, and Power as functions of channel conditions is very challenging, since these functions are nondeterministic and nonlinear. Only worst-case or average values can be determined. Also, there are some dependencies between some of the strategies PHY$_i$, MAC$_i$, and APP$_i$.

The algorithms and protocols at the various layers are often designed to optimize each layer independently and often have different objectives. Various layers operate

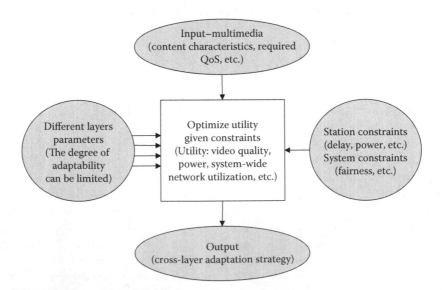

FIGURE 7.7   The block scheme of the cross-layer optimization.

on different units of multimedia traffic and take as input different types of information. For example, the physical layer is concerned with symbols and depends on the channel characteristics. On the other hand, the application layer is concerned with semantics and dependence between flows and the multimedia content. The wireless channel conditions and multimedia content characteristics may change continuously, requiring constant updating of parameters. Finally, different practical considerations for the deployed wireless standard must be taken into account to perform the cross-layer optimization.

The previously formulated cross-layer optimization problem can be solved using iterative optimization or decision tree approaches, where a *group of strategies* are optimized while keeping all other strategies fixed, and this process is repeated until convergence. For the optimization of each group of strategies, one can use derivative and nonderivative methods (e.g., linear and nonlinear programming). Because this is a complex multivariate optimization with inherent dependencies (across layers and among strategies), an important aspect of this optimization is determining the best procedure for obtaining the optimal strategy $S^{opt}(x)$. This involves determining the *initialization*, grouping of strategies at different stages, a suitable order in which the strategies should be optimized, and even which parameters, strategies, and layers should be considered based on their impact on multimedia quality, delay, or power. The *selected procedure* determines the rate of convergence and the values at convergence. The rate of convergence is extremely important, since the dynamic nature of wireless channels requires rapidly converging solutions (this is illustrated in the example later). Depending on the multimedia application, wireless infrastructure, and flexibility of the adopted WLAN standards, different approaches can lead to optimal performance. A classification of the possible solutions is given in the next subsection.

CROSS-LAYER SOLUTIONS

To obtain further insights into the principles that guide cross-layer design, the following cross-layer solutions have been proposed in Reference 4.

Top-down approach. The higher layer optimizes the parameters and the strategies at the next lower layer. This cross-layer solution has been deployed in most existing systems, wherein the application dictates the MAC parameters and strategies, while the MAC selects the optimal PHY layer modulation scheme.

Bottom-up approach. In this architecture the lower layer isolates the higher layers from losses and bandwidth variations. This cross-layer solution is not optimal for multimedia transmission, due to the incurred delays and unnecessary throughput reduction.

Application-centric approach. The application layer optimizes the lower-layer parameters one at a time in either a bottom-up (starting from the PHY layer) or top-down manner, based on its requirements. However, this approach is not always efficient, as the application operates at slower timescales and coarser data granularities (multimedia flows or group of packets), and hence is not able to instantaneously adapt its performance to achieve an optimal level.

MAC-centric approach. In this cross-layer technique the application layer passes its traffic information and requirements to the MAC, which decides which application layer packets/flows should be transmitted and at what QoS level. The disadvantage of this approach resides in the inability of the MAC layer to perform adaptive source channel coding trade-offs given the time-varying channel conditions and multimedia requirements.

Integrated approach. The strategies to design a cross-layer architecture are determined jointly by all the OSI layers. Unfortunately, exhaustively trying all the possible strategies and their parameters in order to choose the composite strategy leading to the best quality performance is impractical due to the associated complexity. A possible solution to solve this complex cross-layer optimization problem in an integrated manner is to use learning and classification techniques.

7.4 CROSS-LAYER DESIGN APPROACHES FOR RESOURCE ALLOCATION AND MANAGEMENT

It is well known that 3G wireless networks, also called IMT2000, aim at providing multimedia mobile services and achieving a maximum bit rate of 2 Mbps. Researchers have been proposing how 3G networks will evolve to beyond 3G or fourth generation (4G) networks. To achieve a successful and profitable commercial market, network service designers and providers need to pay much attention to efficient utilization of radio resources. Although the available bandwidth is much larger in 3G and beyond networks (compared to 2G networks), it is still critical to efficiently utilize radio resources due to fast growth of the wireless subscriber population, increasing demand for new mobile multimedia services over wireless networks, and more stringent QoS requirements in terms of transmission accuracy, delay, jitter, throughput, and so on.

To meet the *anywhere and anytime* concept, the future wireless network architecture is expected to converge into a heterogeneous, all-IP (Internet Protocol) architecture that includes different wireless access networks such as cellular networks, WLANs, and personal area networks (PANs, e.g., Bluetooth and ultra-wideband networks). It is well known that the success of today's Internet has been based on independent and transparent protocol design in different layers, a traditional network design approach that defines a stack of protocol layers (OSI protocol stack). By using the services provided by the lower layer, each protocol layer deals with a specific task and provides transparent service to the layer above it. Such architecture allows the flexibility to modify or change the techniques in a protocol layer without significant impact on overall system design. However, this strict layering architecture may not be efficient for wireless networks when heterogeneous traffic is served over a wireless channel with limited and time-varying capacity and high BER. Efficiently utilizing the scarce radio resources with QoS provisioning requires a cross-layer joint design and optimization approach. Better performance can be obtained from information exchanges across protocol layers.

CDMA RESOURCE ALLOCATION

One major challenge in multimedia services over CDMA cellular networks is QoS provisioning with efficient resource utilization. Compared to circuit-switched voice service in the 2G CDMA systems (i.e., IS-95), heterogeneous multimedia applications in future IP-based CDMA networks require a more complex QoS model and more sophisticated management of scarce radio resources. QoS can be classified according to its implementation in the networks, based on a hierarchy of four different levels: bit, packet, call, and application. Transmission accuracy, transmission rate (i.e., throughput), timeliness (i.e., delay and jitter), fairness, and user perceived quality are the main considerations in this classification. This classification also reflects the principle of QoS categories from the customer point of view.

- **Bit-level QoS.** To ensure some degree of transmission accuracy, a maximum BER for each user is required.
- **Packet-level QoS.** As real-time applications, such as voice over IP (VoIP) and videoconferencing, are delay-sensitive, each packet should be transmitted within a delay bound. On the other hand, data applications can tolerate delay to a certain degree, and throughput is a better QoS criterion. Each traffic type can also have a packet loss rate (PLR) requirement.
- **Call-level QoS.** In a cellular system, a new (or handoff) call will be blocked (or dropped) if there is insufficient capacity. From the user's point of view, handoff call dropping is more disturbing than new call blocking. Effective call admission control (CAC) is necessary to guarantee a blocking probability bound and a smaller dropping probability bound.
- **Application-level QoS.** Bit- and packet-level QoS may not directly reflect service quality perceived by the end user. On the other hand, application layer perceived QoS parameters are more suitable to represent the servie seen by the end user, for example, the peak signal to noise ratio (PSNR) for video applications, and the end-to-end throughput for data applications provided by the responsive Transmission Control Protocol (TCP).

To guarantee the bit- and packet-level QoS requirements of mobile stations (MSs), an effective link layer packet scheduling scheme with appropriate power allocation is necessary. Specifically, the power levels of all the MSs should be managed in such a way that each MS achieves the required bit energy to interference-plus-noise density ratio, and the transmissions from/to all the MSs should be controlled to meet the delay, jitter, throughput, and PLR requirements. A centralized scheduler at the base station (BS) benefits from more processing power and more available information than a distributed one. For the downlink, the BS has information on the traffic status of each MS. For the uplink, each MS needs to send a transmission request upon new packet arrivals and update its link status to the BS, as shown in Figure 7.8. The request and update information can be transmitted in a request channel, or piggybacked in the transmitted uplink packets to avoid possible contention in the request channel, and can be stored in the MS profile at the BS. The BS responds by broadcasting transmission decisions to MSs.

To efficiently utilize scarce radio resources and achieve overall QoS satisfaction, cross-layer information is necessary. In traditional layering architecture, the link layer has statistical knowledge of the lower physical layer, such as the average channel capacity. However, to exploit the CDMA time-varying channel, it is better for

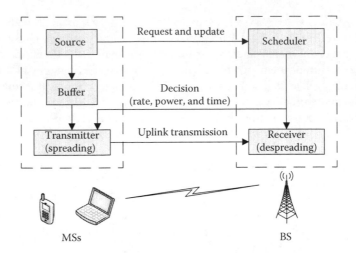

FIGURE 7.8   Scheduler for the uplink transmission. Reproduced with permission from H. Jiang and X. Shen. "Cross-layer design for resource allocation in 3G wireless networks and beyond," *IEEE Commun. Magazine* 43 (December 2005): 120–26.

the link layer to have knowledge of instantaneous channel status. Also, to guarantee the application-level QoS, such as an acceptable visual quality of video services or a guaranteed TCP throughput of data services, the application or transport layer should be jointly designed with the link layer. In a five-layer reference model, Figure 7.9 shows three possible cross-layer information directions, from physical to link layer; from link to transport layer, and vice versa; and from link to application layer, and vice versa. This leads to three cross-layer design approaches: channel-aware scheduling, TCP over CDMA wireless links, and joint video source/channel coding and power allocation, as discussed in the following.

Channel-Aware Scheduling

In a multiple access wireless network, the radio channel is normally characterized by time-varying fading. To exploit the time-varying characteristic, a kind of diversity (multiuser diversity) can be explored to improve system performance. The principle of multiuser diversity is that for a cellular system with multiple MSs having independent time-varying fading channels, it is very likely that there exists an MS with instantaneous received signal power close to its peak value. Overall resource

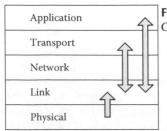

FIGURE 7.9   The cross-layer information for IP-based CDMA resource allocation.

utilization can be maximized by providing service at any time only to the MS with the highest instantaneous channel quality.

With the capability to support simultaneous transmissions in a CDMA system, multiuser diversity can be employed more effectively and flexibly than traditional channel-aware scheduling schemes for a TDMA system. An MS does not need to wait until it has the best channel quality among all MSs, but rather can transmit as long as its channel is good enough.

It should be mentioned that for real-time traffic (e.g., voice or video) with a delay constraint, if an MS is in a bad channel state for a relatively long period, its packets will be discarded when multiuser diversity is employed, as the MS has to wait for a good channel state. Hence, it is challenging to apply multiuser diversity to real-time traffic. An effective way is to incorporate the packet delay in the scheduling decision.

TCP over CDMA Wireless Links

For data services, TCP guarantees error-free delivery. TCP was originally designed for wireline networks with a reliable physical layer, where packet loss mainly results from network congestion. In such networks TCP adjusts its sending rate based on the estimated network congestion status so as to achieve congestion control or avoidance. In a wireless environment, TCP performance can be degraded severely as it interprets losses due to unreliable wireless transmissions as signs of network congestion and invokes unnecessary congestion control. To improve TCP performance over the wireless links, several solutions have been proposed to alleviate the effects of noncongestion-related packet losses,[2] among which snoop TCP and explicit loss notification (ELN) are based on cross-layer design. In snoop TCP, TCP layer knowledge is used by link layer schemes, while in ELN, the network layer takes advantage of cross-layer information from the physical layer.

When a TCP connection is transmitted over CDMA cellular networks, further considerations are needed. First, CDMA capacity is interference limited, and TCP transmission from an MS generates interference to other MSs. It is desired to achieve acceptable TCP performance (e.g., a target throughput) and at the same time introduce minimum interference to other MSs (i.e., to require minimum low-layer resources). Second, power allocation and control in CDMA can lead to a controllable BER, which affects TCP performance.

Joint Video Source/Channel Coding and Power Allocation

Video transmission is an important component of multimedia services. Typical video applications include mobile videoconferencing, video streaming, and distance learning. Due to their real-time nature, video services typically require QoS guarantees such as a relatively large bandwidth and a stringent delay bound.

For video services over a CDMA channel with limited capacity, an effective way is to pass source significance information (SSI) from the source coder in the application layer to the channel coder in the physical layer. More powerful FEC code (and therefore more overhead) can be used to protect more important information, while weaker FEC may be applied to less important information. Such joint source/channel coding is a cross-layer approach called unequal error protection (UEP). UEP can easily be performed with Bose-Chaudhuri-Hocquenghem (BCH) codes,

Reed-Solomon (RS) codes, and rate-compatible punctured convolutional (RCPC) codes with different coding rates for packets with different priorities. UEP can also be implemented by means of power allocation in CDMA systems; for example, transmission power can be managed so that a more important packet experiences a smaller error probability.[19] In case of capacity shortage, UEP schemes can result in more graceful quality degradation (and thus smaller distortion, or higher PSNR) than equal error protection (EEP). Based on channel capacity, the optimal transmission rate and power allocation for packets of each priority can be found to minimize the average distortion of the received video by means of an optimization formulation over CDMA channels.[20] It outperforms uniform power allocation, as it exploits the degree of freedom added by CDMA power allocation.

Ability of a Video Codec to Adjust Its Source Coding Rates

This flexibility can also be exploited to improve system performance. Consequently, it is desirable to employ a joint source/channel coding scheme that allocates bits for source and channel coders to minimize the end-to-end distortion under a given bandwidth constraint.[21]

With interference-limited capacity, it is important to take into account the power management in CDMA systems when designing the source and channel coding. More flexibility can be obtained when power allocation is considered jointly with source and/or channel coding. A large source coding rate can lead to low quantization distortion, and a high transmission accuracy level can achieve low channel-error-induced distortion.

Transmission power consumption minimization can also be achieved in joint source/channel coding and power allocation schemes subject to acceptable video distortion.[22] Apparently, when transmission power consumption is reduced, CDMA system capacity can be enlarged. However, the above optimization is complicated to achieve. The case is worse when time scheduling for multiplexed video traffic is implemented. Further investigation is necessary.

INTEGRATED VOICE-DATA

A cross-layer design is often used to provide QoS for voice and data traffic in wireless cellular networks with a differentiated services (DiffServ) backbone. Optimal resource allocation problems for voice and data flows are formulated to guarantee prespecified QoS with minimal required resources.

In the past decade, the Internet has started to penetrate into the wireless domain. It is now widely recognized that the 3G (and beyond) wireless mobile CDMA cellular networks are evolving into an all-IP architecture in order to provide broadband and seamless global access for various IP multimedia services. A major task in the establishment of such an all-IP platform is provisioning of QoS to different Internet applications.

Recently, the DiffServ[23] approach has emerged as an efficient and scalable solution to ensure QoS in future IP networks. As a class-based traffic management mechanism, DiffServ does not use per-flow resource reservation and per-flow signaling in core routers, which makes DiffServ scalable. Current research on DiffServ is mainly focused on the wireline network. The bottleneck for an end-to-end application across a hybrid wireless/wireline domain is usually the link between the BS and the MS due to the limited radio resources and the varying characteristics of the radio channel. On the other hand, current MAC schemes[24] in CDMA wireless systems usually provide priority to voice users. Voice traffic flows are scheduled for transmission first, while data traffic flows use the residual system capacity and are then differentiated from each other. So far, research on QoS support for data traffic is very limited. In References 25 and 26, two packet-switching scheduling schemes are proposed for wireless CDMA communications. Both are based on per-packet information, thus increasing the scheduling burden and system overhead. Furthermore, the QoS provisioning for data traffic in these two schemes is limited up to the link layer; that is, only physical layer QoS and link layer QoS are considered.

Concerning the above issues, a cross-layer design scheme for wireless cellular networks with a DiffServ backbone to provide QoS to MBs is proposed in Reference 27. The proposed scheme combines the transport layer protocols and link layer resource allocation to both guarantee QoS requirements in the transport layer and achieve efficient resource utilization in the link layer.

In what follows we consider a hybrid wireless/wireline IP-based network for providing multimedia traffic to MSs. The Internet backbone is DiffServ-based, and the wireless subnet is a wideband time-division (TD)/CDMA cellular system with frequency division duplexing (FDD). Multicode CDMA (MC-CDMA) is considered in the code domain. Figure 7.10 shows the stack architecture.[27] It consists of an MS, BS, DiffServ Internet backbone, and correspondence node.

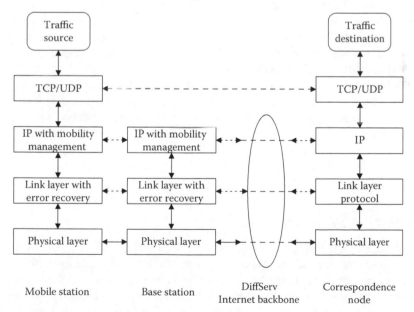

FIGURE 7.10 Protocol stack architecture in the hybrid wireless/wireline IP-based network. Reproduced with permission from H. Jiang and W. Zhuang. "Cross-layer resource allocation for integrated voice/data traffic in wireless cellular networks," *IEEE Trans. Wireless Commun.* 5 (February 2006): 457–68.

In the DiffServ core network, packets are aggregately differentiated by different per-hop behaviors: expedited forwarding (EF) which is aimed at providing a connection of low loss, low delay and low jitter; and assured forwarding (AF),[29] to provide a reliable connection with a target transmission rate, referred to as committed information rate (CIR), denoted by $V_{CI}R$. Generally, in core networks, EF can be applied to voice/video conversation traffic with a stringent delay requirement, while data traffic can be serviced by AF.

UDP is used for voice traffic in the system which does not use retransmissions to guarantee reliable delivery. UDP itself does not provide mechanisms to ensure timely delivery or other QoS guarantees, but relies on lower layer services to do so. When a voice user is on talk spurt, the UDP packets will be generated periodically. On the other hand, TCP can provide reliable end-to-end transmission over unreliable IP service which is suitable for the data traffic. Each transport layer (TCP or UDP) packet is segmented into a number of link layer units for transmission over the error-prone wireless link, and then reassembled at the BS.

To interwork with the DiffServ backbone, the QoS requirement for voice traffic in the CDMA cellular network is guaranteed delay bound. If a UDP packet cannot be delivered to the BS within this bound, it will be dropped, and the dropping probability is bounded by PU,m (e.g., PU,m = 1%, a typical value for voice service). The QoS requirement for data service is the guaranteed transport layer throughput V (>VCIR) with a reliable end-to-end transmission.

In traditional network models, the transport layer and link layer are designed separately and independently. This works well in wireline networks because of the highly reliable transmission provided by the optical fiber channels. However, in a wireless network, for various applications with different QoS requirements, different transport layer protocols have different impacts on lower layers. Hence, an independent link layer resource allocation strategy will not work well. More specifically, for data traffic over a wireless link, TCP dynamically adjusts the sending rate of TCP packets (which will be fed into the link layer transmission queue) according to the network congestion status (e.g., packet loss events and round trip delay). On the other hand, the wireless link layer resource allocation also affects the TCP performance, as it ultimately determines the TCP packet loss and transmission delay over the wireless link. That is, the TCP protocol and wireless link layer resource allocation interact with each other.

OFDM-Based Wireless Networks

The allocation and management of resources are crucial for wireless networks, in which the scarce wireless spectral resources are shared by multiple users. In the current dominant layered networking architecture, each layer is designed and operated independently. However, wireless channels suffer from time-varying multipath fading; moreover, the statistical channel characteristics of different users are different. The suboptimality and inflexibility of this architecture result in inefficient resource use in wireless networks. We need an integrated adaptive design across different layers.

Recently, the principles of multiuser downlink and MAC designs have been changed from the traditional point-to-point view to a multiuser network view. For instance, channel-aware scheduling strategies are proposed to adaptively transmit data and dynamically assign wireless resources based on channel state information (CSI). The key idea is to choose one user with good channel conditions to transmit packets.[30] Taking advantage of the independent channel variation across users, channel-aware scheduling can substantially improve network performance through multiuser diversity, the gain of which increases with the number of users. From a user point of view, packets are transmitted in a stochastic way in the system using channel-aware scheduling, which is also called opportunistic communications.[31] Based on this concept, channel-aware dynamic packet scheduling is applied in 1x evolution (1xEV) for CDMA2000, and high-speed downlink packet access (HSDPA) for wideband CDMA.

The growth of Internet data and multimedia applications requires high-speed transmission and efficient resource management. To avoid intersymbol interference (ISI), OFDM is desirable for high-speed wireless communications. OFDM-based systems are traditionally used to combat frequency-selective fading. From a resource allocation point of view, however, an OFDM system naturally has a potential for more efficient MAC since subcarriers can be assigned to different users.[32,33] Another advantage of OFDM is that adaptive power allocation can be applied for further performance improvement.

A cross-layer framework for the downlink resource management of IP-based OFDM wireless broadband networks is able to effectively enhance spectral efficiency and guarantee QoS built on a utility-optimization-based architecture. In this architecture, exploiting knowledge of the CSI and the characteristics of traffic, the network aims to maximize the total utility, which is used to capture the satisfaction levels of users.

OFDM-based networks offer more degrees of flexibility resource management than do single-carrier networks. Taking advantage of knowledge of the CSI at the transmitter (BS), OFDM-based systems can employ the following adaptive resource allocation techniques:[34]

- **Adaptive modulation and coding (AMC).** The transmitter can send higher transmission rates over the subcarriers with better conditions to improve throughput and simultaneously ensure an acceptable BER at each subcarrier. Despite the use of AMC, deep fading at some subcarriers still leads to low channel capacity.
- **Dynamic subcarrier assignment (DSA).** The base station dynamically assigns subcarriers according to CSI and/or QoS requirements. Channel characteristics for different users are almost mutually independent in multiuser environments—subcarriers experiencing deep fading for one user may not be in a deep fade for other users—therefore, each subcarrier could be in good condition for some users in a multiuser OFDM wireless network. In addition, frequency multiplexing provides fine granularity for resource allocation.
- **Adaptive power allocation (APA).** The BS allocates different power levels to improve the performance of OFDM-based networks, which is called multiuser water filling.

Employing these adaptive techniques at each subcarrier results in large control overhead. In practice, subcarriers can be grouped into a cluster (subchannel) in which we apply adaptive techniques. The size of a cluster determines the resource granularity. Obviously, the resource allocation schemes or algorithms designed for a subcarrier-based adaptive OFDM system can be directly used in a cluster-based system.

The major issue is how to effectively assign subcarriers and allocate power on the downlink of OFDM-based networks by exploiting knowledge of the CSI and the characteristics of traffic to improve spectral efficiency and guarantee diverse QoS.

There are three main challenges for cross-layer design for resource management in OFDM-based networks:

- DSA belongs to the matching or bin packing problems in discrete optimization.
- Unlike a single-carrier network, a multicarrier network can serve multiple users at the same time; hence, the design of multicarrier scheduling for bursty traffic is a new and interesting problem.
- The general relationships among spectral efficiency, fairness, and the stability property of wireless scheduling are not clear for wireless networks with time-varying fading.

Addressing all the above problems is crucial for establishing high-speed efficient wireless Internet networks.

7.5 CONCLUDING REMARKS

A cross-layer design for real-time video streaming maintains a general layered structure and identifies the key parameters to be exchanged between adjacent layers. In this context, adaptive link layer techniques that adjust packet size, symbol rate, and constellation size according to channel conditions are used to improve link throughput, which in turn improves the achievable capacity region of the network.

At the MAC and network layers joint allocation of capacity and flow optimize the supportable traffic rate significantly, and consequently can improve the end-to-end video quality by a wide margin. Smart scheduling at the transport layer further protects the video stream from packet losses and ensures excessive network congestion. Knowledge of the video rate-distortion trade-off and latency requirement at the application layer is used to select the most appropriate source rate for video delivery.

There is always a tendency, and in fact a need, to optimize performance in any system. This generally creates tension between performance and architecture. In the case of wireless networks, we currently see this tension manifesting itself in the current interest in cross-layer design.

Cross-layer design creates interactions—some intended, others unintended. Dependency relations may need to be examined, and timescale separation may need to be enforced. The consequences of all such interactions need to be well understood, and theorems establishing stability may be needed. Proposers of cross-layer design must therefore consider totality of the design, including interactions with other layers, and also what other potential suggestions might be barred because they would interact with the particular proposal being made. The long-term architectural value of the suggestions must also be considered.

For a cross-layer QoS mapping architecture for video delivery in a wireless environment, there are several components that must be taken into account. These include a proposal of an adaptive QoS model that allows QoS parameters to be adaptively adjusted according to the time-varying wireless channel condition, interaction mechanisms between the priority network and video applications to provide proper QoS selection, and a resource management scheme to assign resources based on the QoS guarantee for each priority class under the time-varying wireless channel.

Realistic integrated models for the delay, multimedia quality, and consumed power of various transmission strategies/protocols need to be developed. Moreover, in terms of multimedia quality, the benefits, of employing a cross-layer optimized framework for different multimedia applications with different delay sensitivities and loss tolerances still need to be quantified.

In cross-layer design, the overall system performance can be improved by taking advantage of the available information across different layers. To achieve this, an appropriate signaling method is necessary.

Recent research has provided preliminary results for cross-layer design over all-IP CDMA networks. Further research efforts should include:

- Joint source/channel coding and power allocation for multiplexed video services with time scheduling
- Cross-layer design for DiffServ-based QoS
- Cross-layer design for heterogeneous voice/video/data traffic (a cross-layer design approach usually focuses on a specific traffic type). For a CDMA network supporting heterogeneous voice/video/data traffic with different QoS requirements, it is critical to consider the trade-off among cross-layer approaches for different traffic types, and to achieve desired overall system performance with efficient resource utilization.

In a multicell environment due to the intercell interference, the schedulers in different cells should not act independently, thus making the resource allocation much more complex.

Resource management and scheduling in a wireless OFDM-based downlink can be used to serve multiple users and to support various applications based on a cross-layer approach.

8 Mobile Internet

The goal of this chapter is to find a large audience and help stimulate further interest and research in the mobile Internet and related technologies. With the advance of wireless/mobile communication technologies, the market for mobile Internet is growing dramatically. For example, cellular networks have greatly increased the link bandwidth with recent standardization efforts on the Internet Protocol (IP) multimedia subsystem (IMS). Also, WiFi hotspot services-based IEEE 802.11 wireless local area networks (WLANs) have been widely deployed in public spaces (airports, convention centers, cafes, etc). Furthermore, new wireless technologies (e.g., IEEE 802.16/20) are emerging. All of these techniques will accelerate the growth of the mobile Internet market.

This chapter is organized as follows. First, we present and analyze related protocols for mobile Internet. Then, we review IP mobility and wireless LANs and wide area networks (WANs). Next, we describe IP mobility for cellular and heterogeneous mobile networks. After that, we continue with quantitative analysis of enhanced mobile Internet, together with scalable application-layer mobility protocol. We also review mobility and quality of service (QoS). A network architecture analysis for seamless mobility services concludes the chapter.

8.1 INTRODUCTION

The emerging mobile Internet will facilitate convergence of various wireless networks and protocols including WLANs and voice over IP (VoIP). New innovation will drive the incorporation of mechanisms that will allow seamless handoff over disparate networks, manage user mobility across heterogeneous networks and service providers, ensure such portability, and guarantee quality of service. A number of standards bodies such as the Internet Engineering Task Force (IETF), *WiMAX*, Third Generation Partnership Project (3GPP)/3GPP2, IEEE, Open Mobile Alliance (OMA), Unlicensed Mobile Access (UMA), and Fixed Mobile Wireless Convergence Forum (FMWC) are in the forefront of the evolution of the next-generation wireless IP network architectures, protocols, and standards. Their efforts have led to the specification of infrastructure and radio access networks that are fundamentally different and must interoperate if service delivery is to be perceived as seamless by the end user. Besides the ever-changing technology, mergers and acquisitions have forced carriers to deal with the expense and complexity of operating heterogeneous wireless networks in different parts of the world. At the same time, traditional operators are faced with the challenges introduced by a new set of service providers, unburdened by legacy networks that carriers have inherited and developed over the years. The newcomers are exploiting the power of IP-based protocols to offer new services such as VoIP, data, video, messaging, push-to-talk, conferencing, and others using peer-to-peer structures that overlie existing network structures and reduce them to bit pipe connectivity.

In this chapter we explore and address numerous issues relevant to the evolving mobile Internet: cost, complexity, scalability, and design trade-offs. This topic also increases general understanding of the value of Internet-based solutions and evaluates how these solutions facilitate convergence of the mobile Internet to an appropriate Internet-based open mobile wireless architecture. Such an architecture will be flexible, capable of seamlessly delivering services to end devices over disparate wireless and wireline networks, and able to support all types of end user devices. The mobile Internet must allow for user mobility while maintaining IP connectivity between two mobile hosts with a minimum of latency. Protocols that facilitate user mobility are mobile IP (MIP) or mobile IP version 4 (MIPv4) and mobile IP version 6 (MIPv6). Besides catering to user mobility, the mobile Internet must incorporate handoff mechanisms at the IP layer that also work in cooperation with link layer handoff mechanisms. MIPv6 improves upon MIPv4 by eliminating triangular routing through incorporation of route optimization.

Scalable mobility management is an integral part of the mobile Internet. While MIP protocols are designed to address mobility management in the mobile Internet, the Session Initiation Protocol (SIP) offers an alternative mobility management strategy.

With the proliferation of WiFi spots (WLANs) and implementation of wireless metropolitan area networks (WMANs), it is not surprising that activities have sprung up that propose integrating these new access networks with existing cellular networks (2G/2.5G/3G). Efforts under UMA and voice call continuity (VCC) are aimed at hybrid networks that seek to provide access to wireless networks via WiFi/WLAN. The main idea is to converge what are considered traditional wireline services (Internet access, call forwarding, etc.).

We will start with an overview of a set of IP-based mobility protocols because mobility support plays an important role in IP-based wireless networks for providing multimedia applications.

8.2   RELATED PROTOCOLS

Deployment of International Mobile Telephony (IMT) 2000 standards for 3G wireless networks gave existing 1G, 2G, and 2.5G operators the flexibility to evolve their networks (primarily designed for circuit-switched voice communications) to support skeleton multimedia transmissions with a nominal bit rate of 384 kbps (fast movers) to 2 Mbps (slow movers).[1] Major efforts are under way to deliver applications and services to mobile nodes (MNs) over a packet switched access network that is heterogeneous with the Internet. So the current trend in mobile wireless network evaluation is directed toward an all-IP network.[2] An end-to-end IP-based solution, referred to as fourth generation (4G) systems, combines mobility with multimedia rich content, high bit rate, and IP transport with support for QoS management and authentication, authorization, and accounting (AAA) security.[3] Standards and related technologies are being developed to help early deployment of such systems and ensure interoperability between equipment from different manufacturers, thereby providing significant investment reductions compared to today's third generation (3G) technologies.

In addition, there will be licensing costs as well, since 4G will utilize frequencies believed to be in the public domain. Realizing commercially viable IP mobility support over the current wireless infrastructure remains a challenging research area.

TRADITIONAL MOBILITY MANAGEMENT

Terminal mobility management in cellular networks consists of two components: location management and handoff management.[4] Classification of mobility management is shown in Figure 8.1.

Location management consists of two complementary operation—registration or location update (LU) and paging—to enable a network to discover the current point of attachment of an MN for information delivery.

Handoff management enables a network to maintain a connection as an MN continues to move and change its point of attachment to the network.

Tracking an MN is performed through registration/LU procedures in which an MN informs the network of its location at times triggered by movement, timer expiration, and so on. Locating an MN is performed through search procedures, when the network pages the MN. There is a trade-off between how closely the network tracks the current location of an MN versus the time and complexity required to locate an MN whose position is not precisely known.

Handoff management in a cellular environment is normally performed in three steps: initiation, connection generation, and data flow control. Whenever an MN changes its point of attachment with a base station (BS), it sends a request to the current BS for handoff to the target BS for initiation. After initiation, control is handed over to the target BS by the current BS. The IP address of the MN also changes as it changes its point of attachment. This is connection generation. After obtaining a new address, data may be sent to that address, completing the task of data flow control.

In a cellular environment, there are two kinds of handoff: intracell and intercell. Intracell handoff occurs when a user moving within a cell changes radio channels to minimize interchannel interference under the same BS. On the other hand, intercell handoff occurs when an MN moves into an adjacent cell and all of the MN connections are transformed to the new BS. Intercell handoff may be performed in two

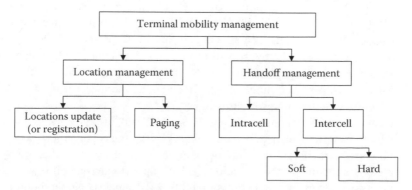

FIGURE 8.1 Mobile management classification.

ways: soft and hard. If two BSs simultaneously handle the interchange between them while performing the handoff, it is a soft handoff with no discontinuity of connection. Soft handoff is achieved by proactively notifying the new BS before actual handoff. Thus, soft handoff minimizes packet loss, but delay incurred may be more. In hard handoff, one BS takes over from another in a relay mode, so delay and signaling are both minimized, but zero packet loss is not guaranteed. Connection may be off for a very small period during the take over.

We can now move on to classifying the IP mobility protocols based on their level of operation from the architectural point of view.

IP Mobility Protocol Classification

Mobility of an MN in a network may be broadly classified into three categories:

- Micromobolity (intrasubnet mobility)—movement within a subnet
- Macromobolity (intradomain mobility)—movement across different subnets within a single domain
- Global mobility (interdomain mobility)—movement across different domains in various geographical regions

In general, the primary goal of mobility management is to ensure continuous and seamless connectivity between micro- and macromobility when occurring on a short timescale. Global mobility involves longer timescales, where the goal is to ensure that MNs can reestablish communication after a move rather than provide continuous connectivity.

Mobile classification of protocols is shown in Figure 8.2. It gives a clear idea of which class of mobility each of the existing protocols aims to support.

Because MobileIP (MIP) is generally targeted for global mobility, it introduces significant network overhead in terms of increased delay, packet loss, and signaling when MNs change their point of attachment very frequently within small geographical areas. To overcome these performance penalties, micro- and macromobility protocols offer fast and seamless handoff control and IP paging support for scalability and power saving. A complete overview of three such protocols, HAWAII, CellularIP (CIP), and Hierarchical MIP (HMIP), is given in Reference 5. Despite many apparent differences, the operational principle of the protocols is quite similar in complementing base MIP by providing local handoff control. Obviously, they are inefficient in handling interdomain LUs, and thus are unable to handle global reachability perfectly.[6] Accordingly, TeleMIP and dynamic mobility agent (DMA)[7] architecture are proposed to resolve this issue.

Although targeted for micromobility, CIP uses MIP support for providing intradomain mobility management. So CIP,[8] along with TIMIP,[9] falls under micro- as well as macromobility. TeleMIP is strictly intradomain, as it cannot support either micromobility or global mobility. HAWAII and DMA can support macromobility as well as global mobility but cannot handle micromobility. MIP supports global mobility but fails to handle micro- or macromobility. HMIP and TR45.6 are two minor extensions

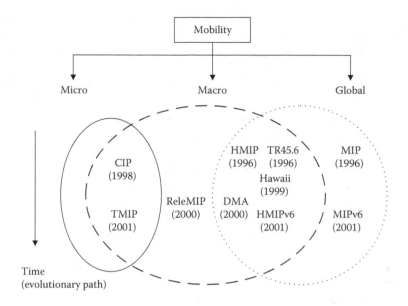

FIGURE 8.2 Mobile classification of protocols. Reproduced with permission from D. Saha et al. "Mobility support in IP: A survey of related protocols," *IEEE Network* 18 (November/December 2004): 34–40.

of MIP to support macromobility as well. As MIPv6 is now replacing MIP, HMIP is being augmented to HMIPv6, and they belong to their parent classes.

GLOBAL MOBILITY PROTOCOLS

The function of MIP lies in the retention of a permanently assigned IP address (known as a home address) by MNs for application transparency. This is achieved by providing a care-of address (CoA) to an MN when it moves out of its home network (HN) to visit a foreign network (FN). While in an HN, the location of an MN is captured by its CoA assigned by a foreign agent (FA) in the FN. A home agent (HA) in the HN maintains a record of the current mobility binding (i.e., the association of the MN home address with its CoA during the remaining lifetime of that association). The HA intercepts every packet addressed to the MN home address and tunnels them to the MN at its current CoA. This is known as triangular routing.

Once a correspondent node (CN) has learned the MN CoA, it may cache it and route its own packets to the MN directly, bypassing the HA completely. This is mainly done for route optimization as triangular routing suffers from various problems due to poor route selection, including increased impact of possible network partitions, increased load on the network, and increased delay in delivering packets. Route optimization can improve service quality but cannot eliminate poor performance when an MN moves while communicating with a distant CN. In this case the registration/LU delay contributes significantly to the handoff delay, leading to reduction in throughput. Also, frequent LUs incur extensive overhead for location cache management in route optimization.

The current IPv4 has only a 32-bit address size, which is not enough to support the addressing needed on the Internet. Since 1994, the IETF has been working on IPv6 to solve the limitations inherent in IPv4 in terms of addressing, routing, mobility support, and QoS provisioning. In IPv6, 128-bit addressing is used instead of the 32-bit addressing in IPv4. The increased IP address size allows the Internet to support more levels of addressing hierarchy, a much greater number of addressable nodes, and simpler autoconfiguration of addresses. IPv6 is considered the core protocol for next-generation IP networks.[10,11]

The network entitles of IPv6 for mobility support, MIPv6[12,13] is similar to that in MIPv4, except that MIPv6 does not have the concept of an FA. MIPv6, unlike MIPv4, uses an extensible packet header including both home address and CoA, along with the authentication header to simplify routing to the MN and perform route optimization in a secure manner. While discovery of a CoA is still required, an MN uses the stateless address autoconfiguration and neighbor discovery functions defined in IPv6 to acquire a colocated CoA of a foreign network in MIPv6. MIPv6 also uses the IP-within-IP tunneling approach to deliver data packets to an MN. If a CN knows the MN CoA, the CN could send data packets to the MN directly (i.e., source routing) using an IPv6 routing header. Otherwise, the data packets are routed to the associated HA, and then tunneled to the MN CoA (i.e., tunneling).

Although MIPv6 uses almost the same terminologies as MIPv4 except for the absence of the FA, security management represents a big difference between them. In MIPv6 all nodes are expected to implement strong authentication and encryption functions.

MIPv6 routing is shown in Figure 8.3. MIPv6 uses both tunneling and source routing to deliver data packets to destination MNs.[14] Tunneling is the only option for MIPv4.

With careful security management, optimized routing could be a solution for MIPv6. The increased number of MNs will cause increased signaling traffic due to mobility support. To improve the efficiency of mobility management, Hierarchical MIPv6 has been proposed.[15] The basic idea is to use regional registration to reduce

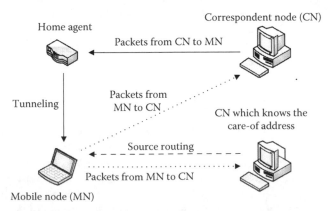

FIGURE 8.3 MIPv6 routing principle. Reproduced with permission from J. Li and H. H. Chen. "Mobility support for IP-based networks," *IEEE Commun. Magazine* 43 (October 2005): 127–32.

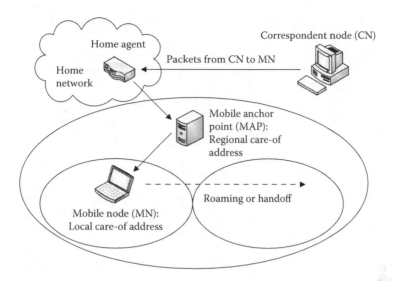

FIGURE 8.4 Hierarchical IPv6. Reproduced with permission from J. Li and H. H. Chen. "Mobility support for IP-based networks," *IEEE Commun. Magazine* 43 (October 2005): 127–32.

the overall registration signaling overhead and improve the QoS in handovers for mobile communications. Hierarchical MIPv6 is shown in Figure 8.4.

With hierarchical MIPv6, the concept of a mobile anchor point (MAP) is introduced. An MAP handles mobility management for MNs within a network domain, including registration and handover. The MAP functions like a local HA. It receives data packets on behalf of MNs and tunnels the packets to the MN CoAs. An MN is assigned two CoAs, a regional one and a local one. The regional CoA is local to the MAP covered region. The local CoA is the same as an MN MIPv6 CoA; it is the address local to the node's link. An MN communicates with its correspondent nodes via its regional CoA. It sends binding updates to CNs (assuming route optimization) only when the MN moves outside the MAP region. Otherwise, it only sends a binding update to the MAP to update its regional CoA (local CoA binding). All packets received by CNs will have the MN regional CoA as the source address; they will thus respond directly to this address. The MAP, on detecting packets sent to this CoA, will encapsulate and forward these packets to the MN at its local CoA.

With the new addressing, hierarchical extensions, and security management, MIPv6 could make MNs available for peer-to-peer (P2P) services, a promising information service (e.g., easier file sharing), since MIPv6 lets MNs have static addresses even as they move around. Without static addressing, MNs have to communicate through a server, which violates the P2P technology's specifications.

In MIP and MIPv6, registration of the CoA naturally requires authentication. The registration in an HA must be initiated by the right MN and not by some other malicious node pretending to be the MN. A malicious node could cause the HA to alter its routing table with erroneous CoA information, and the MN would be unreachable to all incoming communications from the Internet. To secure the registration process, authentication is mandatory and performed by all HAs, MNs, and FAs for MIP and MIPv6.

Generally, the mobility of MNs in IP-based networks is supported by dynamic network architectures and location management. The security consideration due to mobility for IP-based networks rises significantly.

For traditional secure access of services from remote locations, such as Point-to-Point Protocol (PPP) connections, AAA exists. The concept of AAA was introduced by the IETF to support secure roaming of MNs in the wireless Internet.

GLOBAL MOBILITY/MACROMOBILITY PROTOCOLS

For global mobility/macromobility protocol presentations, we will start with hierarchical MIP. We will then continue with wireless IP network architectures including TR45.6, HAWAII, dynamic mobility agent, and hierarchical mobile IPv6.

As an extension of MIP, it employs a hierarchy of FAs to locally handle MIP registrations during macromobility. Registration messages establish tunnels between neighboring FSs along the path from the MN to a gateway FA (GFA). Packets addressed to the MN travel through this network tunnel.

It defines a new node, called a packet data serving node (which contains an FA). Network access identifiers identify MNs in an FN. MNs send registration messages to FAs, which in turn interact with AAA servers residing in the FN (or use a broker network) for authentication with the HN. For macromobility, the scheme proposes to use dynamic HAs (DHAs) that reside in the serving network and are dynamically assigned by visited AAA servers. DHAs allow MNs to receive services from local access service providers while avoiding unnecessarily long routing.

On top of using MIP for interdomain mobility, it supports a separate binding protocol to handle intradomain mobility. Four alternative setup schemes control handoff between access points. An appropriate scheme is selected depending on the service level agreement (or operator's priorities among QoS parameters, e.g., eliminating packet loss, minimizing handoff latency, and maintaining packet ordering). It also uses IP multicasting to page idle MNs when incoming data packets arrive at an access network and no recent routing information is available. Path setup messages generate host-based routing information in tabulated form for MNs within a domain in some specific intermediate routers. The HA sends the encapsulated packets (after intercepting them) to the current border router of the MN. The border router, after decapsulating the packet, again encapsulates and sends it to a nearby intermediate router. This router then decapsulates and finally sends the packet to the MN.

Dynamic mobile agent (DMA) architecture uses the Intradomain Mobility Management Protocol (IDMP) to manage macromobility and allows the use of multiple global binding protocols for maintaining global reachability. A new node called a mobility agent (MA), introduced at network-layer granularity, reduces the generation of global LUs. The MA is similar to the FA of MIP, except that it resides higher in the hierarchy (than individual subnets) and provides an MN with a stable point of attachment throughout the domain. Each FA must be associated with at least one MA in that domain. It also uses subnet agents (SAs) which interact with appropriate MAs to provide authentication. Here, an MN is associated with two current CoAs:

- Global CoA (GCoA) resolves the location of the domain and remains unchanged as long as the MN stays in the current domain.
- Local CoA (LCoA) identifies the MN's present subnet of attachment (similar to CoA of MIP). LCoA has only local scope; an MN notifies the assigned MA of any change in its LCoA.

When an MN first moves into a domain, it is given an LCoA and assigned to an MA. It registers with the designated MA for a GCoA. The MN can then use different global binding protocols to inform the appropriate CNs about this GCoA. Packets from a remote CN, tunneled (or directly transmitted) to the GCoA, are intercepted by the MA and then forwarded (by reencapsulation) to the MN LCoA.

It is the same hierarchical extension to MIPv6 for locally handling MIPv6 registrations as hierarchical MIP is to MIP.

MACROMOBILITY PROTOCOLS

TeleMIP two-level architecture uses the concept of the MA, and it is derived from the registration area-based location management scheme currently employed in cellular networks. An FN is divided into several subnets depending on its geographical location. Each subnet has at least one FA. Whenever an MN changes subnets, it contains a new local CoA (obtained from the FA using conventional MIP techniques) and subsequently informs the MA of this new local address binding. Under a load balancing scenario, MNs in a single subnet may be assigned to different MAs (using different hashing schemes). An MN will be assigned two CoAs:

- A domain-specific CoA (similar to GCoA) from the public space that is unchanged as long as the MN stays within a specific domain or region. This is typically the address associated with the MA.
- A subnet-specific CoA (similar to LCoA) for roaming in a partial subnet. This address may have only local scope and can be either the CoA of the FA or a locally valid colocated address.

This address changes every time an MN changes its foreign subnet. When an MN enters a new domain, it will register the MA CoA with the HA during the initial LU process. The MA is thus aware of the exact (subnet-level) location of the MN and can consequently route the packet to the MN using a domain-specific routing protocol (without requiring source specific routing). As long as the MN is under the control of a single MA, the MN does not transmit any LUs to the HA. The architecture thus ensures the localization of all intradomain-mobility update messages within the domain.

MACRO/MICROMOBILITY PROTOCOLS

Cellular IP (CIP) and terminal independent MIP (TIMIP) belong to macro/micromobility protocols.

CIP supports local mobility (i.e., macro/micromobility) in a cellular network that consists of interconnected CIP nodes. Location management and handoff support are integrated with routing in CIP networks. An MN communicates to its HA with the local gateway's address as the CoA. Consequently, after intercepting the packets from a CN, the HA sends them in encapsulated form to the MN gateway. The gateway decapsulates the packet and forwards it to the MN to minimize control messaging; regular data packets transmitted by MNs are used to refresh host location information. CIP monitors mobile-originated packets and maintains a distributed, hop-by-hop reverse path database used to route packets back to MNs. The loss of downlink packets when an MN roams between APs is reduced by a set of new handoff techniques. CIP tracks idle MNs in an efficient manner, so MNs do not have to update their location after each handoff. This extends battery life and reduces air interface traffic. It supports a fast security model based on special session keys, where BSs independently calculate keys. This eliminates the need for signaling in support of session key management, which would otherwise add additional delay to the handoff process.

TIMIP is a combination of the principles of CIP, HAWAII, and MIP for micro/macromobility scenarios. Here, the IP layer is coupled with layer 2 handoff mechanisms at the APs by means of a suitable interface that eliminates the need for special signaling between MNs and APs. Thus, MNs with legacy IP stacks have the same degree of mobility as MNs with mobility-aware IP stacks. Like CIP, refreshing of routing paths is performed only in the absence of any traffic. Like HAWAII, routing reconfiguration during handoff within a domain changes the routing tables of the access routers located in the shortest path between the new and old APs only. However, in order to support seamless handoff, TIMIP uses context transfer mechanisms compatible with those currently under discussion within the IETF SeaMoby group.

8.3   IP MOBILITY AND WIRELESS NETWORKS

WIRELESS LANs

Wireless fidelity (WiFi) is a technology dominating all WLANs.[16] The term WiFi has been used in general to refer to any type of 802.11 network (802.11a, 802.11b, 802.11g, and so on). The term has been promulgated by the WiFi Alliance. Any products tested and approved as WiFi Certified® by the WiFi Alliance are certified as interoperable with each other, even if they are from different manufacturers. This feature of WiFi technology allows device-level multivendor interoperability to support mobility. A user with WiFi Certified products can use any brand of AP with any other brand of client hardware that is also certified. Typically, any WiFi products using the same radio frequency (e.g., 2.4 GHz for 802.11b or 802.11g, 5 GHz for 802.11a) will work with one another, even if not WiFi Certified.

Initially, WiFi networks were designed to extend enterprise networks. Nowadays, it is a requirement to extend WiFi broadband access to many public places such as universities, hotels, and conference centers. MIP can be a good solution for providing roaming services to WiFi networks, in which WiFi is viewed as a visited network. In this section we address IP mobility support for WLANs, including WiFi and Bluetooth.

Traditional IP considers an entire WLAN as a single subnet, where IP addresses of all hosts have the same address prefix. Over a WLAN, mobility support is implemented through the use of dynamic IP address allocation provided by Dynamic Host Configuration Protocol (DHCP). When an MN remains in a WLAN that could be considered an FN (or subnet), it requests an IP address (i.e., CoA) for some period of time. The server returns an available IP address from a pool of addresses. If the IP address is configured successfully, the MN can communicate within the WLAN. DHCP-based mobility support is simple to implement. However, it cannot provide for the MN roaming across different WLANs, since the network connection can be achieved only within WLAN boundaries. For multimedia applications to MNs, it is important to implement multiple IP subnets across a common WLAN in order to make network management easier, to facilitate location-dependent services, and to decrease the spread of broadcast packets throughout the network.

Therefore, with multiple subnets, MNs must be able to seamlessly roam from one subnet to another while traversing a network. WLAN APs provide support for roaming at the data link layer (ISO/OSI layer 2). Users automatically associate and reassociate with different APs as they move through a network. As MNs roam across subnets, though, there must be a mechanism at the IP/network layer (ISO/OSI layer 3) to ensure that a user device configured with a specific IP address can continue communications within the applications.

Both MIPv4 and MIPv6 provide solutions to the problem by treating a WLAN as a subnet that may include several APs, as shown in Figure 8.5.

To implement MIPv4 or MIPv6, two major components are needed: an MIP server and MIP client software. The MIP server will fully implement the MIP HA functionality, providing mobility management for MNs. The MIP server can generally also keep track of where, when, and for how long users utilize the roaming services. That data can then become the basis for accounting and billing. The registration of an MN CoA when the MN moves is implemented by MIP operation. When an MN moves across the boundary to another subnet during communication, the MIP network-level handoff process is performed. It initiates a handshake between the HA and the new FA (or the MN with colocated address for MIP, or the MN for MIPv6) at the medium access control (MAC) level. After completion of handoff, the data packets destined for the MN are tunneled by the HA to the new subnet, and then to the MN.

Recently, the basic technical concept of WLANs has been extended to wireless WANs.

WIRELESS WANs

Among the current wireless broadband WAN technologies, *WiMAX* is the most promising. It is based on the maturing IEEE 802.16 standard, which specifies the radio frequency technology for MANs and point-to-multipoint wireless networking. IEEE 802.16 divides its MAC layer into sublayers that support different transport

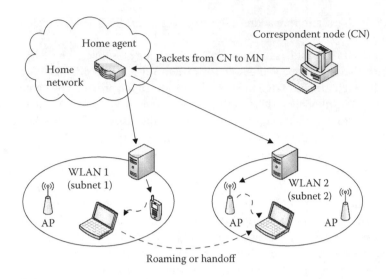

FIGURE 8.5 IP mobility support for WLANs. Reproduced with permission from J. Li and H. H. Chen. "Mobility support for IP-based networks," *IEEE Commun. Magazine* 43 (October 2005): 127–32.

technologies, including IPv4 and IPv6, Ethernet, and asynchronous transfer mode, which lets vendors use *WiMAX* no matter which transport technology they support.

WiMAX extends the area coverage of WiFi. *WiMAX* provides fixed and mobile wireless broadband connectivity without the need for direct line of sight with a BS. In a typical cell radius deployment of 3 to 10 km, *WiMAX* systems are expected to deliver a capacity of up to 40 Mbps/channel for fixed and portable access applications. This provides enough bandwidth to simultaneously support hundreds of business clients with speedy connectivity, and thousands of residences with digital subscriber line (DSL) speed connectivity. Mobile network deployments are expected to provide up to 15 Mbps of capacity within a typical cell radius deployment of up to 3 km.

Compared to WiFi, *WiMAX* networks are relatively large and can support more MNs. Although the early versions of IEEE 802.16a and 802.16d do not support inter-domain mobility, IEEE 802.16e supports mobility at pedestrian speeds. *WiMAX* has begun adding IP mobility support in the IP/network layer.

CELLULAR AND HETEROGENEOUS MOBILE NETWORKS

Cellular networks have been developed for voice telephony service using circuit-switched technology. They are usually complex and large in terms of their network scale and operational features, high-speed mobility, low data rate, and wide area coverage. Cellular networks are in the process of evolution. The aim of the process is to have an all-IP network architecture to provide high-bit-rate multimedia services including voice, video, and data. Multimedia services require multiple sessions over one physical channel which could be provided by packet-switched networks. The common protocol set for packet-switched networks is IP.

The 3G cellular technologies include Universal Mobile Telecommunications System (UMTS) and code-division multiple access 2000 (CDMA2000). The UMTS evolved from the Global System for Mobile (GSM) network in Europe, and CDMA2000 evolved from the CDMA One network which originated in the United States. Both CDMA2000 and UMTS were defined by the International Telecommunication Union (ITU) in the IMT-2000 framework. Based on the combination of circuit and packet switching, both CDMA2000 and UMTS combine mobile and IP technologies to provide personal communications and personalized content. A data session is established to carry IP packets between the network access server and the MN in both CDMA2000 and UMTS networks. Both networks use tunnels to support user mobility. However, the 3G networks including CDMA2000 and UMTS currently solve their mobility problems at the link layer (layer 2) only, not in the IP layer (layer 3). Several overlaid wireless networks including 3G networks, WLANs, and WWANs may exist over the same geographical area.

Figure 8.6 shows IP mobility support with the provision of AAA and QoS control services.

MIPv6 and its hierarchical mobility management extensions may provide a solution for internetwork mobility as well as intranetwork mobility.

With hierarchical MIPv6, the MIPv6 protocols may manage global mobility while the MAP may handle local mobility. QoS and security considerations arise with mobility support.

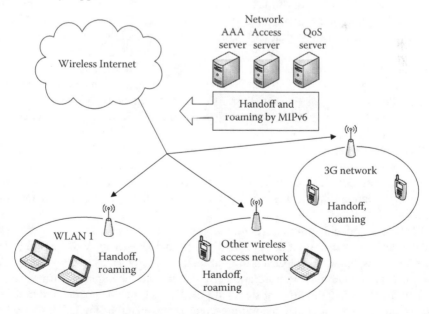

FIGURE 8.6 IP mobility support for next generation heterogeneous mobile networks. Reproduced with permission from J. Li and H. H. Chen. "Mobility support for IP-based networks," *IEEE Commun. Magazine* 43 (October 2005): 127–32.

As the popularity of mobile computing grows, the associated protocols and their scalability are subject to match closer research. Namely, MIP relies heavily on the use of IP-to-IP tunneling, requiring 20 bytes of overhead for every packet routed to or from an MN, assuming inverse tunneling is enabled. Enhanced MIP (EMIP) was developed to eliminate the need for tunneling when providing services to mobile nodes.[18] In what follows, we deal with qualitative analysis of EMIP.

8.4   QUANTITATIVE ANALYSIS OF ENHANCED MOBILE INTERNET

As the popularity of real-time traffic such as voice and video grows, mobile computing is also being used to facilitate these applications. MIP was developed to allow nodes to change location while maintaining network connectivity.[19] Tunneling was an existing networking concept adopted as part of MIP for packet redirection. Although reusing existing networking technologies has benefits, tunneling has several drawbacks. The added overhead required for tunneling packets reduces the bandwidth available in the wired network. Scalability becomes a concern as the number of MNs increases and the tunneling overhead consumes available network bandwidth.

MIP defines the HA and FA in order to facilitate an MN that maintains connectivity as it changes location. The FA assigns the MN a CoA while it resides in the FN. The MN then registers its CoA with the HA. After registration takes place, a tunnel is created between the HA and the FA. When a packet is sent to the MN from a CN, the packet is routed to the home network of the MN. The HA intercepts the packets and sends it through the tunnel to the CoA, which is typically the FA. When the FA receives the packet, it forwards the packet to the MN. IP-in-IP encapsulation must be supported by HAs and FAs for tunneling datagrams in MIP and is used as the tunneling mechanism in this research.[20] For IP-in-IP encapsulation, the encapsulating IP header adds 20 bytes to the size of each packet in the tunnel. Often packets sent from the MN to a CN are routed back through the tunnel to the HA before reaching their destination. This process is called reverse tunneling, and can be used to prevent packet filtering and provide accounting information to the home network.

Providing QoS with MIP has been researched using both integrated services and differentiated services techniques.

Since the Resource Reservation Protocol (RSVP), commonly used to provide QoS in wired networks, requires added overhead and complexity to work across a tunnel, several modifications have been proposed for RSVP with MIP.[21-23] Network address translation (NAT) when applied at the edge of foreign tunnels does not work well with NAT.[24]

EMIP, developed to eliminate tunneling, uses the HA and the FA defined by MIP, and the same mechanisms for discovering the CoA and registering with the HA. EMIP differs from MIP in the way packets are redirected from the HA to the FA. A concept built on network address port translation (NAPT) is used the in place of tunnels.[25]

With EMIP, when the MN registers with the HA, a tunnel is not created between the HA and the FA. Instead, a mapping is created between the HA and the FA when each connection the MN communicates across is established. Mappings are created by intercepting packets to and from the MN at the HA and the FA. The HA and FA then exchange mapping requests and mapping reply messages containing the source

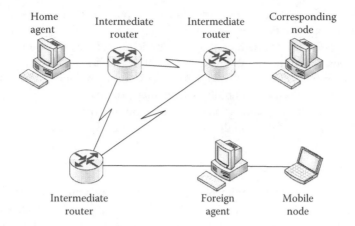

FIGURE 8.7 Test network. Reproduced with permission from P. K. Bestand and R. Pendse. "Quantitative analysis of enhanced mobile IP," *IEEE Commun. Magazine* 44 (June 2006): 66–72.

and destination IP and port addresses of the MN and the CN. The mobility agent that intercepts the packet also supplies a care-of port (CoP) that is used by the HA and the FA to identify the mapping. Once the mapping between the HA and the FA is established for a communication session, the mobility agents can redirect packets to and from the MN by modifying the IP and Transmission Control Protocol (TCP) or User Datagram Protocol (UDP) packet headers instead of using a tunnel.

EMIP eliminates the tunneling overhead, replacing it with a one-time bandwidth overhead to exchange the mapping request and mapping reply packets. However, a delay is introduced with EMIP to buffer the first packet of each new communication in order to establish a mapping between the HA and FA. One example of the test network setup is shown in Figure 8.7.

As the wireless/mobile technologies evolve, the market for mobile Internet is rapidly growing. To proliferate the mobile Internet market, scalable mobility support is a key question. In what follows, we introduce a Scalable Application-Layer Mobility Protocol (SAMP) that is based on P2P overlay networking and SIP.

8.5   SCALABLE APPLICATION-LAYER MOBILITY PROTOCOL

To address the scalability problem, a number of schemes, such as a dynamic HA assignment mechanism, have been proposed.[26] These approaches result in additional signaling overhead to learn the current end condition and to synchronize among multiple mobility agents. Furthermore, in the current mobile Internet architecture, it is not easy to design a self-organized and well-balanced mechanism among multiple mobility agents scattered across different network domains.

Mobility management is one of the key issues to proliferate mobile Internet services, together with the scalability aspect of mobility management in mobile Internet services. A few types of mobility agents are employed for mobility management: an FA in MIPv4, an HA in MIPv4 and MIPv6, and a MAP in HMIPv6. These mobility agents play important roles in mobility management and packet routing. However, if a high burden of these tasks is concentrated on a single mobility agent, the mobility agent will suffer from the increased processing load. This results in a long response time and even a system failure. The overhead at the mobility agent may lead to service unavailability. Consequently, how to provide a scalable service by distributing the network traffic load (for mobility management and packet routing) among multiple mobility agents is an important design issue in mobile Internet services.

SAMP is based on SIP and P2P overlay networking. These are briefly described.

SESSION INITIATION PROTOCOL

SIP is an Internet standard protocol for initiating, modifying, and terminating an interactive multimedia session. The multimedia session involves various applications such as video, voice, instant messaging, and online games. Moreover, SIP is accepted as a call-control protocol in IMS. SIP can also be employed to support mobility at the application layer. SIP is an appropriate mobility solution especially for interactive multimedia applications that need an explicit signaling for session management. In addition, SIP allows users to maintain access to their services while moving (i.e., service mobility), and to maintain sessions while changing terminals (i.e., session mobility).

A typical SIP architecture consists of SIP servers and user agents. SIP servers are classified into proxy, redirect, and registrar servers, depending on their functions. A proxy server relays received SIP messages to another SIP server or SIP user agents, whereas a redirect server performs redirection of received SIP messages. A registrar maintains location information to support mobility. On the other hand, user agents are classified into user agent client (UAC) and user agent server (UAS). Each user agent is identified by a SIP universal resource identifier (URI) that follows a form similar to an e-mail address. The UAC initiates a SIP session by sending an INVITE message, while the UAS responds with SIP reply messages that contain suitable status codes. Basically, the user agent registers its location at the registrar before establishing a SIP session.

P2P OVERLAY NETWORKING

A P2P overlay network is a distributed network that relies on the computing power and bandwidth of peer nodes in the network. Unlike the client/server model, each node participates in the P2P overlay network as a peer with equal responsibility. In P2P overlay networks, since there is no central entity to control overall tasks, it has no role in these networks. Also, P2P overlay networks provide self-organization and load-balancing functions in a distributed manner. Another attractive feature is the technique used for locating and retrieving a desired item (e.g., a file in file-sharing applications). For more efficient locating/retrieving operations, a distributed hash table (DHT) has been introduced. The DHT is a decentralized and distributed system where all items and peer nodes are identified by unique keys. In the DHT, the ownership of keys is distributed among participating peer nodes and hence the peer nodes can efficiently route messages to the owner of any given key. Therefore, the

DHT is scalable to a large number of nodes and can handle continual node arrivals and failures. These features enable the DHT to be widely accepted for large-scale P2P networking.

SCALABLE APPLICATION-LAYER MOBILITY ARCHITECTURE

In SAMP architecture, each SIP server participates to form a P2P network. The SIP server acts as a peer node in a DHT. As previously mentioned, the DHT is a decentralized system where multiple keys are distributed among peer nodes. SAMP employs the DHT for location management in mobile Internet. A key in the DHT corresponds to the location information of an MN in SAMP and the keys are maintained at SIP servers. To join the DHT-based overlay network, the SIP servers should perform the functions of a registrar as well as proxy/redirect services. Hence, we define two SIP server modes: registrar and proxy (RP) mode and registrar and redirect (RR) mode. As its name implies, a SIP server in the RP mode handles SIP messages as a proxy server and also maintains location information as a registrar. Similarly, a SIP server in the RR mode redirects the received SIP messages and keeps track of the MN location information. In this section, we describe the SAMP operations by means of RP-mode SIP servers.

Figure 8.8 shows an example of a session setup procedure in SAMP. First, a CN sends an INVITE message to SIP server 20, which is the corresponding anchor SIP server. Then, the invited message is forwarded to SIP server 4, that is, the home SIP server of the MN.[27]

Next, SIP server 4 relays the INVITE message to the MN anchor SIP server 12 and the MN receives the INVITE message from its anchor SIP server. Because SAMP is based on the P2P overlay network, a session setup procedure requires a

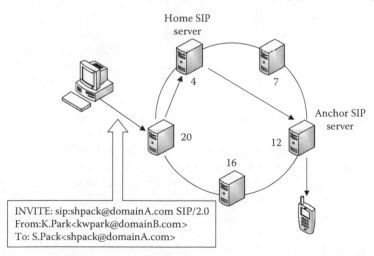

FIGURE 8.8   An example of a session establishment procedure in SAMP. Reproduced with permission from S. Pack and K. Park. "SAMP: Scalable application-layer mobility protocol," *IEEE Commun. Magazine* 44 (June 2006): 86–92.

number of lookups in peer SIP servers. This may lead to a long session setup latency. To reduce the session setup latency, SAMP can employ a two-tier scheme.

The forthcoming convergence of cellular and wireless data networks is often manifested in an "all" approach in which all communications are based on an end-to-end IP protocol framework. In what follows, we focus on major technical highlights of mobility and QoS management subsystems for converged networks. An enhanced IPv6 mobility platform architecture that provides mobility and QoS, as key drivers of all-IP-based networks, is also analyzed.

8.6   MOBILITY AND QOS

Previous works[28,29] have proposed solutions that support seamless mobility based on IPv6. Although these works have shown that the basic concepts are viable, the DAIDALOS project architecture proposes an enhanced IPv6 mobility platform that provides mobility and QoS, as key drivers of the future all-IP-based 4G networks. Fast intra-end intertechnology handovers are a solution to the requirements of seamlessness. For next generation integrated systems, additional requirements are the optimization of resource usage, scalability for an increasing number of customers, and increased network flexibility.

The DAIDALOS mobility architecture aims to provide an efficient and scalable integration of multiple network technologies with sustained QoS support. The simplified general view of the architecture is shown in Figure 8.9.

The architectural design has a hierarchical structure. The network of each mobile operator consists of a core network (two such networks, from different operators, are represented in the figure) and a set of access networks. The access networks contain multiple access routers (ARs), with multiple radio APs each. The architecture supports multiple access technologies, including WLAN, *WiMAX*, TD-CDMA, and digital video broadcasting (DVB).

Each access network is called a region. Resources in each region are independently managed by an access network QoS broker (QoSB-AN), thus providing a first scalability step. Resources in the core are managed by the core network QoS broker (QoSB-CN), which communicates for end-to-end QoS with the QoSB-ANs of the mobile operator's networks as well as with the QoSB-CNs of the other operator's networks. The architecture is based on widely accepted standards for mobility and QoS. Mobility is implemented by means of the MIPv6 protocol, with fast handover extensions,[30] and QoS is based on the differentiated services (DiffServ) architecture.[31] However, additional mechanisms that integrate and complement MIPv6 and DiffServ are needed to achieve the objective of providing QoS to mobile users while optimizing the overall performance. Such mechanisms have been designed in the architecture.

Handover decisions in this architecture are sustained both by measurements of signal quality as well as QoS measures (such as load and resource availability). Handovers can be started either by the terminal or by the network. We refer to the former as a mobile-initiated handover (MIHO), and to the latter as a network-initiated handover (NIHO). Handover execution is improved with functions for maintaining quality during handovers, along with tight coupling with QoS functions.

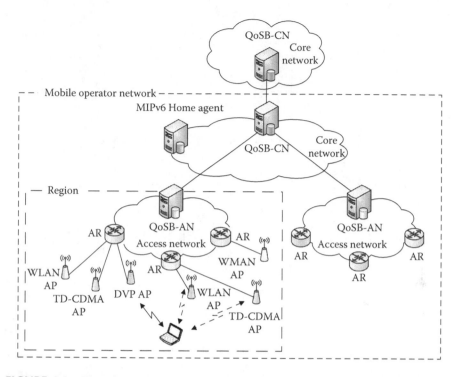

FIGURE 8.9   Hierarchical structure of mobility and QoS architecture. Reproduced with permission from R. L. Aguiar et al. "Scalable QoS-aware mobility for future mobile operators," *IEEE Commun. Magazine* 44 (June 2006): 95–102.

The enhanced functionalities of the architecture modules are described in Reference 29. The needed modules for functionalities are shown in Figure 8.10. They are organized according to physical location.

Enhanced mobile-initiated handover decisions. Handover decisions in the case of an MIHO are enhanced with the objective of ensuring that, from all the possible AP candidates, the best one is chosen. The module responsible for the handover decision at the MT is intelligent interface selection (IIS). This module relies on the mobile terminal controller (MTC) to obtain the information it uses to make a decision. This includes signal quality measurements, obtained from the mobility abstraction layer (MAL), as well as QoS measures, such as the load of the APs, retrieved from the candidate APs. The latter information is obtained from the QoS abstraction layer (QAL) in the neighboring ARs, and conveyed by means of the candidate access router discovery (CARD) protocol[32] to the MT. With this information, the target AP for the handover is chosen so that both signal strength and QoS requirements are met in the new AP, thus guaranteeing appropriate operation and service quality after the handover.

Network-initiated handover functionality. The enhanced MIHO functionality ensures that handover decisions are taken optimally according to local information, but does not guarantee that the overall distribution of resources will be optimal from an operator perspective—which is essential for a realistic network. To achieve this,

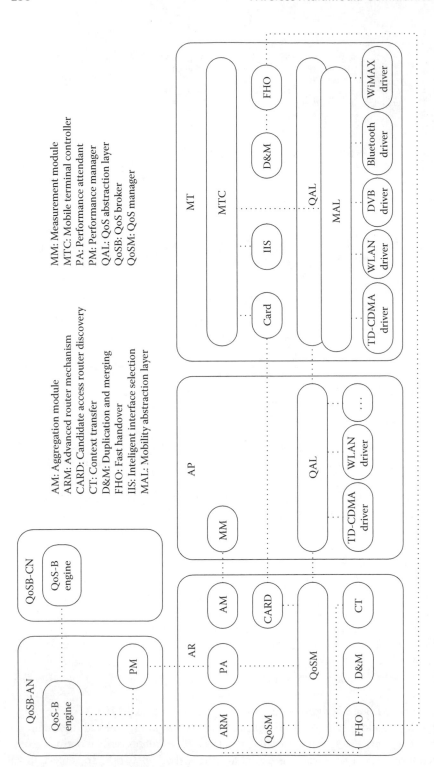

MM: Measurement module
MTC: Mobile terminal controller
PA: Performance attendant
PM: Performance manager
QAL: QoS abstraction layer
QoSB: QoS broker
QoSM: QoS manager

AM: Aggregation module
ARM: Advanced router mechanism
CARD: Candidate access router discovery
CT: Context transfer
D&M: Duplication and merging
FHO: Fast handover
IIS: Inteligent interface selection
MAL: Mobility abstraction layer

FIGURE 8.10 Architecture modules functionalities. Reproduced with permission from R. L. Aguiar et al. "Scalable QoS-aware mobility for future mobile operators," *IEEE Commun. Magazine* 44 (June 2006): 95–102.

NIHO support is required to allow for the optimization of the overall capacity by properly balancing load among the various APs of a region. For this purpose, the performance management (PM) module at the QoSB-AN collects information about the load of the different APs and the radio link quality between the MT and its candidate APs, and based on this information (eventually) reorganizes the wireless connections. Information on the load of the APs is obtained by the performance attendant (PA) modules at the APs, from their interface with the QAL, and delivered to the PM. Signal strength measurements are taken by the measurement modules, filtered out and aggregated by the aggregation module (AM), provided to the PA at the AR, and from there conveyed to the PM. Based on all this data, the PM then reorganizes the connections of all MTs for achieving optimized global performance. This reorganization takes into consideration QoS beyond the wireless access by means of the interaction between the PM and the QoSB engine at the QoSB-AN. The NIHO execution is then triggered by the communication between the QoSB and the fast handover (FHO) execution module at the AR, through the advanced router mechanism (ARM).

Seamless handover execution. In the execution of a handover involving the old AR (oAR) and the new AR (nAR), it is required that continuity of communication be maintained. To perform a low-latency lossless handover, the fast handovers for MIPv6 protocol are enhanced with duplication and merging (D&M) functions. These functions improve performance by duplicating the packets addressed to the MT at the old AR to avoid packet loss. To set up the MT context in the nAR, the context transfer (CT) function is used to transfer the mobility-related state (including security information).

Quality of service. QoS is based on the DiffServ architecture. Admission control and resource reservations are handled by the QoSBs, which act jointly to perform QoS reservations over an end-to-end path. QoS reservations at the routers are performed through the interaction between the QoSB engine at the QoSB-AN and the ARM module at the AR, which performs the reservation via the QoS manager (QoSM). Similarly, reservations in the wireless access part are also performed through the interaction between the QoSB engine at the QoSB-AN and the ARM module at the corresponding AR. The latter communicates with the QoSM, which communicates with the QAL at the AR. QoS reservations in the wireless access are then performed by the QAL modules at the AP and MT.

Multiple technology support. The support of multiple technologies in the architecture is provided by means of a modular design based on the use of abstraction layers (ALs): the MAL and the QAL. These ALs interface with drivers of the different technologies and offer a unique interface to the upper-layer modules of the architecture, while hiding the specifics of the underlying technologies. The QAL offers a technology-independent interface for QoS functions, such as the setup of a QoS connection or the measurement of available resources in an AP. Similarly, the MAL offers a technology-independent interface for mobility-related functions such as the execution of a handover or measurement of signal strength received at the MT.

8.7   SEAMLESS MOBILITY SERVICES: A NETWORK ARCHITECTURE

Seamless mobility implies that the infrastructure transparently manages connectivity so that the use is always best connected. The user's definition of best connected depends on cost, performance, location, or other factors. Several key technical and social drivers are contributing to the interest in seamless mobility. Device convergence, the consolidation of multiple functions into small portable devices with integrated cellular and WiFi-based WLAN interfaces, is a trend that is of interest to users in both the business and consumer environments.

The goals of the network architecture are as follows:

- Support services on existing devices, as well as next generation, multifunction, multiradio devices.
- Align the work with the emerging IMS standards, allowing the solution to leverage industry investment in IMS.[33]
- Define a unified interface to the cellular network that provides a scalable, standards-based solution for seamless mobility. This interface should enable new services such as location-based services to also be supported within the same framework.
- Use network intelligence and create standardized clients to take advantage of the architecture.
- Provide operations, administration, maintenance, and provisioning (OAM&P) systems to support solution deployment.

Seamless mobility is especially well suited to services delivered over IP. The project is challenging because it rides on top of several complex building blocks, each of which must be integrated.[34] Figure 8.11 shows the building blocks for seamless mobility. A key decision is to treat an IMS-based infrastructure as the common service delivery platform to handle all call control and new service delivery. This approach is corrected because innovation in multimedia services is happening in areas of VoIP applications based on SIP applications. Interworking with the cellular network uses existing signaling system 7 (SS7) interfaces[35] and is configured so that

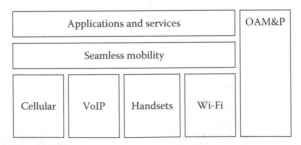

FIGURE 8.11   Building blocks for seamless mobility. Reproduced with permission from Ch. Kalmanek et al. "A network-based architecture for seamless mobility services," *IEEE Commun. Magazine* 44 (June 2006): 103–9.

all cells to and from cellular devices are routed to the IMS infrastructure for call control.

Seamless mobility sits as a middleware layer on top of these building blocks with application services such as IP Centrex (central exchange), IP-PBX (private branch exchange), and so on riding on the top of the seamless mobility layer.

There are a number of different approaches to provide seamless mobility services. These solutions can be grouped into three categories: enterprise or PBX-based solutions, cellular network-based solutions (such as UMA) and IP-network-based solutions.

PBX-based seamless approaches utilize several techniques, such as virtual numbers (a solution for incoming calls that allows caller to reach a subscriber using *find-me follow-me* techniques), various two-stage dialing approaches, or software loaded on handsets to control their operations. The handset is an extension on the IP-PBX within the corporate WiFi network, allowing the user to roam the building or campus with full IP-PBX services. The advantage of this solution is that it leverages existing infrastructure on the IP-PBX. The dial plan and feature are familiar to enterprise users. However, IP-PBX solutions may not address the full range of seamless mobility requirements, and face challenges in scaled deployments. Also, this solution does not extend easily outside of the individual enterprise domain.

One of the cellular network-based solutions is UMA. It delivers GSM and GPRS mobile services over unlicensed spectrum technologies like WiFi essentially by encapsulating GSM signaling in IP, and routing it to a gateway in the cellular network. UMA enables subscribers to roam, and supports handover between cellular networks and public/private WiFi networks using bimode handsets. One of the advantages of UMA is that subscribers receive the same user experience for both their cellular and WiFi mobile voice services as they transition between networks, but limited to the cellular experience. Another advantage for UMA is its standardization, which could help to promote widespread adoption and interoperability. The limitations are that many enterprise customers prefer enhanced features such as corporate dialing plans, coverage plans, voice mail, and other features. Additionally, UMA-based solutions may not integrate easily with other IP services or advancements in VoIP. UMA can be viewed as a near-term opportunity for cellular operators to bring WiFi-based access solutions to market.

Another cellular-based solution emulates a traditional GSM visitor location register (VLR) for WiFi-based VoIP calls. The objective is to enable roaming between WiFi networks. A VoIP gateway emulates a serving mobile switching center (MSC) to register the IP interface of the phone and perform handoffs from/to the WiFi access domain using standard VLR techniques.

In the long term, cellular providers will migrate to a 3G air interface capable of end-to-end VoIP with services enabled on the next generation service platform defined by the 3GPP, referred to as the IMS. This platform will support all users and provide a new generation of services in the future.

8.8 CONCLUDING REMARKS

The mobile Internet has emerged as a powerful paradigm promising to transform the way services are delivered to end users. The widespread growth of mobile wireless networks, applications, and services has ushered in the era of mobile computing, where handheld computing devices or terminals have become the predominant choice for users.

MIP and MIPv6 are basic technologies for mobile networks. To improve the performance of mobile IPv6, the HMIPv6 can be used. IP-based wireless networks will become the core for future cellular networks. MIPv6 and its hierarchical mobility management extensions offer internetwork as well as intranetwork mobility.

SAMP is based on P2P, overlay networking. SAMP is highly scalable compared to MIPv6 and HMIPv6. Moreover, because SAMP is based on an Internet Standard Protocol (ISP), SAMP can be easily implemented and extended from the existing infrastructures. SAMP is expected to play a key role in proliferating the mobile Internet as a scalable mobility solution.

The IP-based architecture which integrates multiple technologies in a seamless environment is very flexible in terms of handover possibilities—that is, MIHO and NIHO, intra- and intertechnology—and is fully integrated with QoS support. The architecture is highly scalable both at mobility and QoS-support levels.

Given the tremendous growth in mobile voice and data usage and the emergence of new multifunctions, there is an emerging interest in seamless mobility applications and services that provide service portability and application persistence across multiple network connections. Seamless mobility ultimately allows users to transparently access all of their data and services in a consistent method.

9 Evolution toward 4G Networks

Fourth generation (4G) seems to be a very promising generation of wireless communications that will change people's lives in the wireless world. There are many striking attractive features proposed for 4G which ensure a very high data rate, global roaming, and so on. New ideas are being introduced by researchers throughout the world, but new ideas introduce new challenges. There are several issues yet to be solved such as incorporating the mobile world into the Internet Protocol (IP)-based core network, establishing an efficient billing system, and perfecting smooth hand off mechanisms. The world is looking forward to the most intelligent technology that will connect the entire globe. 4G networks are expected to deliver more advanced visions, including improvements over the third generation (3G). These improvements include enhanced multimedia, smooth streaming video, universal access, and portability across all types of devices. Also, 4G enhancements are expected to include worldwide roaming capability. After an introductory discussion including migration to 4G mobile systems, as well as beyond 3G and toward 4G networks, we then discuss 4G technologies from the user's perspective. The emphasis is on heterogeneous system integration and services. After that, we present all-IP 4G network architecture. Then, we outline the issues concerning quality of service (QoS) for 4G networks, QoS and end-to-end QoS support are included, as well. Next, we continue with security in 4G networks together with infrastructure security and secure handover between heterogeneous networks. Network operators' security requirements conclude the chapter.

9.1 INTRODUCTION

4G is the latest in a series of wireless network technology families for personal communication systems that over the past four decades have produced movable, portable, and finally truly mobile personal communication (productivity, entertainment, etc.) companions and a rich collection of wireless services, as well. Each successive generation offers progressively higher quality and richer services (analog voice, digital voice, packetized data, text, audio, images, video, mobility, seamless roaming, etc.). A quick search on the Web reveals that over the years there have been many more service offerings announced than the four—three to be exact, as 4G is still in the definition/design phase—technology families.

Explosive growth is expected in mobile communications over the next decade with higher speeds and larger capacities than those provided by 3G communications mobile systems. These features must be made available to meet the requirements for faster speeds and more diverse usage formats. Accordingly, studies are now being carried out to develop 4G mobile systems. 4G mobile communications involves a blend of concepts and technologies in the making. Some can be recognized as having been derived from 3G, whereas others involve new approaches to wireless mobile networks.

Although second generation (2G) and 3G wireless technologies are well defined with mature and established standards efforts and strong industry backing guiding their deployment, 4G networks are not yet defined with respect to either underlying technology or network architecture. Nevertheless, by extrapolating from the previous generations, it is anticipated that 4G technologies will be all-IP based, able to support speeds of several tens of megabits per second (with some projections pointing at over 100 Mbps), which will enable high-quality multimedia services, and allow provision of seamless roaming across multiple access technologies. Quite often, 4G technologies are considered the high-speed IP-based glue that will integrate personal area network (PAN), local area network (LAN), metropolitan area network (MAN), and wide area network (WAN) wireless technologies under a common umbrella. This is certainly an ambitious objective, and to achieve it many technology-oriented as well as business-oriented (e.g., billing and subscriber ownership and management) hurdles need to be overcome.

Interest in the 4G area is on the upswing with an increasing number of technical meetings and publications devoted entirely or in part to the topic.

Driven by the need to address the aforementioned limitations of 3G wireless systems and support context-rich multimedia services and applications, 4G wireless systems are envisioned to provide high data rates in the downlink as well as the uplink direction. While 3G downlink data rates at the present time do not exceed 2 Mbps and 384 kbps at pedestrian and vehicular speeds, respectively, with a potential upgrade to 10 Mbps, 4G systems are expected to attain data rates of 20 Mbps or higher even at high speeds. Recent advances pertaining to the underlying networking, multiplexing, scheduling, and physical layer technologies have been making this realizable at a steady pace. In addition to providing high data rates, supporting global roaming and multiple classes of service with variable end-to-end QoS requirements across heterogeneous wireless systems, including cellular, satellite-based networks, and wireless LANs, are the key features of these emerging global systems. In 4G systems, the radio access network (RAN), as well as the core network, will be packet switched, and a pure end-to-end IP architecture is conceivable.[1]

In the presence of new stringent quantitative and qualitative QoS requirements, there are numerous challenges yet to be addressed before 4G networks come to full fruition. These challenges pertain not only to the (radio) access and serving strata, but also to the home and transport strata in current Universal Mobile Telecommunications System (UMTS)/3G wireless systems, those based on UMTS Release 6 and beyond. Honoring an end-to-end service level agreement across heterogeneous wireless networks, with variable physical, networking, and architectural characteristics and constraints, is worth investigating, but is extremely intricate and challenging.

Devising novel schemes for distributed dynamic channel allocation (DDCA), call/connection admission control (CAC), and cross-layer adaptation, to mention a few, is of utmost importance. In terms of CAC for circuit-switched connections, and packet scheduling for packet-switched services, for example, limited interference/power-based and threshold-based solutions have been adopted by most 2.5G and

3G operators. New, adaptive, and possibly hybrid schemes are sought to optimally, or near optimally, control the radio access bearers for real-time and non-real-time traffic. Developing and utilizing sophisticated traffic engineering schemes to handle bursty traffic, random arrival and reading times, and the random number of packets per session in the case of packet connections will be instrumental to the success of 4G network technologies. Maintaining the QoS of the accepted real-time and non-real-time sessions, without jeopardizing cellular and network coverage and capacity, is also essential to the success of such systems.

We also anticipate that the emerging 4G-compliant systems will encompass novel multihop network architectures to enhance capacity, reduce power consumption, and enhance coverage and throughput. Consequently, near-optimal computationally inexpensive distributed schemes for channel assignment, resource allocation, and CAC ought to be developed.

All access network technologies and stakeholders have their own security requirements and mechanisms, making the integration of multiple access technologies for simple networks a challenge.

Security is one of the major technical challenges in vertical handover. Each network may deploy its own security mechanisms that are incompatible with others. Apart from that, seamless mobility sets time constraints on handovers. To provide seamless handover while maintaining the security level, it is preferable to transfer security context information from one network to another in a timely fashion. Security context contains the state of authentication and authorization, cryptography keys, and the like. It is used to support trust relations and to provide communication security for network/entities. The context is typically shared between a pair of network nodes or networks.

To provide seamless mobility it is understandable that handover between networks either of heterogeneous technologies (vertical handover) or of different administrative domains must be seamless.

Just as 3G networks are starting to establish a strong foothold in the industry, we are already witnessing unprecedented interest in the next generation of wireless technologies. Building on current trends in cellular-centric wireless technologies and seamlessly incorporating high-speed local and personal area networks, 4G technologies hold the promise of even more exciting personalized wireless broadband services. They will seamlessly integrate multiple access technologies and end-user devices, and offer true anywhere, anytime, and through-any-device services to highly mobile professionals and residential customers. To achieve these promises, 4G technologies have plenty of challenges to address.

9.2   MIGRATION TO 4G MOBILE SYSTEMS

To migrate from current systems to 4G with the main features intact, we face a number of challenges. The main desired features of 4G are as follows:

- **High usability and global roaming.** The end user terminals should be compatible with any technology, at any time, anywhere in the world. The basic idea is that the user should be able to take his mobile to any place; for example, from a place that uses code-division multiple access (CDMA) to another place that employs Global System for Mobile (GSM).

- **Multimedia support.** The user should be able to receive high data rate multimedia services. This demands higher bandwidth and higher data rates.
- **Personalization.** This means that any person should be able to access the service. The service providers should be able to provide customized services to different types of users.

To achieve the desired features listed above researches have to solve some of the main challenges that 4G is facing. The main challenges are described below.

- **Multimode user terminal.** To access different kinds of services and technologies, the user terminals should be able to configure themselves in different modes. This eliminates the need for multiple terminals. Adaptive techniques like smart antennas and software radio have been proposed for achieving terminal mobility.
- **Wireless system discovery and selection.** The main idea behind this is that the user terminal should be able to select the desired wireless system. The system could be LAN, Global Positioning System (GPS), GSM, and so forth. One proposed solution for this is to use a software radio approach where the terminal scans for the best available network and then downloads the required software.
- **Terminal mobility.** This is one of the biggest issues researchers are facing. Terminal mobility allows the user to roam across different geographical areas that use different technologies. There are two important issues related to terminal mobility. One is location management where the system has to locate the position of the mobile for providing service. Another important issue is handoff management. In the traditional mobile systems only horizontal handoff has to be performed, while in 4G systems both horizontal and vertical handoffs should be performed.
- **Personal mobility.** Personal mobility deals with the mobility of the user rather than the user terminals. The idea behind this is, no matter where users are located and what device they are using, they should be able to access their messages.
- **Security and privacy.** The existing security measures for wireless systems are inadequate for 4G systems. The existing security systems are designed for specific services. This does not provide flexibility for the users and, as flexibility is one of the main concerns for 4G, new security systems have to be introduced.
- **Fault tolerance.** As we all know, fault tolerant systems are becoming more popular throughout the world. The existing wireless system structure has a treelike topology, and hence if one of the components suffers damage, the whole system goes down. This is not desirable in the case of 4G. Hence, one of the main issues is to design a fault tolerant system for 4G.
- **Billing system.** 3G mostly follows a flat rate billing system where users are charged by a single operator for their usage according to call duration,

transferred data, and so on. But in 4G wireless systems, the user might switch between different service providers and may use different services. In this case it is hard for both the users and service providers to deal with separate bills. Hence the operators have to design a billing architecture that provides a single bill to the user for all the services used. Moreover, the bill should be fair to all the users.

It is convenient to discuss the challenges in the migration to 4G mobile systems and their proposed solutions by grouping them into three different aspects: mobile stations, system, and service. Table 9.1 shows a summary of key challenges and their proposed solutions.

9.3   BEYOND 3G AND TOWARD 4G NETWORKS

In the last few decades, the telecommunications industry has become especially responsive to market demands for new services and capabilities. This has been particularly true of the wireless segment of the industry, which has seen vigorous growth from cordless phones and first-generation analog cellular networks through 2G digital networks, low-speed mobile data networks, paging systems, and now 3G technologies that provide improved voice quality and integration of data and voice services. It has always been difficult to predict the future of the wireless communications industry, but there are certain trends that one can discern and try to project.[10] In what follows, we present some actual technologies.

WAN AND WLAN INTEGRATION

With respect to the critical issues of spectrum allocation, 3G systems are operating in licensed bands where service providers must make investments to secure access to those licenses. On the other hand, WLANs and WPANs operate in unlicensed bands, where one does not need to purchase the spectrum, and where the user is unencumbered by regulatory rules and regulations. However, there is also no regulatory control of signal interface in the unlicensed bands, and thus connectivity and link performance can often be problematic. The last several years have witnessed a renewed interest and vigorous growth in the use of unlicensed-band systems. One possible migration path is the eventual integration of WANs with WLANs in unlicensed bands.

AD HOC NETWORKING

Another important evolving technology is ad hoc networking, which uses a distributed network topology and has the capability for network reconfiguration without the need for a geographically fixed infrastructure. This technology was developed for military networking requirements but has found some applications in commercial voice and data services. The ad hoc networking topology is suitable, as an example, for rapid deployment of any wireless network in a mobile or fixed environment.

In ad hoc networking, the network is reconfigurable and can operate without the need for a fixed infrastructure. This is sometimes referred to as distributed-network

TABLE 9.1
Summary of Key Challenges and Their Proposed Solutions

	Key Challenges	Proposed Solutions
Mobile Station		
Multimode user terminals	To design a single user terminal that can operate in different wireless networks, and overcome the design problems such as limitations in device size, cost, power consumption, and backward compatibilities to systems	A software radio approach can be used: the user terminal adapts itself to the wireless interfaces of the networks[3].
Wireless system discovery	To discover available wireless systems by processing the signals sent from different wireless systems (with different access protocols that are incompatible with each other)	User- or system-initiated discoveries with automatic download of software modules for different wireless systems
Wireless system selection	Every wireless system has its unique characteristic and role; the proliferation of wireless technologies complicates the selection of the most suitable technology for a particular service at a particular time and place	The wireless system can be selected according to the best possible fit of user QoS requirements, available network resources, or user preferences[4]
System		
Terminal mobility	To locate and update the locations of the terminals in various systems; also, to perform horizontal and vertical handoffs as required with minimum handover latency and packet loss	Signaling schemes and fast handoff mechanisms are proposed in Ref. 5
Network infrastructure and QoS support	To integrate the existing non-IP-based and IP-based systems, and to provide QoS guarantee for end-to-end services that involve different systems	A clear and comprehensive QoS scheme for UMTS system has been proposed[6]; this scheme also supports interworking with other common QoS technologies
Security	The heterogeneity of wireless networks complicates the security issue; dynamic reconfigurable, adaptive, and lightweight security mechanisms should be developed	Modifications in existing security schemes may be applicable to heterogeneous systems; security handoff support for application sessions is also proposed[4]
Fault tolerance and survivability	To minimize the failures and their potential impacts in any level of treelike topology in wireless networks	Fault-tolerant architectures for heterogeneous networks and failure recovery protocols are proposed in Ref. 7

TABLE 9.1 (CONTINUED)
Summary of Key Challenges and Their Proposed Solutions

	Key Challenges	Proposed Solutions
	Service	
Multioperators and billing system	To collect, manage, and store the customers' accounting information from multiple service providers; also, to bill the customers with simple but detailed information	Various billing and accounting frameworks are proposed in Ref. 8
Personal mobility	To provide seamless personal mobility to users without modifying the existing servers in heterogeneous systems	Personal mobility frameworks are proposed; most of them use mobile agents. but some do not[9]

Reproduced with permission from S. Y. Hui and K. H. Yeung, "Challenges in the migration to 4G mobile systems," *IEEE Commun. Magazine* 41 (December 2003): 54–59.

topology. Such networks are used primarily in military communications, but have also found application in some commercial networks for voice and data transmission. Ad hoc networks may employ either single-hop (peer-to-peer) or multihop connectivity. By way of example, the 802.11 WLAN standards support single-hop peer-to-peer ad hoc networking. When an 802.11 terminal is powered up, it first searches for a beacon signal transmitted by an access point or another terminal announcing the existence of an ad hoc network. If no beacon is detected, the terminal takes the responsibility of announcing the existence of an ad hoc network. Also, several other wireless technologies, such as the Personal Handyphone System (PHS) and the NEXTEL satellite network, utilize peer-to-peer push-to-talk communication to establish a connection between pairs of voice terminals.

Important emerging areas for application of ad hoc networking technology include wireless PANs (WPANs). At present, the wireless industry differentiates WPANs from WLANs by their smaller signal coverage area, ad-hoc-only topology, low power consumption, plug-and-play architecture, and support of both voice and data devices. The earliest WPANs were BodyLANs, developed by the U.S. Department of Defense to connect sensors and communications devices carried by a soldier or attached to a soldier's clothing. Commercial applications of the same technology can provide connectivity among laptops, notepads, and cellular phones carried by the business traveler.

INFRASTRUCTURE-BASED ACCESS TECHNOLOGIES

In infrastructure-based broadband access, the network includes a fixed (wired) infrastructure that supports communication between mobile terminals and between mobile and fixed terminals. A typical example is a WLAN employing one or multiple access points (APs), with APs connected by a wired (typically cabled) backbone.

Two mobile stations in the same AP coverage area will communicate through that
AP, and wider area connectivity is supported by AP-to-AP communication over the
wired backbone. A common example of infrastructure-based broadband access is a
WLAN based on the popular IEEE 802.11b standard, operating in the 2.4 to 2.497
GHz band, providing broadband access to the Internet at data rates of 1, 2, 5.5, and
11 Mbps.

UWB AND S-T CODING

It is clear that CDMA is emerging as the preferred transmission technology for 3G
systems, providing enhanced voice quality and increased network capacity relative
to 2G systems, while orthogonal frequency division multiplexing (OFDM) has been
adopted in WLANs operating at 5 GHz. It is safe to project that OFDM will continue
to play an important role in the future of broadband wireless access. Other important
emerging technologies include ultrawideband (UWB) communication and space-
time (S-T) coding. The UWB concept uses transmission of narrow noiselike pulses
with spectrum extending over several gigahertz, and offers promise of supporting
very large numbers of simultaneous users. The S-T coding concept was devised to
improve performance and increase spectrum utilization efficiency on band-limited
wireless channels by combining channel coding, modulation, transmitter diversity,
and optional receiver antenna diversity.

LOCATION AWARENESS

Another evolving technology is position location awareness, and there is particular
interest now in indoor applications. Examples of how this technology can be beneficial include location of patients, medical professionals, and instrumentation in a hos-
pital; location tracking of merchandise in a large warehouse; and tracking of systems
and components in a large factory. Other potential applications include personnel
location in military, firefighting, and disaster-recovery situations. It is expected that
this technology will become an integral part of future wireless networks.

9.4   4G TECHNOLOGIES FROM THE USER'S POINT OF VIEW

It was originally expected that 4G would follow sequentially after 3G and emerge as
an ultra-high-speed broadband wireless network. For example in Asia, the Japanese
operator NTT DoCoMo defines 4G by introducing the concept of mobile multimedia
anytime, anywhere, to anyone; global mobility support; integrated wireless solution;
and customized personal service (MAGIC), which mainly concentrates on public systems and envisions 4G as the extension of 3G cellular service. This view is referred
to as the linear 4G vision and, in essence, focuses on a future 4G network that will
generally have a cellular structure and will provide very high data rates (exceeding
100 Mbps). In general, the latter is also the main tendency in China and South Korea.
Nevertheless, even if 4G is named as the successor of the previous generations, the
future is not limited to cellular systems, and 4G should not be seen exclusively as a
linear extension of 3G. In Europe, for example, the European Commission envisions

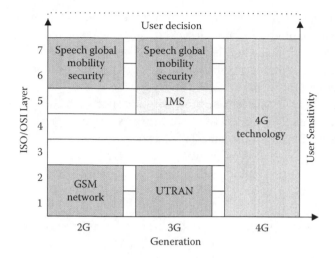

FIGURE 9.1   Protocol layer innovations versus wireless generations. Reproduced with permission from S. Frattasi et al. "Defining 4G technology from the user's perspective," *IEEE Network* 20 (January/February 2006): 35–41.

that 4G will ensure seamless service provisioning across a multitude of wireless systems, and provide optimum delivery via the most appropriate (i.e., efficient) network available. This view is referred to as the concurrent 4G vision. However, it does not give us the underlying methodology that could justify such a broad definition.[11]

To define and solve relevant technical problems, the system-level perspective has to be envisioned and understood with a broader view, taking the user as its departing point. This user-centric approach can result in a beneficial method for identifying innovation topics at all the different protocol layers, and avoiding a potential in terms of service provisioning and user expectations. Protocol layer innovations versus wireless generations are shown in Figure 9.1.

Novel technologies may have a significant and unpredictable impact on the user's behavior and, consequently, their usage may change the emerging products. So, understanding users in general means understanding how they change as the society around them changes, and specifically, how they change through the interaction with the products that are introduced. In particular, if technological developers start from understanding human needs, they are more likely to accelerate the evolutionary development of useful technology. The payoff from technological innovations is that they support some human needs while minimizing the downside risks. Therefore, responsible analysis of technology opportunities will consider positive and negative outcomes, thus amplifying the potential benefits for society. Clearly, there is a need for a new approach, there is a need for contextual understanding, and there is a major methodological challenge in the design of the next generation of wireless mobile communication technologies.

The top-down approach methodology focuses on a user-centric vision of the wireless world and consists of the following four stages:

- Consideration of the user as a sociocultural person with subjective preferences and motivations, cultural background, customs, and habits. This leads to the identification of the user's functional needs and expectations in terms of services and products. However, to interrelate sociocultural values and habits with functional needs is a sociological problem.
- Reflection about the functional needs and expectations derived from step 1 in everyday life situations, where new services are significant assets for the user. In this way, fundamental but exemplary user scenarios are derived from sketches of people's everyday lives.
- Extrapolation and interrelation of the key features of 4G from the user scenarios assessed in step 2. They represent the basic pillars for a very relevant and pragmatic definition of the forthcoming technology.
- Identification of the real technical step-up of 4G with respect to 3G by mapping the key features described in step 3 into advances in terms of system design, services, and devices. These technological developments are necessary to support the requirements of the different user scenarios defined in Step 2.

USER SCENARIOS

User scenarios can be elaborated and listed as some sketches like: business on the move, smart shopping, mobile tourist guide, personalization transfer, and so forth. All the key features are referred to as the user-centric system, which is illustrated in Figure 9.2. Inspired by the heliocentric Copernican theory, the user is located in

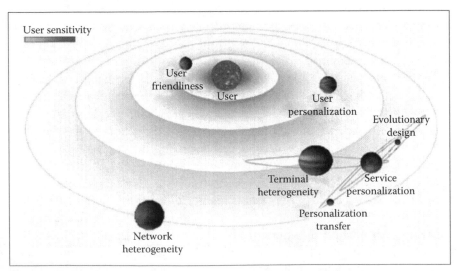

FIGURE 9.2   The user-centric system. Reproduced with permission from S. Frattasi et al. "Defining 4G technology from the user's perspective," *IEEE Network* 20 (January/February 2006): 35–41.

the center of the system, and the different key features defining 4G rotate around the user on orbits with a distance dependent on a user-sensitive scale.

Therefore, the farther the planet is from the center of the system, the less sensitive to it the user is. The decrease of user sensitivity leads to a translation toward the technocentric system, where network heterogeneity has a much stronger impact than user friendliness. Furthermore, this kind of representation also shows the interdependency between key features; for example, service personalization is a satellite of terminal heterogeneity. The user-centric system demonstrates that it is mandatory in the design of 4G to focus on the upper layers (maximum user sensitivity) before improving or developing the lower ones. If a device is not user friendly, for example, the user cannot exploit it and have access to other features, such as user personalization.

User Friendliness

User friendliness exemplifies and minimizes the interaction between applications and users thanks to a well-designed transparency that allows the users and the terminals to naturally interact (e.g., the integration of new speech interfaces is a great step for achieving this goal). For instance, users can get traveling information in the most user-friendly way: text, audio, or video format.

User Personalization

User personalization refers to the way users can configure the operational mode of their device and preselect the content of the services chosen according to their preferences. Because every new technology is designed keeping in mind the principal aim to penetrate the mass market and to have a strong impact on people's lifestyles, the new concepts introduced by 4G are based on the assumption that each user wants to be considered a distinct, valued customer who demands special treatment for his or her exclusive needs. Therefore, to embrace a large spectrum of customers, user personalization must be provided with high granularity, so that the huge amount of information is filtered according to the users' choices. For example, users can receive targeted pop-up advertisements. The combination between user personalization and user friendliness provides users with easy management of the overall features of their devices and maximum exploitation of all the possible applications, thus conferring the right value to their expense.

Terminal and Network Heterogeneity

To be a step ahead of 3G, 4G must provide not only higher data rates but also a clear and tangible advantage in people's everyday lives. Therefore, the success of 4G will depend on a combination of terminal heterogeneity and network heterogeneity. Terminal heterogeneity refers to the different types of terminals in terms of display size, energy consumption, portability/weight, complexity, and so forth. Network heterogeneity is related to the increasing heterogeneity of wireless networks due to the proliferation in the number of access technologies available (e.g., UMTS, *WiMAX*, WiFi, Bluetooth). These heterogeneous wireless access networks, shown in Figure 9.3, typically differ in terms of coverage, data rate, latency, and loss rate. Therefore, each of them is practically designed to support a different set of specific services and devices.

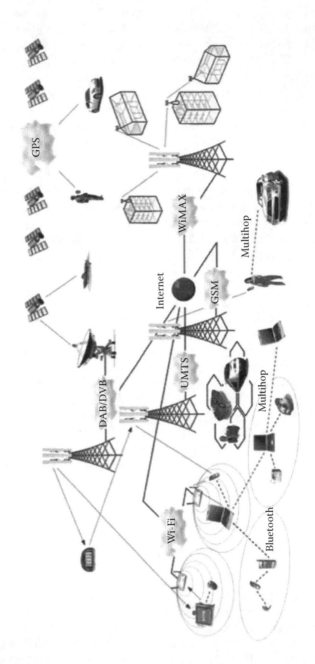

FIGURE 9.3 Heterogeneous wireless networks. Reproduced with permission from S. Frattasi et al. "Defining 4G technology from the user's perspective," *IEEE Network* 20 (January/February 2006): 35–41.

Service Personalization

4G will encompass various types of terminals, which may have to provide services independently of their capabilities. To optimize the service presentation, tailoring content for end-user devices will be necessary. The capabilities of the terminal in use will determine whether or not new services are to be provisioned, so as to offer the best enjoyment to the user and prevent declining interest and elimination of a service offering. This concept is referred to as service personalization. It implicitly constrains the number of access technologies supportable by the user's personal device. However, this limitation may be solved in two ways: by the development of devices with evolutionary design, and by means of a personalization transfer.

Having the most adaptable device in terms of design can provide customers with the most complete application package, thus maximizing the number of services supported.

The advantage for the customers is to buy a device on which they have the potential to get the right presentation for each service, freeing the device from its intrinsic restrictions. In a private environment, users can optimize the service presentation, thus exploiting the multiple terminals they have at their disposal.

The several levels of dependency highlighted by the user-centric system definitely stress the fact that it is not feasible to design 4G starting from the access technology in order to satisfy the user's requirements.

HETEROGENEOUS SYSTEMS INTEGRATION

The real step-up of 4G with respect to 3G can be summarized with seamless integration of already existing and new networks, services, and terminals. The final goal is to satisfy ever-increasing user demands. The 4G coming generation will be able to allow complete interoperability among heterogeneous networks and associated technologies. Generational evolution from 2G to 4G is demonstrated in Figure 9.4.

Although 2G has focused on full coverage for cellular systems offering only one technology, and 3G provides its services only in dedicated areas and introduces the concept of vertical handover through the coupling with WLAN systems, 4G will be a convergence platform extended to all the network layers. Hence, the user will be connected almost anywhere thanks to widespread coverage due to the exploiting of the various networks available. In particular, service provision will be granted with at least the same level of QoS when passing from one network's support area to that of another.

On the other hand, resource sharing among the various networks available will smooth the problem related to spectrum limitations relative to 3G.

HETEROGENEOUS SERVICES

Apart from some soft additional emerging services (e.g., fast Internet connection, pop-up advertisements, etc.), there is still a lack of really new and distinct services that will enable new applications with tangible benefits for their users. Therefore, we envision the real advantage in terms of services that 4G will bring will be based on the integration of technologies designed to match the needs of different market segments.

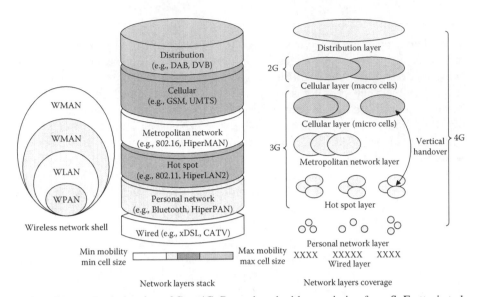

FIGURE 9.4   Evolution from 2G to 4G. Reproduced with permission from S. Frattasi et al. "Defining 4G technology from the user's perspective," *IEEE Network* 20 (January/February 2006): 35–41.

Short-range wireless technologies, such as WiFi and Bluetooth, will enable machine-to-machine (M2M) communications. As an example, users sign up online to the waiting list, which sends them back the approximate waiting time. Also, users can transfer content to a publicly available larger display. In particular, from the sociological point of view, in the latter case, the private and public spheres are mixed. This recombination can result in the enhancement of public access such that access to displays will be as common as access to public telephone booths is nowadays. Short-range wireless technologies also open the possibility for cooperative communication strategies, which can provide better services at lower costs, thus maximizing the users' profit. In this way, the technology increases the social cooperative behavior and empowers the consumer to make clever use of it. Hence, the user's personal device is no longer a mere medium for transferring information, but a social medium that helps to build groups and friendships.

Because 3G networks are not able to deliver multicast services efficiently or at a decent level of quality, the synergy of UMTS and digital audio/video broadcasting (DAB/DVB) will open the possibility to provide to mobile users interactive or on-demand services—so-called data casting—and audio and video streaming in a much more efficient way than using the point-to-point switched network.

The embedding in the user terminal of a GPS receiver will offer the essential feature of location awareness, which is necessary to provide users with the most comprehensive and extensive level of information, thus bringing about a real revolution in terms of personalized services. The user terminal can hence provide not only location-based information, such as maps and directions to follow to reach a specific place, but also useful information relevant in time and space, such as pop-up advertisements concerning offers in shops nearby. However, GPS technology can only support outdoor localization. Indoor localization requires the cooperation of short-range wireless technologies.

Finally, it is worth highlighting that although users are attracted by high data rates, they would certainly be even more attracted by useful services exploiting high data rates. The support of imaging and video, as well as high-quality audio, gives service providers (SPs) a myriad of possibilities for developing appealing applications. These features, blended with the support of high data rates, result in a particularly attractive combination. Indeed, in addition to an explosive increase in data traffic, we can expect changes to the typically assumed downlink–uplink traffic imbalance. Data transfer in the uplink direction is expected to increase considerably and, as a result of these trends, the mobile user will ultimately become a content provider (CP). In future wireless networks, the CP concept will broaden to encompass not only the conventional small- or middle-size business-oriented service companies, but also any single user, or group of users. Mobile CP will open a new chapter in service provision.

INTERWORKING DEVICES

Because 4G is based on the integration of heterogeneous systems, the future trend of wireless devices will move toward multimode/reconfigurable devices and exploitation of interworking devices.

The user terminal is able to access the core network by choosing one of the several access networks available, and to initiate the handoff between them without the need for network modification or interworking devices. This leads to the integration of different access technologies in the same device (multimodality) or to the use of the software-defined radio (SDR) (reconfigurability).[14] For example, whereas the integration of Bluetooth in the user terminal will enable a personalization-transfer service, a built-in GPS receiver will allow users to utilize their personal devices as navigators just by plugging them in their cars, thereby lightening the number of needed devices. However, the reconfigurability of the user terminal could be a key aspect that would make the future of 4G technology as highly adaptable as possible to the various worldwide markets.

To reduce the hardware embedded in the user terminal and the software complexity, the use of interworking devices is exploited. For example, consider the case of integrated AP performing the interworking between a WMAN technology and a WLAN technology, such as *WiMAX* and WiFi, respectively[15]: the WMAN is considered the backbone and the WLAN the distribution network; therefore, instead of integrating both technologies, the user terminal will only incorporate the WiFi card. The price to be paid for this relief is, therefore, increased system (infrastructure) complexity.

9.5 ALL-IP NETWORK

All-IP network (AIPN) describes a longer-term vision for the 3GPP networks.[16] For this work item, a feasibility study including user scenarios as well as the service requirements was created.[17]

AIPN is a common IP-based network that provides IP-based network control and transport across and within multiple access systems. This includes the provision of IP-based mobility with a performance comparable to other cellular mobility mechanisms, independent of specific access or transport technologies. It is the aim of the AIPN to provide a seamless user experience for all services within and across the various access systems.

AIPN will support a wide range of networking scenarios. These scenarios include a moving PAN receiving seamless service while handing over from a home WLAN to a city-owned hotspot service to the AIPN. Scenarios also include a virtual secure personal network where the AIPN provides the connection between different (sub) networks owned by the same user, providing access to the AIPN for the group of users that move together. The system architecture evolution (SAE) work item targets the timeframe between today's 3G Partnership Project (3GPP) networks and AIPN, and a subset of the AIPN scenarios. The objective is the evolution of the 3GPP network to a higher-data-rate, lower-latency, packet-optimized system that supports multiple radio access networks and mobility between them.[18,19]

SAE covers the networking aspects of the 3GPP system. It is synchronized with a work item on the 3GPP RAN, which includes a new air interface technology targeting peak bit rates of up to 100 Mbps.

In SAE, interworking with non-3GPP RANs is supported by a fundamental redesign of the 3GPP architecture, which includes an intersystem mobility anchor point. Mobility in 3GPP today is achieved by a 3GPP-specific protocol GTP (GPRS Tunneling Protocol). For intra-3GPP mobility, GTP will be maintained. However, to achieve mobility across heterogeneous access networks, the usage of native IP-based protocols (e.g., mobile IP) has been agreed upon. The challenge is to provide seamless service continuity and to maintain and support the same security, privacy, and charging capabilities available in today's 3GPP system when moving between the different network types.

Because a strong focus in the evolution of the 3GPP system is to also interwork with non-3GPP systems, for which it might become much harder to establish long-lived interworking agreements, the need for a dynamic mechanism or automatism such as network composition to enable interworking across heterogeneous networks arises.

In the AIPN work item, PANs and personal networks (PNs) have been identified as new types of user-owned networks that need to be supported by and interwork with the AIPN network of the 3GPP operator. A separate work item has been initiated for the study of personal network management (PNM). The objective is specifying the service requirements on how to manage and connect devices of a single user that are forming a PN or PAN. As this is 3GPP, the focus of course is on management support by the 3GPP network.[20]

NETWORK COMPOSITION PROCEDURE

Although the network composition concept is applicable to any kind of network, for this study item these results are considered in a 3GPP context and focus on user cases relevant to 3GPP operators. Feedback from 3GPP meanwhile resulted in updates and refinements of the continuing work in the ambient networks project.

Today, 3GPP networks interwork with other networks. This means that user-plane traffic is exchanged between these networks after configuring control-plane interworking. A concrete example of control-plane interworking is a roaming agreement between 3GPP networks. Another example is the WLAN–3GPP interworking defined in Reference 21, where a WLAN access network provides networks access to mobile nodes based on their 3GPP subscription.

The concept of network composition enables dynamic interworking between networks, whereby interworking is achieved through a uniform procedure that enables interworking between the composing networks at the control plane level. As a result, network composition is a mechanism through which all kinds of interworking scenarios are independently involved so that control functionalities can be achieved.

Network composition furthermore is a uniform procedure that allows dynamic establishment of interworking and interworking agreements among different networks. Network composition is also a uniform procedure. The composing networks may be of a rather heterogeneous nature, ranging from (for example) operator-owned 3GPP networks over PANs to third-party-operated access networks. The interworking enabled through network composition can be quite loose, as in the case of a dynamic roaming agreement. It can also be very tight, as in the case of the dynamic integration of a non-3GPP access network into the evolved 3GPP network.

For 3GPP network operators, network composition is interesting for a number of reasons. For example, a dynamic, automated procedure for establishing network interworking saves costs compared to off-line configuration. Composition also allows operators' networks to react quickly to changing resource demands by automatically extending network resources. Furthermore, the uniformity of the composition procedure is thought to also facilitate interworking with future and emerging network types and network technologies, with reduced standardization effort.

The procedure for composing networks includes a number of phases denoted media sense, discovery/advertisement, security and internetworking establishment, composition agreement negotiation, and composition agreement realization. Some of these phases might be optional depending on the composition scenario, and they are not necessarily passed in a one-way fashion. User interaction in all of these phases is minimized. We therefore assume the network is configured with policies that determine how and when to compose. Figure 9.5 shows the basic flow diagram.[22]

In the first step, a network willing to compose must sense the physical or logical medium. Depending on the particular scenario, media sense may be performed in different ways. For instance, a new AP detects a beacon of the operator network to which it should attach, or a user device is switched on and searches for networks in its vicinity (i.e., a RAN or a PAN).

The *sensing* also includes the case of discovering the link to a specific remote network (no physical vicinity); for example, when two operator networks that are to be connected discover each other, facilitating what is known as *virtual composition*.

Depending on the situation, media sense is followed by either an advertisement or a discovery phase, or they could also be combined. These messages can be broadcasted or they can be sent as targeted composition queries.

With active advertisements, a network can offer resources and control services to other networks. The advertisement message includes an identifier, possibly based

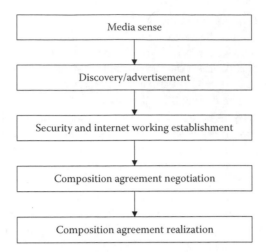

FIGURE 9.5 The basic flow diagram for the composition procedure. Reproduced with permission from C. Koppler et al. "Dynamic network composition for beyond 3G networks: A 3GPP viewpoint," *IEEE Network* 21 (January–February 2007): 47–52.

on cryptographic techniques used by a network, which is included to bind the advertisement to a particular network, and may be authenticated and/or authorized at a later phase. Alternatively, the network may discover a particular resource or control service by either actively asking particular neighbors, or by listening to advertisements by other networks. The discovery/advertisement phase allows setting up a list of candidate networks and selecting one for composition.

Two networks intending to compose need to establish basic security and internetworking connectivity. The identities of the networks might be authenticated and authorized using a trusted third party. Alternatively, the required trust relationship may be based on a preestablished shared secret or may even be opportunistic (e.g., the networks only make sure they keep communicating with the same network). At some point during this message exchange or immediately afterward, internetworking connectivity between the two networks is established.

The next step of the composition procedure is the negotiation of a composition agreement (CA). The CA includes the policies to be followed in the composed network; the identifier of the composed network; how logical and physical resources are accessed, controlled, and/or shared between the composing networks; and so forth. Together with the framework agreement, the CA specifies the rights and duties of each composing party. For example, in the case of a moving network, the CA may settle that a mobile router[23] performs mobility control toward the outside world on behalf of all mobile nodes in the moving network. In case of 3GPP–WLAN interworking, the CA may settle that the WLAN offers access to authentication, authorization, and accounting (AAA) services in the 3GPP network. Also, the CA should determine whether and how IP addresses may be reassigned.

To speed the process, in certain environments the use of CA templates as well as the reuse of previously established CAs are envisioned.

Where the CA includes commercial factors, the CA should be digitally signed by both networks to provide nonrepudiation. It is possible that the process of establishing a CA may involve increasing levels of authorization; for example, negotiation of certain resources and services may be authorized only once the two networks have agreed on the commercial aspects of the CA.

The composition agreement realization phase represents the completion of the composition. During this phase, network elements are configured to reflect the CA, and each of the composing networks is configured to reflect the CA. Each of the composing networks must carry out the configuration of its own resources by updating their policies and control functions. In practice, for mobile nodes, this may mean in a moving network that they switch off their own 3GPP-based mobility control such that the mobile router can handle it on their behalf.

The result of the network composition procedure is either a new network, an enlarged network (i.e., one network is absorbed into the other), or two interworking networks.

Inside a composed network, if one or more of the networks decide to discontinue their interworking (which could be due, for example, to switch-off of one network, a node leaving coverage, etc.), decomposition takes place, which then also leads to the invalidation of the composition agreement.

ALL-IP 4G NETWORK ARCHITECTURE

The overall 4G architecture is IP version 6 (IPv6)-based supporting seamless mobility between different access technologies. Mobility is a substantial problem in such an environment, because intertechnology handovers have to be supported. For example, Ethernet IEEE 802.3 can be targeted for wired access, WiFi IEEE 802.11b for WLAN access, and WCDMA, the radio interface of UMTS, for cellular access. With this diversity, mobility cannot be simply handled by the lower layers, but needs to be implemented at the network layer. An IPv6-based mechanism has to be used for interworking, and no technology-internal mechanisms for handover, either on the WLAN or on other technology, can be used. IPv6 mobility will handle handover between cells.

The key entities of the general network architecture are as follows:

- **A user.** a person or company with a service level agreement (SLA) contracted with an operator for a specific set of services.
- **A mobile terminal (MT).** a terminal from which the user accesses services.
- **Access router (AR).** the point of attachment to the network, which takes the name of radio gateway (RG) for wireless access (WCDMA or IEEE 802.11).
- **Paging agent (PA).** entity responsible for locating the MT when it is in idle mode while there are packets to be delivered to it.
- **The AAA and charging (AAAC) system.** entity responsible for service level management including accounting and charging.
- **Network management system (NMS).** the entity responsible for managing and guaranteeing availability of resources in the core network, as well as overall network management and control.

This network is capable of supporting multiple functions:

- Interoperator information interchange for multiple-operator scenarios
- Confidentiality both of user traffic and of the network control information
- Mobility of users across multiple terminals
- Mobility of terminals across multiple technologies
- QoS level guarantees to traffic flows (aggregates)
- Monitoring and measurement functions, to collect information about network and service usage
- Paging across multiple networks to ensure continuous accessibilty of users

9.6 QOS ISSUES FOR 4G NETWORKS

Wireless communications are currently experiencing a fast integration toward the beyond 3G (B3G)/4G era. This represents a generational change in wireless systems; they not only will be involved in noteworthy changes in technologies to augment their communication capability, but also will be characterized by demonstrating a keen interest in users more than technologies. New terms are rapidly coming into use[24] that try to describe this novel approach to communications: individual-centric, user-centered, or ambient-aware communications. The common idea behind these terms is surely represented by a clear focus on (multimedia) services tailored to user needs, and personalized and ubiquitous access. Last but not least, B3G systems have been envisioned as an evolution and convergence of mobile communications systems and Internet technologies to offer a multitude of services over a variety of access technologies.[25]

Fundamental assumptions and requirements driving B3G/4G design are being tackled by the Wireless World Research Forum (WWRF), which is working on a series of white papers outlining B3G visions and roadmaps, architectural principles, research challenges, and candidate approaches. From the WWRF, a novel vision of a user always connected to the global communication infrastructure emerges.

Future 4G mobile systems are envisioned to offer wireless services to a wide variety of mobile terminals ranging from cellular phones and personal digital assistants (PDAs) to laptops. This wide variety of mobile terminals is referred to as heterogeneous terminals. Heterogeneous terminals have various processing power, memory, storage space, battery life, and data rate capabilities.

To use the spectrum efficiently, heterogeneous terminals in 4G should use the same spectrum in case the users are interested in the same devices. One solution is the use of multiple description coding (MDC), where the service information is split into multiple streams. MDC has the capability to split the information stream into multiple substreams, where each of the substreams can be decoded without the information carried by the neighboring substreams, and therefore have no dependencies to other substreams such as layered video coding. In a multicast scenario, high class terminals would receive a large number of streams, while low class terminals would go for a smaller number. Note that the substreams of the low class terminal are also received by the high class terminal. Therefore, the spectrum is used more efficiently. The quality at the receiver in terms of video size, frame rate, and so forth increases as the number of received descriptors increases. The flexibility of the bandwidth assigned to each descriptor, and the number of descriptors assigned to end users, makes MDC a very attractive coding scheme for 4G networks.

The advantage of MDC is achieved at the expense of higher bandwidth usage due to the smaller video compression of the encoding process. Therefore, existing video traffic characterizations such as single and multiple layer coding, as presented in Reference 27, cannot be used for the evaluation of future wireless communication systems as they underestimate the bandwidth required.

4G wireless systems are expected to extend wireless service to high data rates in high-mobility environments.

Developers need to do much more work to address end-to-end QoS. They may need to modify many existing QoS schemes, including admission control, dynamic resource reservation, and QoS renegotiation to support 4G users' diverse QoS requirements. The overhead of implementing these QoS schemes at different levels requires careful evaluation.

A wireless network could make its current QoS information available to all other wireless networks in either a distributed or centralized fashion so they can effectively use the available network resources. Additionally, deploying a global QoS scheme may support the diverse requirements of users with different mobility patterns. The effect of implementing a single QoS scheme across the networks instead of relying on each network's QoS scheme requires study.

QoS provisioning comprises data plane (mainly traffic control, e.g., classification and scheduling) and control plane (mainly admission control and QoS signaling) functions. Following the above exploration of mobility problems, we can identify the fundamental difference of QoS provisioning in all-IP 4G mobile networks from traditional, wired or wireless IP networks: whereas its resource control mechanisms can be similar to that of traditional networks, changing a location during the lifetime of a data flow introduces changed data path, thus requiring identifying the new path and installing new resource control parameters via path-coupled QoS signaling. Hence, the problem is how to apply any QoS signaling mechanism to achieve end-to-end resource setup in mobility scenarios. The current QoS signaling protocol, Resource Reservation Protocol (RSVP), exhibits lack of intrinsic architectural flexibility in adapting to mobility requirements.[28] Difficulties arise, for example, because of its inability to adapt to the introduction of mobility routing in the data plane encountered in 4G networks, which results in either solutions that are too complicated, or simply being unable to satisfy the needs.[29,30]

Availability of the network services anywhere, at anytime can be one of the key factors that attract individuals and institutions to the new network infrastructures, stimulate the development of telecommunications, and propel economies. This bold idea has already made its way into the telecommunication community, bringing new requirements for network design, and envisioning a change of the current model of providing services to customers. The emerging new communications paradigm assumes that a user will be able to access services independently of his or her location, in an almost transparent way, with the terminal able to pick the preferred access technology at the current location (ad hoc, wired, WLAN, or cellular), and move between technologies seamlessly, that is, without noticeable disruption.

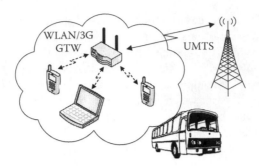

FIGURE 9.6 Cellular coverage connecting a moving WLAN to the Internet. Reproduced with permission from S. Pack et al. "Performance analysis of mobile hotspots with heterogeneous wireless links," A. Lera et al. "End-to-end QoS provisioning in 4G with mobile hot spots," *IEEE Network* (Sept.-Oct. 2005): 26–34.

Unified, secure, multiservice, and multiple-operator network architectures are now being developed in a context commonly referred to as B3G networks or, alternatively, 4G networks. The 4G concept supports the provisioning of multiple types of services, ranging from simple network access to complex multimedia virtual reality, including voice communication services, which are themselves a challenge in packet-based mobile communications environments.

As a result of the heterogeneity of the access technologies, IPv6 is targeted as the common denominator across multiple access technologies, and makes the solution basically independent of the underlying technology—and therefore future proof. However, fitting such important concepts as support for QoS, AAAC, and mobility into the native Internet architecture poses numerous difficulties and is a real challenge.

Networks in motion (NEMO) is one of the most interesting concepts emerging from the 3G/4G scenario. A NEMO consists of one or more mobile routers with a number of devices connected to it. It can change its point of attachment to other networks as it physically moves or changes in topology. Among NEMOs, so-called moving WLANs (m-WLANs) consist of a collection of wireless terminals carried by a platform in motion. Each m-WLAN communicates via wireless links with a fixed or wireless backbone through an anchor or master device. An example of cellular coverage connecting a moving WLAN to the Internet is presented in Figure 9.6.[25]

A group of users are traveling on public transport (a bus). Each user owns a terminal device equipped with a WLAN network interface card; only some terminals are equipped with a dual 3G-WLAN network card and act as gateways toward the 3G/B3G network. During their wait at the bus station the travelers access the Internet through a public WLAN hotspot; when they get on the bus and leave the hotspot, they continue accessing the Internet through a multimode gateway provided in the mobile hotspot.

PROVIDING QUALITY OF SERVICE

The design principle for QoS architecture was to have a structure which allows for a potentially scalable system that can maintain contracted levels of QoS. Eventually, especially if able to provide an equivalent to the Universal Telephone Service, it could possibly replace today's telecommunications networks. Therefore, no specific network services should be presumed nor precluded, though the architecture should be optimized for a representative set of network services. Also, no special charging models should be imposed by the AAAC system, and the overall architecture must be able to support very restrictive network resource usage.

In terms of services, applications that use voice over IP (VoIP), video streaming, web, e-mail access, and file transfer have completely different prerequisites, and the network should be able to differentiate their service. The scalability concerns favor a differentiated service (DiffServ) approach.[32,33] This approach is based on the assumption that end-to-end QoS assurance is achieved by a concatenation of multiple managed entities. With such requirements, network resource control must be under the control of the network service provider. It has to be able to control every resource, and to grant or deny user and service access. This requirement calls for flexible and robust explicit connections admission control (CAC) mechanisms at the network edge, and the ability to take fast decisions on user requests.

Service provisioning for 4G networks is based on separation of service and network management entities. We can define a service layer, which has its own interoperation mechanisms across different administrative domains (and can be mapped to the service provider concept), and a network layer, which has its own interoperation mechanism between network domains. An administrative domain may be composed of one or more technology domains. Service definitions are handled inside administrative domains, and service translation is done between administrative domains.[34]

Each domain has an entity responsible for handling user service aspects (the AAAC system), and at least one entity handling the network resource management aspects at the access level (the QoS broker). The AAAC system is the central point for AAA. When a mobile user enters the network, the AAAC is supposed to authenticate the user. Upon successful authentication, the AAAC sends to the QoS broker the relevant QoS policy information based on the SLA of the user, derived from his or her profile. From then, it is assumed that the AAAC has delegated resource-related management tied to a particular user to the QoS broker. However, two different network types have to be considered in terms of QoS: the core and the access. In the DiffServ approach, the core is basically managed per aggregate based on the network services, and not by user services. In that sense, core management is decoupled from the access.

Service will be offered by the network operator independently of the user applications, but will be flexible enough to support them. The services may be unidirectional or bidirectional. In fact, the QoS architecture can support any type of network service, where the only limit is the level of management complexity expressed in terms of complexity of interaction between the QoS brokers, the AAAC systems, and the AR that the network provider is willing to support. Users will then subscribe to SLAs consisting of different offerings.

END-TO-END QoS SUPPORT

Three distinct situations arise in the QoS architecture:

- Registration, when a user may only use network resources after authentication and authorization
- Service authorization, when the user has to be authorized to use specific services
- Handover, when there is a need to reallocate resources from one AR to another[35]

The registration process is initiated after a care-of address (CoA) is acquired by the MT via stateless autoconfiguration, avoiding duplicate address detection (DAD) by using unique layer-2 identifiers to create the interface identifier part of the IPv6 address. However, getting a CoA does not entitle the user to use resources, except for registration messages and emergency calls. The MT has to start the authentication process by exchanging the authentication information with the AAAC through the AR. Upon a successful authentication, the AAAC system will push the NVUP (network view of user profile) to both the QoS broker and the MT.

One of the specific problems of IP mobility is assuring a constant level of QoS. User mobility can be assured by means of fast handover techniques in conjunction with context transfer between network elements (ARs–old and new QoS brokers).

Building an all-IP architecture based on DiffServ introduces a problem of how to create per-domain services for transport of traffic aggregates with a given QoS. Per-domain services support data exchange by mixing traffic of different applications; therefore different aggregates are required to support delay-sensitive traffic and delay tolerant traffic, as well as inelastic, elastic, and network maintenance traffic.

As applications generate traffic of different characteristics in terms of data rates, level of burstiness, packet size distribution, and because the operator needs to protect the infrastructure against congestion, it is very important that aggregate scheduling will be accompanied by:

- Per-user rate limitation performed in the ingress routers (ARs) based on user profiles
- Dimensioning and configuration of network resources to allow for a wide range of user needs and services
- Resource management for edge-to-edge QoS

To maintain resource utilization in the entire domain, the QoS broker is expected to know the demand, current utilization factors of all links based on incoming call parameters or on measurements, and on additional information such as a traffic load matrix. The real data traffic is provided by monitoring functions in the network, while traffic matrixes are induced on historical profiling (and with varying degrees of complexity). The QoS broker will then use this knowledge for admission control and resource provisioning. The mathematical formulations have the disadvantage of relying on the worst-case scenario, which leads to substantial overdimensioning.

An architecture for supporting end-to-end QoS is able to support multiservice, multioperator environments handling complex multimedia services, with per user and per service differentiation, integrating mobility and AAAC aspects. The main elements are the MT, the ARs, and the QoS brokers.

9.7 SECURITY IN 4G NETWORKS

Security requirements of 3G networks have been widely studied in the literature. Different standards implement their security for their unique security requirements. For example, GSM provides highly secure voice communications among users. However, the existing security schemes for wireless systems are inadequate for 4G networks. The key concern in security designs for 4G networks is flexibility. As the existing security schemes are mainly designed for specific services, such as voice service, they may not be applicable to 4G environments that will consist of many heterogeneous systems. Moreover, the key sizes and encryption and decryption algorithms of existing schemes are also fixed. They become inflexible when applied to different technologies and devices (with varied capabilities, processing powers, and security needs). To design flexible security systems, some researchers are starting to consider reconfigurable security mechanisms.

Security in 4G networks mainly involves authentication, confidentiality, integrity, and authorization for the access of network connectivity, and QoS resources for the MN flows. First, the MN needs to prove authorization and authenticate itself while roaming to a new provider's networks. AAA protocols (such as Radius, COPS, or Diameter)[36] provide a framework for such support, especially for control plane functions (including key establishment between the MN and AR, authenticating the MN with AAA server(s), and installing security policies in the MN or the AR data plane such as encryption, decryption, and filtering), but they are not well suited for mobility scenarios. There needs to be an efficient, scalable approach to address this. The Extensible Authentication Protocol (EAP)[37] provides a flexible framework for extensible network access authentication and potentially could be useful. Second, when QoS is concerned, QoS requests need to be integrity protected, and moreover, before allocating QoS resources for an MN flow, authorization needs to be performed to avoid denial of service attacks. This requires a hop-by-hop way of dynamic key establishment between QoS-aware entities to be signaled on. Finally, most security concerns lie in network layer functions. Although security can also be provided by higher layers, the network layer provides privacy and data integrity between two communicating applications.[38]

INFRASTRUCTURE SECURITY FOR FUTURE MOBILE NETWORKS

The MNs in future will be open to different services and SPs. This will mean that the MN operator (MNO) will have trust relations with different networks and SPs. An MNO should not limit the services it provides to its users, whereas a LAN administrator can. A LAN administrator supports a limited set of users.

Security features groups are defined by 3GPP. Each of these accomplishes certain security objectives.[39–42] There are five security features groups:

- **Network access security.** This comprises encryption of the data and the signaling data integrity. The serving network (SN) verifies the validity of the UMTS Subscriber Identity Module (USIM) and its entitlement to receive UMTS services, while the USIM verifies the authenticity of the network. When the SN requests security data from the home environment, the latter should verify that the former is a trusted network that can receive the requested data. Finally, the access network and the mobile equipment (ME) can communicate. This is known as authentication and key agreement (AKA).
- **Network domain security (NDS).** This is a set of security features that enable nodes in the network provider domain to securely exchange the signaling data.
- **User domain security.** Includes features within the USIM so that only authorized users, i.e., these who know the personal identification number (PIN), should be able to access the USIM. Some USIM data should be protected from being accessed by the user.
- **Application domain security.** Includes security mechanisms for accessing the user profile data and IP security mechanisms to provide secure messaging between the network and the USIM.
- **Visibility and configurability of security.** This feature enables users to inform themselves whether a security feature is in operation and whether the use and operation of a certain service should depend on the security feature.

3GPP also defines network access services for the IP multimedia core network subsystem, which is essentially an overlay over the packet-switched domain. A separate security association is required between the multimedia client and the IP multimedia core network subsystem before access is guaranteed to the multimedia services.

There are four types of security domains that can be identified:

- The user, which is also the subscriber of external parties or related parties
- The MNO
- Parties with which the MNO has a trust relation
- External parties with which the MNO has no trust relation

The purpose of introducing a domain model is to

- Identify types of domains based on their security policy plus their relations and dependencies
- Add an atomicity that helps maximize the number of possible combinations
- Identify reference points (RPs) between the domains which can lead to implementations of interfaces and might be used as points of conformance or policy enforcement
- Build a model that is simple but helpful to in understanding the interworking between domains

Infrastructure Security Definition

Mobile communication systems have a number of distinct characteristics and properties related to architecture and technologies. Some of these present challenges to developing network infrastructure security:[43]

- **Flexibility/multifunctionality.** Future mobile systems will provide not only voice, but also data and multimedia communications with seamless mobility over heterogeneous access technologies. New function modules and elements have to be installed in the network. Most of the subsystems will define their security architecture, but to be posed within one network, requirements on the infrastructure security must be soundly concluded.
- **Use standard technology.** Unlike the present and former systems, the future mobile communication systems will adopt IP as the basic network technology, which is well studied and understood among a rather large and open engineering society—the Internet world. Therefore, many IP-based security threats in the Internet could be easily imported here.
- **Critical Services.** More and more services that ease our daily lives and serve some critical missions will be created and provided over the future mobile communication systems. Users' satisfaction and perhaps safety will be very dependent on the mobile systems, which could then be a target of malicious people/groups.

To distinguish from other aspects of security, the goals of infrastructure security of mobile communication networks are identified as

- Physical security of the network infrastructure including the nodes and the cables.
- Access control to the network infrastructure nodes.
- Protection of the infrastructure nodes against unauthorized access, for example, obtain administrator's power by exploiting software flaws; theft of sensitive data, including network configurations and internal structure information that should be kept in secret, and other critical databases.
- Protection of critical data against unintended (user mistakes) or intended changes (tampering), for example, erase/change of the critical network configurations and the databases of routers, servers, and other nodes.
- Availability of the network infrastructure.
- Protection of critical infrastructure nodes against denial of service (DoS) attacks.
- Protection of critical communication media, including wireless link.
- Security of network management signaling; infrastructure could be automatically maintained and managed, network management signaling must be secured (e.g., authentication, confidentiality, integrity, and antireplay, depending on the signaling).
- Amenability to security management; infrastructure, including deployed security mechanisms, is subject to change. For the reliable enforcement of

the security policies, the infrastructure security has to be capable of adapting according to OAM (operations and management) processes which are aiming for adequate security management.

- Support of secure coupling to foreign networks; as the main characteristics of MNO domains is the openness to interoperate with various foreign domains, the network infrastructure security mechanisms have to support interoperability with user, CP/SP, and other MNO domains.

Infrastructure Security Requirements

Based on the security goals, a nonexhaustive list of major requirements on infrastructure security is as follows:

- All the elements of a communication network should be hardened so that it is resistant to various security attacks. Maintaining the latest update levels of network elements and their software is an essential part of any OAM concept applied in a network domain.
- Critical infrastructure elements must be identified and well protected; the necessity of redundant elements should be carefully evaluated, for example, to still allow management during exceptional situations.
- Information on the network's internal structure, including the topology, the platform types, the distribution of functional elements, the capacities, and so on, or the customer data concerning location, service usage, usage pattern, account information, and so on, should be made available only to authorized parties and only to the extent that is actually required. Limited availability of this information reduces the knowledge about potential targets.
- An intrusion detection system (IDS) should be deployed in the network to detect, monitor, and report security attacks, such as DoS and attacks that utilize system flaws, as well as any compromise of system security.
- Fast response to security attacks and automatic recovery of security compromise must be provided to increase the probability of business continuity/ continuous operation, and to mitigate the effects of attacks.
- The network should be easy to manage and realistic and realizable security policies must be developed.
- Rapid couplings of the networks of different administrative domains must be secured (authentication, integrity, confidentiality, availability, and anti-replay protection, etc.) against external attackers.
- Mutual dependencies between the infrastructure security and new services to be created must be minimized.
- Authenticity, confidentiality, integrity, antireplay protections for network management signaling shall be provided.
- Interoperability between the administrative domains shall not compromise the security of any involved domain.

SECURE HANDOVER BETWEEN HETEROGENEOUS NETWORKS

Handover is a basic mobile network capability for dynamic support of terminal migration, while handover management is a process of initiating and ensuring a seamless and lossless handover of a mobile terminal from a region covered by one base station (BS) to another BS, which may belong to a different access network (AN). Handover procedures involve a set of protocols to notify all related entities of a particular connection which has been executed. In data networks, the mobile terminal is usually registered with a particular point of attachment. In mobile networks, an idle mobile terminal would have selected a particular BS that is serving the cell in which it is located. This is for the purpose of routing incoming data packets or voice calls appropriately. When the mobile terminal moves and executes a handover from one point of attachment to another, the old serving point of attachment has to be informed about the change. This is usually called disassociation. The mobile terminal must reassociate itself to the network with the new point of access. Other network entities involved in routing data packets to the mobile terminal or switching voice calls may be informed of the handover to seamlessly continue the ongoing connection or call. Depending on whether or not a new connection is created before breaking the old one, handovers can be classified into hard and seamless.[44] They can be further classified from the technical point of view—change of basic service set (BSS)/AP, change of radio resource, change of technology—as well as from a demonstrative point of view—intradomain handover and inter-domain handover.[45]

Figure 9.7 gives an idea of wireless overlay networks. This is an environment where vertical handover (VHO) can take place.[46]

Considering different communication networks, some VHO senarios could be as follows:

- At an office, between Ethernet and WLAN, both of which are parts of the company's intranet, the session should not be interrupted (seamless handover)

FIGURE 9.7 Wireless overlay networks.

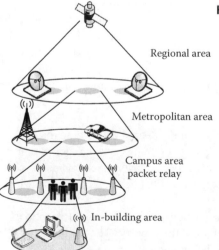

Regional area

Metropolitan area

Campus area
packet relay

In-building area

- At a hotspot, between cellular networks and WLAN of different network operators
- In an urban area, between 3G (UMTS) and 2G (GSM) networks

VHO is an important feature of B3G systems. Handover between UMTS Terrestrial Radio Access Network (UTRAN) and GSM Edge Radio Access Network (GERAN) has been studied by 3GPP. Interworking between cellular network (UMTS) and WLAN has also been studied by 3GPP. The main intention is to extend 3GPP services and functionality to the WLAN access environment, thus leading to an assumption that WLAN is mainly operated as an extension to the 3GPP access network.[47]

Based on the above understanding, some requirements on seamless VHO from the security point of view can be concluded:

- Unified authentication/security and billing
- Ease of access to applications from all locations with acceptable QoS at all times
- VHO must not compromise security of any involved access networks or mobile terminals; as a whole it should not compromise security of B3G systems
- Trade-off between security and issues concerning performance and resource should be considered
- VHO should be executed without user intervention, while also allowing users to configure which network he or she likes to use, perhaps based on the QoS and tariff

Security Context

Security context (SC) is used to support the trust relation of, and to provide communication security for, entities/nodes in distributed networks. Security related information such as authentication state, authorization results, cryptographic (session) keys, and algorithms, comprise the contents of SC. Usually SC is negotiated when creating a communication association and shared by two or more parties.

When creating a new SC, authenticity and integrity of the information must be ensured, and confidentiality of some or all of the information is required.[48]

Trust between access networks (ANs), or an AN and a service provider network (SPN), should also be ensured by an SC that can be created either statistically or dynamically, and may have a comparably long lifetime. Some important parameters of security context are as follows:

- **Authentication state.** Includes identifiers of the peer(s) with whom the context is shared, and the authentication results. In many cases, authorization, accounting (possibly later charging), and other security mechanisms make sense only when the other peer, or peers, are authenticated. This holds true not only for ANs, but also for MTs.
- **Authorization state.** Access to each participant's services and functions should be authorized by either itself or a third party. The authorization results are part of the context and can be dynamically updated.

- **Communication security parameters**. To have confidentiality and/or integrity in a communication association, some cryptographic parameters, like keys and algorithms, must be established by both peers. These security parameters may be established during authentication.

Security Context in Handover

SC shared between an MT and an AN can be retained during a horizontal handover. Similar action can take place during VHO. Another way is to determine old SC with the serving network (SN) and create new SC with the target network (TN), which will be time-consuming. Security context transfer can be studied part by part.

Authentical state includes whether other peers are authenticated. The SN may transfer this state to the TN if they trust each other.

The SN may also transfer the authorization state to the TN. Questions are as follows: authorization may depend on technology of the AN (such as different QoS levels authorized for 3G cellular networks and WLANs), and network domains (e.g., is it a partner of SPN or not). Furthermore, the authorization state needs to be formalized and standardized to avoid misunderstanding between ANs.

Security parameters are also transferable, especially for intrasystem handover; in the case of VHO this issue might be different.

SC can be transferred through a wired connection, which is desirable when both the SN and TN's BSs and APs are permanently installed in the area. Nevertheless, confidentiality and integrity of SC transfer must be ensured; otherwise both ANs and MT are in serious danger.

NETWORK OPERATORS' SECURITY REQUIREMENTS

Within the B3G environment, the role of the MNO is regarded as aggregating services and partner products beyond sole provisioning of connectivity. A successful MNO will play a central role in B3G while having "partnerships" with various participants of the value chain. This role of a mobile network operator is depicted in Figure 9.8. In this figure, "others" can be taken as any other business with which the MNO will have a partnership, this will be mainly to provide services; examples could be a travel agency or a consumer electronics manufacturer. SPs provide services such as roaming, while, as the name suggests, content providers provide content. Site owners will play an important role mainly for hotspot services. Manufacturers, on the other hand, can provide features for networks or terminals. Last but not least, individual users may appear in different roles, for example, in their role as employees, as private persons, or other specific roles.

It is clear that most of the partners provide one type of service or another; in the future even users can provide services and play the role of content or service provider. Thus "service" is an important asset for an MNO. Without users there is no business; thus "users" are also an asset. Without a network and its elements one cannot provide any service to users; thus the "network" itself is also an asset. In other words, one can say that service is a tool of the MNO to generate profits from the customers—users are the (main) source of the value chain and the network is the platform to provide the services.

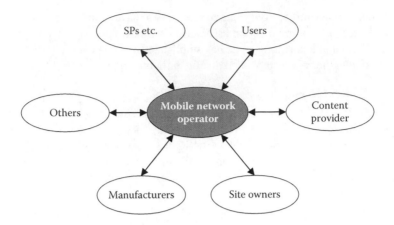

FIGURE 9.8   Role of an MNO in B3G.

Figure 9.9 shows the envisioned role of operators in future communication systems and their assets. Based on this, the assets of the MNO are identified, and security requirements, to protect these assets, can be studied. Because the MNO is the main contact point for users, their security requirements on B3G systems should be carefully considered by the operator with support from other partners. Users' security requirements, such as to protect their terminal and data from possible network attack, are regarded as part of operators' requirements because of the direct influence on operators' business when the requirements are not fulfilled.

MNOs link different parties of the value chain. Thus MNOs have the responsibility to take care of party-specific (security) concerns, while balancing this with suitable security solutions to protect the MNO infrastructures. Only in this connection is there a chance to play a role among all participants of the value chain. To achieve a truly secure B3G system, requirements for overall systems should also be drawn, some of which are touched on in this chapter; however, it is a task for future research.

Requirements from Users' Perspective

Because user satisfaction is crucial to MNOs, users' security requirements must be regarded as one of the most important issues for them. Customers will be happy if they can get the service that they want, almost anywhere at any time at a good price and of the required quality. On the other hand, users' security requirements should

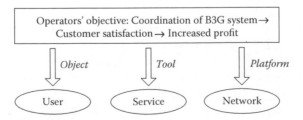

FIGURE 9.9   Network operators' objective in B3G.

be fulfilled—this is a basic prerequisite for MNOs. Any compromise of security that has an effect on users' assets may finally turn out to be a serious problem for the business of an MNO.

Users' requirements of security on systems B3G can be categorized as follows:

- **Terminal security.** Obviously the mobile terminal is an important asset of the user; requirements on terminal security include:
  - Access control (a mobile terminal can be activated and used only by an authorized user)
  - Virus-proof (terminal should be protected against viruses, network worms, etc.)
  - Theft prevention (stolen terminals should be blocked to access networks)
- **Communication and data privacy**, includes:
  - Security of voice and data communications
  - Privacy of location, call setup information, user ID, call pattern, and so on
  - Service usage privacy (unauthorized partners must not know which services are used by any specific user, its usage pattern and volume, etc.).
- **Service provision security:**
  - Service availability should be ensured to prevent or mitigate DoS attack
  - Security against fraudulent service providers
  - E-commerce/m-commerce security

To secure users' assets is part of the goal of security architecture of B3G systems. This is a task of, not only, MNOs, but also manufactures, regulatory bodies, and other participants in the future communication environment.

Because of competition, the MNO will usually enhance its service provision (capacity, quality, variety, etc.) to satisfy its customers, possibly by cooperating with other business partners that can be SPs or other network operators.

Security mechanisms and services should also be intelligible and easy to use. Besides the direct requirements from users, some requirements which are related to the operation stage should also be considered in system design.

Security mechanisms should be either transparent to users, or sufficiently usable, without causing any difficulty or inconvenience for users. Simple operation is more preferable to users; otherwise the mechanism may fail because of users' reluctance to use it.

Security mechanisms should not impair service quality, otherwise business competence will be hindered and, more important, the mechanisms may be bypassed.

Giving users the chance to choose between different levels of security through technical configuration may not be a good idea, because inappropriate configurations, which can happen for a large number of users, may be exploited by attackers, and normally leads to serious consequences for public relations. On the other hand, customer support, like education and consulting, are very expensive.

Requirements from the Network's Perspective

Wireless access networks are different in terms of their general characteristics such as bandwidth, coverage, cost, QoS, and security. In the B3G era, we are talking about the principle of heterogeneity.[51] Heterogeneous networks refer to a combination of different network technologies, and possibly opening the MNO "managed/controlled" network to the Internet, which is not under the control of anyone. This, in addition to the general security requirements for a network, creates several new security issues. In the following we have presented issues related to networks of MNOs.

- Secure attachment and detachment to/from the network must be provided. This is to prevent unauthorized users from accessing the network or making use of the connection of a detaching user.
- Access control to various services or network elements must be provided by the operator. The allowed level/extent of access should also be decided.
- Trust relationships should be built between different networks to which a user might move. This is applicable for both homogeneous and heterogeneous networks.
- The network must have a good accounting mechanism to charge users correctly; this is both for the benefit of the user as well as the MNO.
- In general, the operator should have infrastructure security which prevents tampering of the network and its elements. This goes hand-in-hand with the requirement that the network should be able to identify the tampering, give an alarm, and heal automatically.
- Reconfigurability is also a major issue for B3G networks. If network elements are reconfigurable, then it must happen in a secure fashion; the type of reconfiguring should also be correct, and reconfiguration must be in trusted hands.
- DoS attack is easily possible in wireless medium, although not easy to prevent in current systems; methods should be sought for future systems to prevent such attacks.
- Changes in wireless medium require adaptation in physical and MAC (medium access control) layers; this should not compromise security.
- Rogue BSs are also a threat; to prevent this, the MNO should have mechanisms that will identify such BSs and thus protect the users.
- MNOs should also watch out for service or content providers making illegal use of the network.
- Lawful interception to fulfill the legal obligations.
- Network must be installation and repair fraud resistant.
- Operations and management of security solutions must be possible and relatively simple.
- Extension of network should not lead to weakness in security.
- Network architecture of B3G systems should not compromise the extensibility of security services.

Requirements from the Service Providers' Perspective
Service and contents to the user can be provided by

- MNO
- SPs

- Content providers
- Other businesses
- Site owners

Some of the security requirements that arise here from the MNO point of view are as follows:

- Service should be provided to the specified set of users (authentication), according to the agreed upon contractual obligation (authorization), and the usage should be accountable.
- Rogue service or content providers can appear, and methods must be developed to deter them.
- Secure access to services, from any partner, should be provided.

The operator should take care that the SPs are correctly charged, or if the SP is paying the operator, then the operator should take care that it bills the SP correctly.

Because cooperation with many other SPs is expected in future communication and service provision systems, nonrepudiation will be very important between operators and SPs to prevent and combat fraud. However, appropriate business models may be more efficient than technology based systems.

In the future the number of network operators will increase, and thus a service that can be provided is openness, which will create a positive perception for users. Having openness brings forward several security requirements, level and builds trust among the operators.

9.8   CONCLUDING REMARKS

Migrating current systems to 4G presents enormous challenges. The challenges can be grouped into three aspects: mobile station, system, and service. There is much work to be done in the migration to 4G systems. Current systems must be implemented with a view to facilitate a seamless integration into 4G infrastructure.

A great deal of literature has appeared presenting 4G as the ultimate boundary of wireless mobile communication without any limit to its potential, but this literature has not provided any practical design rules. A pragmatic definition of 4G that considers the user as the cornerstone of the design states that 4G will be a convergence platform providing clear advantages in terms of coverage, bandwidth, and power consumption. It will offer a variety of new heterogeneous services. All these characteristics will be supported by multimode/reconfigurable devices and the implementation of interworking ones.

As a result of the increase in demand for speed, multimedia support, and other resources, the wireless world is looking forward to a new generation technology to replace 3G. This is where the 4G wireless communication comes into play. 4G wireless communication is expected to provide better speed, high capacity, lower cost, and IP-based services. The main aim of 4G wireless is to replace the current core technology with a single universal technology based on IP. Yet there are several challenges that inhibit the progress of 4G, and researchers throughout the world are contributing their ideas to solve these challenges.

The following characteristics can be anticipated to define the 4G and service provision models:

- Open access heterogeneity
- Network access heterogeneity
- Service branding

Keeping in mind the above 4G characteristics, end-to-end service architectures should have the following desirable properties:

- Open service and resource allocation model
- Open capability negotiation and pricing model
- Trust management
- Collaborative service constellations
- Service fault tolerance

4G mobile communication networks are expected to provide all IP-based services for heterogeneous wireless access technologies, assisted by mobile IP to provide seamless Internet access for mobile users. Methodologies for QoS and security support in 4G networks integrate signaling with AAA services to guarantee the user applications, QoS requirements, and achieve efficient AAA. An integrated service and resource management approach is based on the cooperative association among QoS brokers, AAA, and charging systems.

Seamless VHO is an important function of beyond 3G systems. Concept transfer can help support seamless VHO while maintaining the required security level.

To summarize the security requirements, the following points should be carefully treated in research of 4G systems:

- Provision of seamless mobility over heterogeneous networks with sufficient security but no apparent performance compromise
- Mobility versus location privacy
- Anonymity versus accountability
- Assurance that services provided to users are trustworthy, because users will most probably complain to the MNO when they have a problem
- Special terminal features and reconfigurability versus security (user may buy mobile devices directly from vendors instead of from operators—the issue here is security for heterogeneous devices)
- Last but not least, we should not forget that human and software bugs can be the weakest link in security

References

WIRELESS COMMUNICATIONS OVERVIEW

1. A. Goldsmith. *Wireless Communications*. Cambridge: Cambridge University Press, 2005.
2. V. H. McDonald. "The cellular concept," *Bell System Tech. J.* 58 (January 1979): 15–41.
3. F. Abrishamkar and Z. Siveski. "PCS global mobile satellites," *IEEE Commun. Magazine* 34 (September 1996): 132–36.
4. R. Ananasso and F. D. Priscoli. "The role of satellites in personal communication services," *Issues on Mobile Satellite Communications for Seamless PCS, IEEE J. Selected Areas in Commun.* 11 (January 1993): 6–23.
5. Q. Bi, I. Zysman, and H. Menkes. "Wireless mobile communications at the start of the 21st century," *IEEE Commun. Magazine* 39 (January 2001): 110–16.
6. K. R. Rao, Z. S. Bojkovic, and D. A. Milovanovic. *Introduction to Multimedia Communications: Applications, Middleware, Networking.* Hoboken, NJ: John Wiley & Sons, 2006.
7. Z. Bojkovic and D. Milovanovic. "Challenges in mobile multimedia requirements and technologies," *WSEAS Trans. on Signal Processing* 1, no. 1 (October 2005): 37–78.
8. B. Al-Manthari, H. Hassanien, and N. Nasser. "Packet scheduling in 3.5G high-speed downlink packet access networks: breadth and depth," *IEEE Network* 21 (January/February 2007): 41–46.
9. A. J. Goldsmith and L. J. Greenstein. "A measurements-based model for predicting coverage areas of urban microcells," *IEEE Journal Selected Areas in Commun.* 11 (September 1993): 1013–23.
10. A. Mehrotra. *Cellular Radio: Analog and Digital Systems.* Norwood, MA: Artech House, 1994.
11. J. E. Padgett, C. G. Gunther, and T. Hattori. "Overview of wireless personal communications," *IEEE Commun. Magazine* 33 (January 1995): 28–41.
12. J. D. Vriendt et al. "Mobile network evolution: A revolution on the move," *IEEE Commun. Magazine* 40, no.4 April 2002): 104–11.
13. W. Song, W. Zhang, and Y. Cheng. "Load balancing for cellular/WLAN integrated networks," *IEEE Network* 21 (January/February 2007): 27–33.

CHAPTER 1   INTRODUCTION TO WIRELESS NETWORKING

1. M. Shafi et al. "Wireless communications in the 21st century: A perspective," *Proc. of the IEEE* 85 (October 1997): 1622–38.
2. W. Hocharenko et al. "Broadband wireless access," *IEEE Commun. Magazine* 35 January 1997): 20–26.
3. M. Dinis and J. Fernandes. "Provision of sufficient transmission capacity for broadband mobile multimedia: A step toward 4G," *IEEE Commun. Magazine* 39 (August 2001): 46–54.
4. T. Tjelta et al. "Future broadband radio access systems for integrated services with flexible resource management," *IEEE Commun. Magazine* 39 (August 2001): 56–63.

5. T. Robles et al. "QoS support for an all-IP system beyond 3G," *IEEE Commun. Magazine* 39 (August 2001): 64–72.

6. L. Beccheti, P. Mahonen, and L. Munoz. "Enhancing IP service provision over heterogeneous wireless networks: A path toward 4G," *IEEE Commun. Magazine* 39 (August 2001): 74–81.

7. M. Mehta et al. "Reconfigurable terminals: An overview of architectural solutions," *IEEE Commun. Magazine* 39 (August 2001): 82–89.

8. J. M. Pereira. "Balancing public and private in 4G,"*Proc. IEEE PIMRC 2001*, 125–32, San Diego, September/October 2001.

9. ETSI General Description of a GSM PLMN, European Telecommunications Standards Institute. GSM Recommendation 01.02, 1991.

10. W. R. Krenik. "Wireless user perspectives in the United States," *Wireless Personal Communications Journal* 32, no.2 (August 2002): 153–60.

11. N. Nakajima. "Future communication systems in Japan," *Wireless Personal Communications* 17, no.2 (June 2001): 209–23.

12. G. Priggouris, S. Hdjiefthymiades, and L. Merkos. "Supporting IP QoS in the GPRS," *IEEE Networks* 15, no.5 (September/October 2000): 8–17.

13. B. Sarikaya. "Packet mode in wireless networks: Overview of transition to the 3G," *IEEE Commun. Magazine* 38 (September 2000): 164–72.

14. F. Muratore. *UMTS: Mobile Communications for Future.* New York: John Wiley & Sons, 2000.

15. R. van Nohelen et al. "An adaptive radio link protocol with enhanced data rates for GSM evolution," *IEEE Personal Communications* 6, no.1 February 1999): 54–64.

16. A. Furuskar et al. "WDGE: Enhanced data rate for GSM and TDMA/136 evolution," *IEEE Personal Communications*, 6 (June 1999): 56–66.

17. R. Prasad, W. Mohr, and W. Konhauser. *Third-Generation Mobile Communication System.* Norwood, MA: Artech House, 2000.

18. T. Ojanpera and R. Prasad. *WCDMA: Towards IP Mobility and Mobile Internet.* Norwood, MA: Artech House, 2001.

19. R. Prasad. *Towards a Global 3G System: Advanced Mobile Communications in Europe.* Norwood, MA: Artech House, 2001.

20. J. Shiller. *Mobile Communications.* Reading. MA: Addison-Wesley, 2000.

21. R. Prasad and A. Prasad. *WLAN Systems and Wireless IP for Next Generation Communication.* Norwood, MA: Artech House, 2002.

22. R. Prasad. *Wideband CDMA for 3G Mobile Systems.* Norwood, MA: Artech House, 1998.

23. J. Rapeli. "Future directions for mobile communication business, technology and research," *Wireless Personal Communications* 17 (June 2001): 155–73.

24. M. Zeng, A. Annamalai, and V. Barghava. "Recent advances in cellular wireless communications," *IEEE Commun. Magazine* 37 (September 1999): 128–38.

25. H. Holma and A. Toskala. *WCDMA for UMTS.* New York: John Wiley & Sons, 2000.

26. P. Kauffman. "Fast power control for third generation DS-CDMA Mobile Radio System," in *Proc. 2000 International Zurich Seminar on Broadband Communications*, 9–13, Zurich, Switzerland, 2000.

27. R. Prasad and M. Ruggieri. *Technology Trends in Wireless Communications.* Norwood, MA: Artech House, 2003.

28. L. Jorguseski, J. Farseotu, and R. Prasad. "Radio resource allocation in third generation mobile communication systems," *IEEE Commun. Magazine* 39 (February 2001): 117–23.

29. S. Benedetto and G. Montorsi. "Unveiling turbo codes: Some results on parallel concatenated coding schemes," *IEEE Trans. Information Theory* 42 (March 1996): 409–28.

30. L. C. Perez, J. Seghers, and D. J. Costello, Jr. "A distance spectrum interpretation of turbo codes," *IEEE Trans. on Information Theory* 42 (November 1996): 1698–1709.

31. J. Baunister, P. Malter, and S. Coope. *Convergence Technologies for 3G Networks: IP, UMTS, EGPRS and ATM*. New York: John Wiley & Sons, 2004.

32. A. Salkintrizis and N. Passas, eds. *Emerging Wireless Multimedia: Services and Technologies*. New York: John Wiley & Sons, 2005.

33. 3GPP TS 29.060. GPRS Tunneling Protocol (GTP) across the Gn and Gp Interface (Release 5), September 2004.

34. G. Perkins, ed. IP Mobility Support for IPv4, RFC3344, August 2002.

35. Z. Bojkovic, J. Turan, and L. Ovsenik. "Toward multimedia across wireless," *Journal of Electrical Engineering* 56 (January 2005): 9–14.

36. Z. Bojkovic and B. Bakmaz. "QoS architecture over heterogeneous wireless access networks," in *Proc. XL International Scientific Conference on Information, Communication and Energy Systems and Technologies ICEST2005*, 419–22, Nis, Serbia, June 2005.

37. B. Bakmaz and Z. Bojkovic. "Internet Protocol version 6 as backbone of heterogeneous networks," in *Proc. 12th Int. Workshop on Systems, Signal and Image Processing IWSSIP2006*, 255–59, Chalkida, Greece, September 2005.

38. Z. Bojkovic and B. Bakmaz. "Next-generation mobile services," in *Proc. WSEAS International Conference on Communications*, 240–44, Vouliagmeni, Athens, Greece, July 2006.

39. IEEE 802.11. Standard for Wireless LAN Medium Access Control (MAC) and Physical Layer (PHY) Specification, November 1997.

40. IEEE 802.11e/D13.0. Draft Supplement to Standard for Telecommunications and Information Exchange between Systems—LAN/MAN Specific Requirements, Part 11: Wireless Medium Access Control (MAC) and Physical Layer (PHY) Specification: MAC Enhancements for QoS, January 2005.

41. S. Xu and T. Saadani. "Does the IEEE 802.11 MAC Protocol work well in multihop wireless ad hoc networks?" *IEEE Commun. Magazine* 39 (June 2001): 130–37.

42. I. G. Niemegeers and S. M. H. DeGroot. "From personal area networks to personal networks: A user oriented approach," *Wireless Personal Communications* 22 (August 2002): 175–86.

43. C. Bisdikian. "An overview of the *Bluetooth* wireless technology," *IEEE Commun. Magazine* (December 2001): 86–94.

44. 3GPP TS 23.234 V6.2.0, 3GPP System in Wireless Local Area Network (WLAN) Interworking: System Description (Release 6), September 2004.

45. 3GPP TS 22.934 V6.2.0. Feasibility Study on 3GPP System to Wireless Local Area Network (WLAN) Interworking (Release 6), September 2003.

46. S. Ohmou, Y. Yamao, and N. Nakajima. "The future generations of mobile communications based on broadband access methods," *Wireless Personal Communications* 17 (June 2001): 175–90.

47. W. Mohr. "Development of mobile communications systems beyond third generation," *Wireless Personal Communications* 17 (June 2001): 191–207.

CHAPTER 2   CONVERGENCE TECHNOLOGIES

1. J. Bannister, P. Mather, and S. Coope. *Convergence Technologies for 3G Networks: IP, UMTS, EGPRS and ATM*. New York: John Wiley & Sons, 2004.

2. W. W. Lu. "Compact multidimensional broadband wireless: The convergence of wireless mobile and access," *IEEE Communications Magazine* 38 (November 2000): 119–23.

3. S. Maye and A. Umar. "The impact of network convergence on telecommunications software," *IEEE Communications Magazine* 39 (January 2001): 78–84.

4. www.umts-forum.org.
5. K. R. Rao, Z. S. Bojkovic, and D. A. Milovanovic. *Introduction to Multimedia Communications: Applications, Middleware, Networking.* Hoboken, NJ: John Wiley & Sons, 2006.
6. K. R. Rao, Z. S. Bojkovic, and D. A. Milovanovic. *Multimedia Communication Systems: Techniques, Standards and Networks.* Upper Saddle River, NJ: Prentice-Hall, 2002.
7. E. B. Dahlman and Y-Ch. Jou. "Evolving technologies for 3G cellular wireless communications systems," *IEEE Commun. Magazine* 44 (February 2006): 62–64.
8. P. Bender et al. "CDMA/HDR: A bandwidth efficient high-speed data service for nomadic users," *IEEE Commun. Magazine* 38 (July 2000): 75–87.
9. N. Bhushan et al. "CDMA 2000 1xEV-DO revision A: A physical layer and MAC layer overview," *IEEE Commun. Magazine* 44 (February 2006): 75–87.
10. S. Pakval et al. "Evolving 3G mobile systems: Broadband and broadcast services in WCDMA," *IEEE Communications Magazine* 44 (February 2006): 68–74.
11. P. Viswanatah et al. "Opportunistic beamforming using dumb antennas," *IEEE Trans. Inform. Theory* 48, no.6 (June 2002): 1277–94.
12. A. Jolali et al. "Data throughput of CDMA HDR a high efficiency, high data rate personal communication wireless system," in *Proc. IEEE VCT,* 1854–58, Tokyo, Japan, May 2000.
13. J. F. Cheng. "On the coding gain of incremental redundancy over chase combining," in *Proc. IEEE Globecom,* 107–12, San Francisco, CA, December 2003.
14. A. Samunkic. "UMTS universal mobile telecommunications system: Development of standards for the third generation," *IEEE Trans. Vehic. Tech.* 47 (November 1998): 1099–1104.
15. F. Akyyildiz et al. "Medium access control protocols for multimedia traffic in wireless networks," *IEEE Network* 13 (July/August 1999): 39–47.
16. E. Dahlman et al. "UMTS/IMT2000 based on wideband CDMA," *IEEE Commun. Magazine* 36 (September 1998): 70–81.
17. Z. Bojkovic, M. Stojanovic, and B. Milovanovic. "Current developments towards the 4G wireless system," in *Proc. Int. Conf TELSIX,* 229–32, Nis, Serbia, September 2005.
18. Z. Bojkovic and B. Bakmaz. "Quality of service and security as frameworks towards next-generation wireless networks," *WSEAS Trans. Commun.* 4, no. 4 (April 2005): 147–53.
19. D. Milovanovic and Z. Bojkovic. "Trends in multimedia over wireless broadband networks," *WSEAS Trans. on Commun.* 4, no. 11 (November 2005): 1292–97.
20. B. Bakmaz and Z. Bojkovic. "Internet Protocol version 6 as backbone of heterogeneous networks" in *Proc. IWSSIP,* 255–59, Chalcida, Greece, September 2005.
21. S. Y. Hui and K. H. Yeung. "Challenges in the migration to 4G mobile systems," *IEEE Commun. Magazine* 41 (December 2003): 54–59.
22. Q. Zhang et al. "Cooperative opportunistic transmission for wireless ad hoc networks," *IEEE Network* 21 (January/February 2007): 14–20.
23. 3GPP TS 25.308. High Speed Downlink Packet Access (HSDA): Overall Description, Rel. 5, March 2003.
24. 3GPP TS25.214. Physical Layer Procedures, Rel. 5, N.5.5.0, June 2003.
25. B. A. Manthari, H. Hassanien, and N. Nasser. "Packet scheduling in 3.5G high-speed downlink packet access networks: Breadth and depth," *IEEE Network* 21 (January/February 2007): 41–46.
26. T. Kolding et al. "High speed downlink protocol access WCDMA evolution," *IEEE Vehic. Soc. News* 50 (February 2003): 4–10.

CHAPTER 3 WIRELESS VIDEO

1. M. Etoh and T. Yoshimura. "Advances in wireless video delivery," *Proc. of the IEEE* 93 (January 2005): 111–22.
2. M. Etoh and T. Yoshimura. "Wireless video applications in 3G and beyond," *IEEE Wireless Commun.* 12 (August 2005): 66–72.
3. A. Vetro, J. Xin, and H. Sun. "Error resilience video transcoding for wireless communications," *IEEE Wireless Commun.* 12, no. 4 (August 2005): 14–21.
4. A. Katsaggelos et al. "Energy-efficient wireless video coding and delivery," *IEEE Wireless Commun.* 12 (August 2005): 24–30.
5. M. Chen and A. Zakhor. "Rate control for streaming video over wireless," *IEEE Wireless Commun.* 12 (August 2005): 32–41.
6. ITU Rec.H.264 ISO/IEC 14996-10. AVC, Advanced Video Coding for General Audiovisual Services, 2003.
7. T. Stockhammer and M. M. Hannuksela. "H.264/AVC video for wireless transmission," *IEEE Wireless Commun.* 12 (August 2005): 6–13.
8. A. Vetro, C. Christopoulos, and H. Sun. "An overview of video transcoding architectures and techniques," *IEEE Signal Processing* 20, (March 2003): 18–29.
9. K. R. Rao and P. Yip. *Discrete Cosine Transform: Algorithms, Advantages, Applications.* New York: Academic Press, 1990.
10. M. van der Schaar et al. "Adaptive cross-layer protection strategies for robust scalable video transmission over 802.11 WLANs," *IEEE J. Selected Areas in Commun.* 21 (December 2003): 1752–63.
11. W. Tan and A. Zakhor. "Real-time Internet video using error resilient scalable compression and TCP-friendly transport protocol," *IEEE Trans. Multimedia* 1 (June 1999): 172–86.
12. K. Ratham and I. Mata. "WTCP: An efficient mechanism for improving wireless access to TCP services," *Int. Commun. Syst.* 16 (February 2003): 47–62.
13. N. Samarawera. "Non-congestion packet loss detection for TCP error recovery using wireless links," *IEE Proc. Commun.* 146, (August 1999): 222–30.
14. P. Sinha et al. "WTCP: A reliable transport protocol for wireless wide-area networks," *Wireless Networks* 8, no. 2–3 (March–April 2002): 301–16.
15. L. S. Brakmo and L. L. Peterson. "TCP Vegas: End-to-end congestion control avoidance on a Global Internet," *IEEE J. Selected Areas in Commun.* 13, no. 8 (October 1995): 1465–80.
16. K. R. Rao and J. J. Hwang. *Techniques and Standards for Image, Video and Audio Coding.* Upper Saddle River, NJ: Prentice-Hall, 1996.
17. Z. S. Bojkovic and D. A. Milovanovic. "Audiovisual integration in multimedia communications based on MPEG-4 facial animation," *Circuits, Systems and Signal Processing* 20 (May–June 2001): 311–39.
18. R. Talluri. "Error-resilient video coding in the ISO MEPG-4 standard," *IEEE Commun. Magazine* 36, no. 6 (June 1998): 112–19.
19. G. Sullivan and T. Wiegand. "Video compression—From concepts to the H.264/AVC standard," *Proc. of the IEEE* 93 (January 2005): 18–31.
20. T. Wiegand et al. "Overview of the H.264/AVC video coding standard," *IEEE Trans. CSVT* 13 (July 2003): 560–76.
21. S. Wenger. "H.264/AVC over IP," *IEEE Trans. CSVT* 13 (July 2003): 545–56.
22. T. Stockhamer, M. M. Hannuksela, and T. Wiegand. "H.264/AVC in wireless environment," *IEEE Trans. CSVT* 13 (July 2003): 657–73.
23. H. Malwar et al. "Low complexity transform and quantization in H.264/AVC," *IEEE Trans. CSVT* 13, no. 7 (July 2003): 598–603.

24. D. Marfe, H. Schwarz, and T. Wiegand. "Context-adaptive binary arithmetic coding for H.264/AVC," *IEEE Trans. CSVT* 13 (July 2003): 620–36.
25. J. Ribas-Corbera, P. A. Chou, and S. L. Regunathan. "A generalized hypothetical reference decoder for H.264/AVC," *IEEE Trans. CSVT* 13 (July 2003): 674–87.
26. Y. W. Huang et al. "Analysis, fast algorithm and VLSI architecture design for H.264/AVC intra-frame coding," *IEEE Trans. CSVT* 15, no. 3 (March 2005): 378–401.
27. F. Pan et al. "Fast mode decision algorithm for intra-prediction in H.264/AVC," *IEEE Trans. CSVT* 13 (July 2003): 813–22.
28. S. Kwan, A. Tamhankar, and K. R. Rao. "Overview of the H.264/MPEG-4 Part10," *Journal of Visual Commun. and Image Representation* 17 (April 2006): 186–216.
29. S. Vanger et al. RTP Payload Format for H.264 Video, IETF RFC3984, February 2005.
30. T. Wiegand. "Rate-constrained coder control and compression of video coding standards," *IEEE Trans. CSVT* 13 (July 2003): 688–703.
31. I. E. G. Richardson. *H.264 and MPEG-4 Video Compression for Next-Generation Multimedia.* New York: John Wiley & Sons, 2003.
32. K. R. Rao, Z. S. Bojkovic, and D. A. Milovanovic. *Introduction to Multimedia Communications: Applications, Middleware, Networking.* New York: John Wiley & Sons, 2006.

CHAPTER 4   WIRELESS MULTIMEDIA SERVICES AND APPLICATIONS

1. A. Salkintzis and N. Passas, eds. *Emerging Wireless Multimedia: Services and Technologies.* New York: John Wiley & Sons, 2005.
2. D. Wu, Y. T. Hou, and Y.-Q. Zhang. "Transporting real-time video over the Internet: Challenges and approaches," *Proc. of the IEEE* 88, no. 12 (December 2000): 1855–77.
3. Q. Zhang, W. Zhu, and Y.-Q. Zhang. "End-to-end QoS for video delivery over wireless Internet," *Proc. of the IEEE* 93 (January 2005): 123–34.
4. J. Wroclawski. The Use of RSVP with IETF Integrated Services, RFC2210, September 1997.
5. D. Grossman. New Terminology and Classification for DiffServ, RFC3260, April 2002.
6. W. Li. "Overview of fine granularity scalability in MPEG-4 video standard," *IEEE Trans. CSVT* 11, no. 3 (March 2001): 301–17.
7. F. Wu, S. Li, and Y.-Q. Zhang. "A framework for efficient progressive fine granularity scalable video coding," *IEEE Trans. CSVT* 11 (March 2001): 332–44.
8. M. van der Schaar and H. Rodha. "Adaptive motion compensation fine-granular-scalability (AMC-FGS) for wireless video," *IEEE Trans. CSVT* 12 (June 2002): 360–71.
9. K. D. Wong and V. K. Varma. "Supporting real-time IP multimedia services in UMTS," *IEEE Commun. Magazine* 41 (November 2003): 148–55.
10. 3GPP TS22.140 v5.1.0 (2002–03). Stage 1 Multimedia Messaging Service (Release 5).
11. 3GPP TS23.060 v6.0.0 (2002–03). Generalized Packet Radio Service (GPRS), Service Description, Stage 2 (Release 6).
12. G. Camarillo and M.-A. Garcia-Martin. *The 3G IP Multimedia Subsystem (IMS), Merging the Internet and the Cellular Worlds.* New York: John Wiley & Sons, 2004.
13. J. Rosenberg et al. SIP: Session Initiation Protocol, RFC3261, June 2002.
14. K. R. Rao, Z. S. Bojkovic, and D. A. Milovanovic. *Introduction to Multimedia Communication: Applications, Middleware, Networking.* New York: John Wiley & Sons, 2006.
15. M. Hundley and V. Jacobson. SDP: Session Description Protocol, RFC2327, April 1998.

16. K. R. Rao, Z. S. Bojkovic, and D. A. Milovanovic. *Multimedia Communications Systems: Techniques, Standards and Networking.* Upper Saddle River, NJ: Prentice-Hall, 2002.
17. B. Bakmaz and Z. Bojkovic. "Internet Protocol version 6 as backbone of heterogeneous networks," in *Proc. IWSSIP05*, 255–59, Chalcida, Greece, September 2005.
18. 3GPP TS22.250 v.6.0.0. IP Multimedia Subsystem (IMS) Group Management Stage 1, December 2002.
19. 3GPP TS22.228 v.6.5.0. Requirements for the IP Multimedia Core Network Subsystem, January 2004.
20. 3GPP TS23.2228 v.6.0.0. IP Multimedia Subsystem (IMS) Stage 2, January 2004.
21. B. Bakmaz, Z. Bojkovic, and M. Bakmaz. "Internet protocol multimedia subsystem for mobile services," in *Proc. IWSSIP and EC-SIPMCS*, Slovenia, June 2007.
22. A. D.-S. Jun et al. "An IMS-based service platform for next-generation wireless networks," *IEEE Commun. Magazine* 44 (September 2006): 88–95.
23. E. Christensen et al. Web Services Description Language (WSDL) 1.1, March 2001. http://www.w3.org/T3/wsdl.
24. W. Bushnesll. "IMS based converged wireleine-wireless services," in *Proc. Int. Conf. Intelligence in Service Delivery Networks*, 18–21, France, October 2004.
25. T. Magedanz, T. Witaszek, and K. Knuettel. "The IMS playground@Fokus—An open testbed for next generation network multimedia services," in *Proc. Int. Conf. Testbeds and Research Infrastructures for the Comp. Soc.*, 2–11, February 2005.
26. Ch. J. Pavlovski. "Service delivery platform in practice," *IEEE Commun. Magazine* 45, no. 3 (March 2007): 114–21.
27. S. Q. Khan, R. Gaglianello, and M. Luna. "Experiences with blending HTTP, RTSP and IMS," *IEEE Commun. Magazine* 45 (March 2007): 122–28.
28. 3GPP TS23.218 v6.3.0. IP Multimedia (IM) Session Handling, IM Call Model Stage 2, March 2005.
29. G. Camarillo et al. "Towards an imnovation oriented IP multimedia subsystem," *IEEE Commun. Magazine* 45 (March 2007): 130–36.
30. D. Petrie. A Framework for SIP User Agent Profile Delivery, IETF Internet Draft, March 2006.
31. V. Hilt and G. Camarillo. A SIP Event Package for Session-Specific Session Policies, IETF Internet Draft, October 2006.

CHAPTER 5   WIRELESS NETWORKING STANDARDS

1. K. R. Rao, Z. S. Bojkovic, and D. Milovanovic. *Introduction to Multimedia Communications: Applications, Middleware, Networking.* Hoboken, NJ: John Wiley & Sons, 2006.
2. K. R. Rao, Z. S. Bojkovic, and D. Milovanovic. *Multimedia Communication Systems: Techniques, Standards and Networking.* Upper Saddle River, NJ: Prentice-Hall, 2002.
3. Z. Bojkovic, D. Milovanovic, and A. Samcovic. Multimedia communication systems: techniques, standards and networking, in *Proc. Int. Workshop Trends and Recent Achievements in Information Technology*, 19–41, Cluj Napoca, Romania, May 2002.
4. A. K. Salkintzis and N. Pasas, eds. *Emerging Wireless Multimedia Services and Technologies.* Chichester, UK: John Wiley & Sons, 2005.
5. A. Ganz, Z. Ganz, and K. Wongthavarewat. *Multimedia Wireless Network: Technologies, Standards and QoS.* Upper Saddle River, NJ: Prentice-Hall, 2003.
6. A. Santamaria and F. J. Lopez-Hernandez. *Wireless LAN Standards and Applications.* Norwood, MA: Artech House, 2001.
7. WIANA, Wireless Networking Standards and Organizations, Wireless LAN Association, April 2002. http://www.wiana.org.

8. WLANA, Wireless Networking Organizations, Wireless LAN Association, April 2002. http://www.wiana.org.

9. V. Hayers. "Standardization efforts for wireless LANs," *IEEE Network Mag.* 5 (November 1991): 19–20.

10. C. E. Perkins. *Mobile IP, Design Principles and Practices.* Reading, MA: Addison Wesley Longman, 1998.

11. P. Newman. "In search of the all-IP mobile network," *IEEE Commun. Magazine* 42 (December 2004): 53–58.

12. H. Soliman et al. "Hierarchical mobile IPv6 mobility management," IETF Networking Group, October 2004. http://www.ietf.org/internet-drafts/draft-ietf-mip-shop-hmipv6-03.txt.

13. R. B. Marks, and R. E. Hebner. "Government activity to increase benefits from the global standards system," in *Proc. IEEE Conference on Standards and Innovation in Information Technology,* 183–90, Boulder, CO, October 2001.

14. IEEE 802 LAN/MAN Standards Committee, http://grouper.ieee.org/groups/802.

15. IEEE 802.11 WG, International Standard for Information Technology—Local and Metropolitan Area Networks Specific Requirements. Part 11, Wireless LAN MAC and PHY Specifications, 1999.

16. Q. Ni. "Performance analysis and enhancements for IEEE 802.11e wireless networks," *IEEE Network Mag.* 19 (July/August 2005): 21–27.

17. R. B. Marks. "IEEE standardization for the wireless engineer," *IEEE Microwave Magazine* 2 (June 2001): 16–26.

18. R. B. Marks, I. C. Gifford, and B. O'Hara. "Standards in IEEE 802 unleash the wireless Internet," *IEEE Microwave Magazine* 2 (June 2001): 46–56.

19. IEEE 802.11 WG. Wireless LAN Medium Access Control (MAC) and Physical Layer (PHY) Specifications, 1999.

20. IEEE P802.11e/D6.0. Wireless Medium Access Control (MAC) and Physical Layer (PHY) Specifications: Medium Access Control (MAC) Quality of Service (QoS) Enhancements, November 2003.

21. X. Yang. "IEEE 802.11e: QoS provisioning of the MAC layer," *IEEE Wireless Commun. Mag.* 11 (March 2004): 72–79.

22. D. Gao, J. Cai, and K. N. Ngan. "Admission control in IEEE 802.11e wireless LANs," *IEEE Network Magazine* 19 (July/August 2005): 6–13.

23. K. R. Rao and Z. S. Bojkovic. *Packet Video Communications over ATM Networks.* Upper Saddle River, NJ: Prentice-Hall, 2000.

24. Y. Pu et al. "Wireless ATM LAN with and without infrastructure," *IEEE Commun. Mag.* 36 (April 1998): 90–95.

25. N. Morinaga, M. Nakagawa, and R. Kohno. "New concepts and technologies for achieving highly reliable and high-capacity multimedia wireless communication systems," *IEEE Commun. Mag.* 35 (January 1997): 34–40.

26. The ATM Forum, http://www.atmforum.com.

27. P. Mishra and M. Srivasta. Effect of virtual circuit rerouting on application performance, ATM Forum/97-0648/WATM, September 1997.

28. P. Mishra and M. Srivasta. Proposed handover signaling architecture for release 1. 0 WATM Baseline, ATM Forum/97-0845/WATM, September 1997.

29. ETSI TR 101 683 V1.1.1(2002-02), BRAN, HiperLAN Type 2 System Overview, 2002.

30. http://portal.etsi.org/bran/kta/hiperlan/hiperlan2.asp.

31. IrDA standards, http://www.irda.com.

32. IEEE 802.15, http://grouper.ieee.org/groups/802/15.

33. T. M. Stiep et al. "Paving the way for personal area network standards: An overview of the IEEE 802.15 Working Group for wireless personal area networks," *IEEE Personal Communications* 7 (February 2000): 37–43.
34. IEEE, Draft Standard for Telecommunications and Information Exchange between Systems—LAN/MAN Specific Requirements. Part 15. 3: Wireless Medium Access Control (MAC) and Physical Palyer (PHY) Specifications for High Rate Personal Area Networks (WPAN), Draft P802. 15. 3/D17, February 2003.
35. *Bluetooth*, http://www.bluetooth.com.
36. http://www.ieee802.org/15/pub.
37. R. VanNee. "New high-rate wireless LAN standards," *IEEE Commun. Mag.* 37 (December 1999): 82–88.
38. R. Prasad and L. Gavrilovska. "Research challenges for wireless personal area networks," in *Proc. 3rd Int. Conf. on Information Communications and Signal Processing ICICS*, Keynote speech, Singapore, October 2001.
39. Specification of the *Bluetooth* System core v1.0A, July 1999.
40. Specification of the *Bluetooth* System profiles v1.0A, July 1999.
41. J. C. Haartsen. "The Bluetooth radio system," *IEEE Personal Communications* (February 2000): 728–36.
42. ETSI TR 101 178. Digital Enhanced Cordless Telecommunications (DECT), A High Level Guide to the DECT Standardization, March 2000.
43. http://www.dect.ch.
44. http://www.dectweb.com.
45. R. A. Schultz. In *Impulse Radio: TDMA versus SDMA*. ed. S. Glisic, and P. A. Lepanen, 245–63. London: Kluwer, 1997.
46. I. Opperman. "The role of UWB in 4G," *Wireless Personal Communications* 29 (2004): 121–33.
47. Y. D. Taylor, ed. *Introduction to Ultra Wideband Radar Systems*. Boca Raton, FL: CRC Press, 1995.
48. http://www.fcc.gov/bureaus/engineering.technology/news.releases/2002/nret0203.html, FCC press release, February 2002.
49. Federal Communications Commission (FCC). "First report and order in the matter of revision of Part 15 of the Commissions rules regarding ultrawideband transmission systems," ET-Docket98-153, FCC 02-48, released 22 April 2002.
50. Federal Communications Commission (FCC), http://www.fcc.gov.
51. J. Foerster. UWB Theory and Applications Final, IEEE P802.15 Working Group for Wireless Personal Area Networks (WPANs), February 2003.
52. A. Scleh and R. Valenzuela. "A statistical model for indoor multipath propagation," *IEEE J. Selected Areas in Commun.* 5 (February 1987): 128–37.
53. http://www.WirelessMAN.org.
54. http://www.wimaxforum.org.
55. ETSI TS102 178. ETSI Broadband Radio Access Networks (BRAN), HiperMAN Data Link Control (DLC) Layer, 2003.
56. ETSI TS102 177. ETSI Broadband Radio Access Networks (BRAN), HiperMAN Physical (PHY) Layer, 2003.
57. WiMAX Forum, www.wimaxforum.org.
58. CEPT/ERC/REC 13-04,.Preferred Frequency Bands for Fixed Wireless Access in the Frequency Range between 3 and 39.5 GHz, Technical report, European Radiocommunications Committee, 1998.
59. R. VanNee and R. Prasad. *OFDM for Wireless Multimedia Communications*. Norwood, MA: Artech House, 2000.
60. IEEE 802.20 WG PD-03. Mobile Broadband Wireless Access Systems "Five Criteria": Vehicular Mobility, November 2002.

61. IEEE 802.20 WG PD-02. PAR Forum, December 2002.
62. IEEE 802.20 WG PD-06r1. System Requirements for IEEE 802.20 Mobile Broadband Wireless Access Systems—Version 14.
63. IEEE 802.20 WG PD-06. Introduction to IEEE 802.20: Technical and Procedural Orientation, March 2003.
64. W. Bolton, Y. Xiao, M. Guizani. "IEEE 802.20: Mobile broadband wireless access," *IEEE Wireless Communications* 14, no. 2 (February 2007): 84–95.
65. IEEE 802.20 WG. Initial Contribution on a System Meeting MBWA Characteristics, March 2003.
66. IEEE 802.20 WG. Desired Characteristics for an MBWA Air Interface, March 2003.
67. IEEE 802.20 WG. User Data Models for an IP-Based Cellular Network, March 2003.
68. F. Zon, X. Jiang, Z. Lin. "IEEE 802.20 based broadband railroad digital network—The infrastructure of M-commerce on the train," *Proc. ICEB*, 771–76, Beijing, China, December 2004.
69. B. Bakmaz, Z. Bojkovic, D. Milovanovic, and M. Bakmaz. "Mobile broadband networking based on IEEE 802.20 standard," in *Proc. TELSIX 2007*, Nis, Serbia.
70. IEEE 802.11g 2003. Part 11 Wireless LAN Medium Access Control (MAC) and Physical Layer (PHY) Specification Band, Supp. IEEE 802.11, 2003.
71. TgnSync Technical Specification, http://www.tgnsync.org/techdocs/tgn-sync.proposal-technical-specification.pdf.
72. IEEE 802.11e. Wireless LAN Medium Access Control (MAC) Enhancement for Quality of Service (QoS), 802.11e Draft 80.0, 2004.
73. ISO/IEC 14496-2. Coding of Audio-Visual Objects—Part 2: Video, Final Committee Draft, v3, January 2001.
74. A. Kossentini, M. Naimi, and A. Gueroul. "Toward an improvement of H.264 video transmission over IEEE 802.11e through a cross-layer architecture," *IEEE Commun. Mag.* 44 (January 2006): 107–14.
75. ISO/IEC JTC1/SC29/WG11 (MPEG) and ITU-T Joint Video Team (JVT). Doc. VCEG-AC01, Meeting Report, 29 VCEG Meeting, Klagenfurt, Austria, July 2006.
76. ISO/IEC JTC1/SC29/WG11 (MPEG) and ITU-T Joint Video Team (JVT). Doc. VCEG-AE01, VCEG Management, Marrakech, Morocco, January 2007.
77. S.-K. Kwon, A. Tamhankar, and K. R. Rao. "Overview of H.264/MPEG-4 Part 10," *J. VCIR* 17 (April 2006): 186–216.
78. Y. Andreopoulos, M. Van der Schaar, Z. Hu, S. Heo, and S. Suh. "Scalable resource management for video streaming over IEEE 802.11a/e," *Proc. ICASPP2006* 5 (2006): 361–64.

CHAPTER 6 ADVANCES IN WIRELESS VIDEO

1. C. E. Shannon. *The Mathematical Theory of Communication.* Champaign, IL: University of Illinois Press, 1948.
2. B. Girod and N. Farber. "Feedback-based error control for mobile video transmission," *Proc. IEEE* 87 (October 1999): 1707–23.
3. P. Haskell and D. Messerschmitt. "Resynchronization of motion compensated video affected by ATM cell loss," *Proc. IEEE ICASSP* 3, 545–48, San Francisco, CA, March 1992.
4. W. M. Lam, A. R. Reibman, and B. Lin. "Recovery of lost or erroneously received motion vectors," in *Proc. IEEE ICASSP* 5, 417–20, Minneapolis, MN, April 1993.
5. P. Salama, N. Shroff, and E. J. Delp. "A fast suboptimal approach to error concealment in encoded video streams," in *Proc. IEEE ICIP* 2, 101–104, Santa Barbara, CA, October 1997.

6. Y. Wang and Q. F. Zhu. "Error control and concealment for video communication: A review," *Proc. of the IEEE* 86 (May 1998): 947–77.
7. M. Etoh and T. Yoshimura. "Wireless video application in 3G and beyond," *IEEE Wireless Commun.* 12 (August 2005): 66–72.
8. T. Stockhamer, M. M. Hannuksela, and T. Wiegand. "H.264/AVC in wireless environment," *IEEE Trans. CSVT* 13 (July 2003):. 657–73.
9. Y. Wang, Q. Zhu, and L. Shaw. "Maximally smooth image recovery in transform coding," *IEEE Trans. Commun.* 41 (October 1993): 1544–51.
10. H. Sun and W. Kwok. "Concealment of damaged block transform coded images using projection onto convex sets," *IEEE Trans. Image Processing* 4 (April 1995): 470–77.
11. M. Ghanbari. "Cell-loss concealment in ATM video codecs," *IEEE Trans. CSVT* 3 (June 1993): 238–47.
12. J. D. Villasenor, Y-Q Zhang, and J. Wen. "Robust video coding algorithms and systems," *Proc. of the IEEE* 87 (October 1999): 1724–33.
13. J. Wen and J. D. Villasenor. "A class of reversible variable length codes for robust image and video coding," in *Proc. ICIP*, 65–68, Santa Barbara, CA, 1997.
14. ISO/IEC JTC1/SC29/WG11 N1383. Description of Error Resilient Core Experiments, November 1996.
15. C. Boyd et al. "Integrating error detection into arithmetic coding," *IEEE Trans. Commun.* 15 (January 1997): 1–3.
16. G. P. Elmestry. "Joint lossless-source and channel codes over the (0-1) vector space," in *Proc. Inform. Sciences Symp.*, 319–24, Johns Hopkins, Baltimore, MD, March 1997.
17. R. E. VanDyck and D. J. Miller. "Transport of wireless video using separate, concatenated and joint source-channel coding," *Proc. of the IEEE* 87 (October 1999): 1734–50.
18. E. Ayanoglu, P. Pancha, and A. R. Reibman. "Video transport in wireless ATM," in *Proc. ICIP*, 400–403, Washington, DC, 1995.
19. R. Mathew and J. F. Arnold, "Layered coding using bitstream decomposition with drift correction," *IEEE Trans. CSVT* 7 (December 1997): 882–91.
20. P. Cherriman and L. Hanzo. "Robust H.263 video transmission over mobile channels in interference limited environment," in *Proc. Workshop Wireless Image/Video Communications*, 1–7, Loughborough, UK, September 1996.
21. R. Swann and N. Kingsbury. "Error resilient transmission of MPEG-2 over noisy wireless ATM networks," in *Proc. ICIP*, 85–88, Santa Barbara, CA, 1997.
22. D. W. Redmill and N. G. Kingsbury. "The EREC: An error resilient technique for coding variable length blocks of data," *IEEE Trans. Image Processing* 4 (April 1996): 565–74.
23. U. Benzler. "Scalable multiresolution video coding using subband decomposition," in *Proc. Workshop Wireless Image/Video Communications*, 109–14, Loughborough, UK, September 1996.
24. H. Gharavi and W. Y. Ng. "H.263 compatible video coding and transmission," in *Proc. Workshop Wireless Image/Video Communications*, 115–120, Loughborough, UK, September 1996.
25. R. E. VanDyck and H. V. Ganti. "Wavelet video transmission over wireless channels," *Signal Processing: Image Communication, EURASIP Special Issue on Mobile Image/Video Transmission* 12, no. 2 (April 2004): 135–45.
26. K. Sayood, F. Liu, and J. D. Gibson. "A constrained joint source/channel coder design," *IEEE J. Selected Areas in Commun.* 12 (December 1994): 1584–93.
27. F. I. Alajaji, N. C. Phamdo, and T. E. Fuja. "Channel codes that exploit the residual redundancy in CELP-encoded speech," *IEEE Trans. Speech and Audio Processing* 4 (September 1996): 325–36.
28. K. Sayood and J. C. Borkenhagen. "Use of residual redundancy in the design of joint source/channel coders," *IEEE Trans. Commun.* 39 (June 1991): 838–46.

29. A. Vetro, C. Christopoulos, and H. Sun. "An overview of video transcoding architectures and techniques," *IEEE Signal Processing* 20 (March 2003): 18–29.
30. A. Vetro, Y. Xin, and H. Sun. "Error resilience video transcoding for wireless communications," *IEEE Wireless Commun.* 12 (August 2005): 14–21.
31. ITU-T Rec. H264, ISO/IEC 14496-10. Advanced Video Coding, 2003.
32. S. Wegner. "H.264/AVC over IP," *IEEE Trans. CSVT* 13 (July 2003): 645–56.
33. G. de los Reyes et al. "Error-resilient transcoding for video over wireless channels," *IEEE J. Selected Areas in Commun.* 18 (June 2000): 1063–74.
34. S. Dogan et al. "Error-resilient video transcoding for robust internetwork communications using GPRS," *IEEE Trans. CSVT* 12 (June 2002): 453–564.
35. G. Cote, S. Shirami, and F. Kossentini. "Optimal mode selection and synchronization for robust video communications over error-prone networks," *IEEE J. Selected Areas in Commun.* 18 (June 2000): 952–65.
36. R. Zhang, S. L. Regunathan, and K. Rose. "Video coding with optimal inter/intra-mode switching for packet loss resilience," *IEEE J. Selected Areas in Commun.* 18 (June 2000): 966–76.
37. A. Puri et al. "An integrated source transcoding and congestion control paradigm for video streaming in the Internet," *IEEE Trans. Multimedia* 3 (March 2001): 18–32.
38. T. C. Wang, H. C. Wang, and L. G. Chen. "Low delay and error robust wireless video transmission for video communication," *IEEE Trans. CSVT* 12 (December 2002): 1049–58.
39. R. G. Gallagher. Energy Limited Channels: Coding, Multi-Access and Spread Spectrum, MIT LIDS Rep. LIDS-P-1714, November 1987.
40. E. Uysal-Biyikoglu, B. Prabhakar, and A. El Gamal. "Energy-efficient packet transmission over a wireless link," *IEEE Trans. Network* 10 (August 2002): 487–99.
41. S. Zhao, Z. Xiong, and X. Wang. "Joint error control and power allocation for video transmission over CDMA networks with multiuser detection," *IEEE Trans. CSVT* 12 (June 2002): 425–37.
42. Y. S. Chan and J. V. Modestino. "A joint source coding-power control approach for video transmission over CDMA networks," *IEEE J. Selected Areas in Commun.* 21 (December 2003): 1516–25.
43. Y. Eisenber et al. "Joint source coding and transmission power management for energy efficient wireless video communications," *IEEE Trans. CSVT* 12 (June 2002): 411–24, June 2002.
44. A. Katsagelos et al. "Energy efficient wireless video coding and delivery," *IEEE Wireless Commun.* 12 (August 2005): 24–30.
45. C. E. Luna et al. "Joint source coding and data rate adaptation for energy efficient wireless video streaming," *IEEE J. Selected Areas Commun.* 21 (December 2003): 1710–20.
46. E. Gustafsson and G. Karlson. "A literature survey on traffic dispersion," *IEEE Network* 8 (March/April 1997): 28–36.
47. N. Gogate et al. "Supporting image/video applications in a multihop radio environment using route diversity and multiple description coding," *IEEE Trans. CSVT* 12 (September 2002): 777–92.
48. J. G. Apostopoluos. "Reliable video communication over lossy packet networks using multiple state encoding and path diversity," in *Proc. SPIE VCIP*, 392–409, San Jose, CA, January 2001.
49. J. G. Apostopulos et al. "One multiple description streaming in content delivery networks," in *Proc. IEEE INFOCOM*, 1736–45, New York, June 2002.
50. E. Setton, Y. Liang, and B. Girod. "Adaptive multiple description video streaming over multiple channels with active probing," in *Proc. IEEE ICME*, 509–12, Baltimore, MD, July 2003.

51. Y. Chakareski, S. Han, and B. Girod. "Layered coding vs. multiple description for video streaming over multiple paths," in *Proc. ACM Multimedia*, 422–431, Berkeley CA, 2003.

52. A. C. Begen, Y. Altunbasak, and O. Ergun. "Multipath selection multiple description encoded video streaming," *Image Commun.*, EURASIP 20, no. 1 (January 2005): 39–60.

53. S. Mao et al. "Video transport over ad hoc networks: Multistream coding with multipath transport," *IEEE J. Selected Areas Commun.* 21 (December 2003): 1721–37.

54. T. Nguyen and A. Yakhor. "Path diversity with forward error correction (PDF) system for packet switched networks," in *Proc. IEEE INFOCOM*, 663–72, San Francisco, CA, April 2003.

55. R. Stewart et al. Stream Control Transmission Protocol, IETF RFC2960, October 2000.

56. S. Mao et al. "Multipath video transport over ad hoc networks," *IEEE Wireless Commun.* 12 (August 2005): 42–49.

CHAPTER 7 CROSS-LAYER WIRELESS MULTIMEDIA

1. W. Kumviaisek et al. "A cross-layer quality of service mapping architecture for video delivery in wireless networks," *IEEE J. Selected Areas Commun.* 21 (December 2003): 1685–98.

2. A. Ortega and K. Ramchandran. "Rate distortions for image and video compression," *IEEE Signal Processing Magazine* 15 (October 2001): 23–50.

3. K. R. Rao, Z. S. Bojkovic, and D. A. Milovanovic. *Multimedia Communication Systems: Techniques, Standards and Networks.* Upper Saddle River, NJ: Prentice-Hall, 2002.

4. M. VanderShaar and S. Shankar. "Cross-layer wireless multimedia transmission: Challenges, principles and new paradigms," *IEEE Wireless Commun.* 12 (August 2005): 50–58.

5. V. Kawadia and P. R. Kumar. "A cautionary perspective on cross-layer design," *IEEE Wireless Commun.* 12 (February 2006): 3–11.

6. E. Salton et al. "Cross-layer design for ad hoc networks for real-time video streaming," *IEEE Wireless Commun.* 12 (August 2005): 59–65.

7. S. Shakkottai, T. S. Rappaport, and P. C. Karlsson. "Cross-layer design for wireless networks," *IEEE Commun. Magazine* 41 (October 2003): 74–80.

8. Q. Liu, Sh. Zhak, and G. B. Giannakis. "Cross-layer scheduling with perceived QoS guarantees in adaptive wireless networks," *IEEE J. Selected Areas Commun.* 23 (May 2005): 1056–66.

9. H. Wang and N. Moayeri. "Finite state Markov channel—A useful model for radio communication channels," *IEEE Trans. Veh. Tech.* 44 (February 1995): 163–71.

10. G. J. Sullivan and T. Wiegand. "Video compression—From concepts to the H.264/AVC standard," *Proc. of the IEEE* 93 (January 2005): 18–31.

11. M. Johnson and L. Xiao. "Cross-layer optimization of wireless networks using nonlinear column generation," *IEEE Trans. on Wireless Commun.* 5 (February 2006): 435–45.

12. Z. Bojkovic and D. Milovanovic. "H.264 video transmission over IEEE802.11 based wireless networks: QoS cross-layer optimization," *WSEAS Trans. on Communications* 5 (September 2006): 1777–94.

13. Z. Bojkovic and D. Milovanovic. "Cross-layer quality of service for video wireless multimedia delivery: Some challenges and principles," *WSEAS Trans. on Communications* 4 (January 2007): 17–22.

14. Z. Bojkovic and B. Bakmaz. "Need for cross-layer optimization in ad hoc networks for real-time video streaming," in *Proc. IWSSIP*, 361–364, Budapest, Hungary, September 2006.

15. B. Girod et al. "Advances in channel-adaptive video streaming," *Wireless Commun. and Mobile Comput.* 2 (September 2002): 549–52.

16. R. Katz, "Adaptive and mobility in wireless information systems," *IEEE Pers. Commun.* 1 (2nd qtr. 1994): 6–17.

17. D. Majumdar et al. "Multicast and unicast real-time video streaming over wireless LAN's," *IEEE Trans. CSVT* 12 (June 2002): 524–34.

18. H. Jiang and X. Shen. "Cross-layer design for resource allocation in 3G wireless networks and beyond," *IEEE Commun. Mag.* 43 (December 2005): 120–26.

19. I. M. Kim and H. M. Kim. "Efficient power management schemes for video service in CDMA systems," *Electronics Letters* 36 (June 2000): 1149–50.

20. S. Zhao, Z. Xiong, and X. Wang. "Joint error control and power allocation for video transmission over CDMA networks with multi user detection," *IEEE Trans. CSVT* 12 (June 2002): 425–37.

21. L. P. Kondi, F. Ishliak, and A. K. Katsaggelos. "Joint source-channel coding for motion compensated DCT-based SNR scalable video," *IEEE Trans. Image Processing* 11 (September 2002): 1043–52.

22. Q. Zhang et al. "Power-minimized bit allocation for video communication over wireless channels," *IEEE Trans. CSVT* 12 (June 2002): 398–410.

23. S. Blake et al. An Architecture for Differentiated Services, IETF RFC 2475, December 1998.

24. T. K. Liu and J. A. Silvester. "Joint admission/congestion control for wireless CDMA systems supporting integrated services," *IEEE J. Selected Areas in Commun.* 16 (August 1998): 845–57.

25. I. F. Akyildiz, D. A. Levine, and I. Joe. "A slotted CDMA protocol with BER scheduling for wireless multimedia networks," *IEEE/ACM Trans. Networking* 7 (April 1999): 146–58.

26. V. Huang and W. Zhuang. "QoS-oriented packet scheduling for wireless multimedia CDMA communications," *IEEE Trans. Mobile Computing* 3 (January–February 2004): 73–85.

27. H. Jiang and W. Zhuang. "Cross-layer resource allocation for integrated voice/data traffic in wireless cellular networks," *IEEE Trans. Wireless Commun.* 5 (February 2006): 457–68.

28. V. Jacobson, K. Nichols, and K. Poduri. An Expedited Forwarding PHB, IETF RFC 2598, June 1999.

29. J. Heinanen et al. Assured Forwarding PHB Group, IETF RFC 2597, June 1999.

30. P. Viswanath et al. "Opportunistic beamforming using clumb antennas," *IEEE Trans. Information Theory* 48 (June 2002): 1277–94.

31. X. Liu, E. Chong, and N. Schroff. "Opportunistic transmission scheduling with resource-sharing constraints in wireless networks," *IEEE J. Selected Areas in Commun.* 19 (October 2001): 2053–64.

32. C. Y. Wong et al. "Multiuser OFDM with adaptive subcarrier, bit and power allocation," *IEEE J. Selected Areas in Commun.* 17 (October 1999): 1747–58.

33. J. Chuang and N. Sollenberger. "Beyond 3G: Wideband wireless data access based on OFDM and dynamic packet assignment," *IEEE Commun. Magazine* 38 (July 2000): 78–87.

34. G. Song and Y. Li. "Utility-based resource allocation and scheduling in OFDM-based wireless broadband networks," *IEEE Commun. Magazine* 43 (December 2005): 127–34.

CHAPTER 8   MOBILE INTERNET

1. P. Stuckmon, *The GSM Evolution—Mobile Packet Data Services.* New York: John Wiley & Sons, 2003.
2. P. Bhagwat, C. Perkins, and S. Tripathi. "Network layer mobility: An architecture and survey," *IEEE Pers. Commun.* 3 (June 1996): 56–64.
3. I. F. Akylidiz et al. "Mobility management in current and future communications networks," *IEEE Network* 12 (July/August 1998): 34–49.
4. D. Saha et al. "Mobility support in IP: A survey of related protocols," *IEEE Network* 18 (November/December 2004): 34–40.
5. A. T. Campbell et al. "Comparison of IP micromobility protocols," *IEEE Wireless Commun.* 9 (February 2002): 72–82.
6. T. B. Zahariadis et al. "Global roaming in next-generation networks," *IEEE Commun. Magazine* 40 (February 2002): 145–51.
7. A. T. Campbell et al. "Design and performance of cellular IP networks," *IEEE Pers. Commun.* 7 (August 2000): 42–49.
8. S. Dos et al. "TeleMIP: telecommunications-enhanced MIP architecture for fast intradomain mobility," *IEEE Pers. Commun.* 7 (August 2000): 50–58.
9. A. Misra et al. "Autoconfiguration, registration and mobility management for pervasive computing," *IEEE Pers. Commun.* 8 (August 2001): 24–31.
10. J. F. Huber. "Mobile next-generation networks," *IEEE Multimedia* 11 (January–March 2004): 72–83.
11. P. Newman. "In search of the all-IP mobile network," *IEEE Commun. Mag.* 42 (December 2004): 53–58.
12. C. E. Perkins. *MobileIP: Design, Principles and Practices.* Reading, MA: Addison Wesley Longman, 1998.
13. J. D. Solomon. *MobileIP, the Internet Unplugged.* Upper Saddle River, NJ: Prentice-Hall, 1998.
14. J. Li and H. H. Chen. "Mobility support for IP-based networks," *IEEE Commun. Mag.* 43 (October 2005): 127–32.
15. H. Solima et al. Hierarchical Mobile IPv6 Mobility Management, IETF Network working group, October 2004.
16. P. S. Henry and H. Luo. "*WiFi*: What's next?" *IEEE Commun. Mag.* 40 (December 2002): 66–72.
17. I. F. Akyildiz, J. Xie, and S. Mohanty. "A survey of mobility management in next-generation all-IP-based wireless systems," *IEEE Wireless Commun.* 11 (August 2004): 16–28.
18. P. K. Bestand and R. Pendse. "Quantitative analysis of enhanced mobile IP," *IEEE Commun. Mag.* 44 (June 2006): 66–72.
19. C. Perkins. IP Mobility Support for IPv4, RFC3344, August 2002.
20. C. Perkings. IP Encapsulation within IP, RFC2003, October 1996.
21. A. K. Talukdar, B. R. Badrinath, and A. Acharya. "MRSVP: A resource reservation protocol for an integrated service network with mobile hosts," *Wireless Networks* 7 (January–February 2001): 5–19.
22. C. Tseng, G. Lee, and R. Liu. "A hierarchical mobile RSVP protocol," *Wireless Networks* 9 (March–April 2003): 95–102.
23. S. Paskalis et al. "An efficient RSVP-mobile IP interworking scheme," *Mobile Networks and Apps.* 8 (June 2003): 197–207.
24. H. Levkowetz and S. Vaarala. Mobile IP Traversal of Network Address Translation (NAT) Devices, RFC3519, April 2003.
25. P. Srisuresh and K. Egevang. Traditional IP Network Address Translator (Traditional NAT), RFC3022, January 2001.

26. D. Johnson, C. Perkins, and J. Arkko. Mobility Support in IPv6, RFC3775, June 2003.
27. S. Pack and K. Park. "SAMP: Scalable application-layer mobility protocol," *IEEE Commun. Magazine* 44 (June 2006): 86–92.
28. J. Hillebrand et al. "Quality-of-service signaling up for next generation IP-based mobile networks," *IEEE Commun. Mag.* 42 (June 2004): 72–79.
29. R. L. Aguiar et al. "Scalable QoS-aware mobility for future mobile operators," *IEEE Commun. Mag.* 44 (June 2006): 95–102.
30. R. Koodli. Fast Handovers for Mobile IPv6, RFC4068, July 2005.
31. S. Blake et al. An Architecture for Differentiated Services, RFC2475, December 1998.
32. M. Liebsh et al. Candidate Access Router Discovery (CARD), RFC4066, July 2006.
33. G. Camarillo and M. A. Garcia-Martin. *The 3G Multimedia Subsystem—Merging the Internet and the Cellular Worlds.* Hoboken, NJ: John Wiley & Sons, 2004.
34. Ch. Kalmanek et al. "A network-based architecture for seamless mobility services," *IEEE Commun. Mag.* 44 (June 2006): 103–9.
35. L. Dryburgh and J. Hewitt. *Signaling System No. 7 (SS7/C7): Protocol, Architecture and Services.* Cisco Press, 2004.

CHAPTER 9   EVOLUTION TOWARD 4G NETWORKS

1. A. M. Safwat and H. Mouftah, "4G network technologies for mobile telecommunications," *IEEE Network* 19 (September–October 2005): 3–4.
2. S. Y. Hui and K. H. Yeung, "Challenges in the migration to 4G mobile systems," *IEEE Commun. Mag.* 41 (December 2003): 54–59.
3. E. Buracchini. "The software concept," *IEEE Commun. Mag.* 38 (September 2000): 138–43.
4. J. Al-Muhtadi, D. Mickunas, and R. Campbell. "A lightweight reconfigurable security mechanism for 3G/4G mobile devices," *IEEE Wireless Commun.* 9 (April 2002): 60–65.
5. N. Montavont and T. Noel. "Handover management for mobile nodes in IPv6 networks," *IEEE Commun. Mag.* 40 (August 2002): 38–43.
6. 3GPPTS 23.107 v5.9.0. Quality of Service (QoS): Concept and Architecture, August 2002.
7. D. Tipper et al. "Providing fault tolerance in wireless access networks," *IEEE Commun. Mag.* 40 (January 2002): 58–64.
8. F. Ghys and A. Vaaraniemi. "Component-based charging in a next-generation multimedia network," *IEEE Commun. Mag.* 41 (January 2003): 99–102.
9. A. D. Stefano and C. Santoro. "NetChaser: Agent support for personal mobility," *IEEE Internet Comp.* 4 (March/April 2000):74–79.
10. K. Pahlaven and A. H. Levesgne. *Wireless Information Networks.* New York: John Wiley & Sons, 2005.
11. S. Frattasi et al. "Defining 4G technology from the user's perspective," *IEEE Network* 20 (January/February 2006): 35–41.
12. B. Scheiderman. *Lenardo's Laptop—Human Needs and the New Computing Technologies.* Cambridge, MA: MIT Press, 2002.
13. http://kom.acu.ch/project/jade.
14. S. Y. Hui and K. H. Yeung. "Challenges in the migration to 4G mobile systems," *IEEE Commun. Magazine* 41 (December 2003): 54–59.
15. S. Frattasi, E. Chanca, and B. Prasad. "An integrated AP for seamless interworking of existing WMAN and WLAN standards," *Kluwer/Springer WPC*, Special issue in increasing efficiency in broadband fixed wireless access systems: from physical to network layer solutions, January 2006.

16. http://www.3g.org.
17. 3GPP TR22.978, v7.1.0. All-IP Network (AIPN) Feasibility Study (Release 7), June 2005.
18. 3GPP TR23.882, Draft v1.4.2. 3GPP System Architecture Evolution: Report on Technical Options and Conclusions, Stage 1 (Release7), October 2006.
19. 3GPP TR22.278, Draft v0.2.0. Service Requirements for Evolution of the 3GPP System, Stage 1 (Release 8), August 2006.
20. 3GPP TR22.259, Draft v7.0.0. Service Requirements for Personal Network Management (PNM), Stage 1 (Release 7), March 2006.
21. 3GPP TR23.234, Draft v7.1.0. 3GPP System for Wireless Local Area Network (WLAN) Interworking: System Description, (Release 7), March 2006.
22. C. Koppler et al. "Dynamic network composition for beyond 3G networks: A 3GPP viewpoint," *IEEE Network* 21 (January–February 2007): 47–52.
23. V. Devarpalli et al. Network Mobility (NEMO) Basic Support Protocol, RFC3963, January 2005.
24. Z. Bojkovic and B. Bakmaz. "Quality of service and security as frameworks toward next-generation wireless networks," *WSEAS Trans. on Communications* 4 (April 2005): 147–53.
25. A. Lera et al. "End-to-end QoS provisioning in 4G with mobile hotspots," *IEEE Network* 19 (September–October 2005) 26–34.
26. R. Prasad and S. Hara. *Multicarrier Techniques for 4G Mobile Communications: Universal Personal Communications*. Norwood, MA: Artech House, 2003.
27. F. H. P. Fitzek and M. Reisslein. Video traces for network performance evaluation, http://trace.eas.asu.edu.
28. R. Braden et al. Resource Reservation Protocol (RSVP) Version 1 Functional Specification, RFC2205, September 2005.
29. V. Marques et al. "An IP-based QoS architecture for 4G operator scenarios," *IEEE Wireless Commun.* 10 (June 2003): 54–62.
30. S. C. Lo et al. "Architecture for mobility and QoS support in all-IP wireless networks," *IEEE J. Selected Areas in Commun.* 22 (May 2004): 691–705.
31. S. Pack et al. "Performance analysis of mobile hotspots with heterogeneous wireless links," *IEEE Trans. on Wireless Commun.* 6 (October 2007): 3717–27.
32. D. Black et al. An Architecture for Differentiated Services, IETF RFC2475, December 1998.
33. K. R. Rao, Z. S. Bojkovic, and D. A. Milovanovic. *Introduction to Multimedia Communications: Applications, Middleware, Networking*. New York: John Wiley & Sons, 2006.
34. T. M. Trang Nguyen et al. "COPS-SLS: a service level negotiations protocol for the Internet," *IEEE Commun. Magazine* 40 (May 2002):158–65.
35. H. Ensiedler et al. The Moby Dick Project: A Mobile Heterogeneous All-IP Architecture, ATAMS 2001, Krakow, Poland, http://www.ist-mobydick.org.
36. P. Calhoun et al. Diameter Base Protocol, RFC3588, September 2003.
37. B. Abova et al. Extensible Authentication Protocol (EAP), RFC3748, June 2004.
38. T. Dierks and C. Allen, The TLS Protocol Version 1, RFC2246, January 1999.
39. 3GPP TS32.210. IP Network Layer Security v6.0.0, December 2002.
40. 3GPP TS32.203. Access Security for IP-Based Services v5.4.0, December 2002.
41. 3GPP TS32.200. MAP Application Layer Security v5.1.10, December 2002.
42. 3GPP TS23.228. IP Multimedia Subsystem (IMS) Stage 2 v6.0.1, January 2003.
43. A. R. Prasad, H. Wang, and P. Schoo. *Network operator's security requirements on systems beyond 3G*, WWRF 9, Zurich, Switzerland, 1–2 July 2003.
44. K. Pahlvan et al. "Handoff in hybrid mobile data networks," *IEEE Personal Commun.* 7 (April 2000): 34–37.

45. 3GPP TS22.129 v5.2.0. 3GPP Handover Requirements between UTRAN and GERAN or Other Radio System, June 2002.

46. M. Stenim and R. H. Katz. "Vertical handoffs in wireless overlay networks," *ACM Mobile Networks and Applications MONTE*, Special issue on Mobile networking in the Internet 3, no. 4 (1998): 335–530.

47. 3GPP TS23.234 v1.3.0. 3GPP System to Wireless Local Area Network (WLAN) Interworking: System Description, January 2003.

48. P. Nikander and L. Metso. "Policy and trust in open multi-operator networks," in *Proc. IFIP SmartNet*, 419–36, Vienna, Austria, September 2000.

49. W. Keller et al. Systems beyond 3G-Operator's Vision, WWRF Meeting 7, Eindhoven, Netherlands, December 2002.

50. BBC News, *Mobile Disables to Beat Thieves*, March 2003.

51. B. Bakmaz, Z. Bojkovic, and M. Bakmaz. "Network selection for heterogeneous wireless environment," in *Proc. IEEE Int. Symp. Personal, Indoor and Mobile Radio Commun., PIMRC*, Athens, Greece, 2007.

Index